T0137397

Lecture Notes in Computational Science and Engineering

109

More information about this series at http://www.springer.com/series/3527

Jochen Garcke • Dirk Pflüger

Editors

Sparse Grids and Applications – Stuttgart 2014

 Springer

Editors
Jochen Garcke
Institut für Numerische Simulation
Universität Bonn
Bonn, Germany

Dirk Pflüger
Institute for Parallel and Distributed
  Systems
Universität Stuttgart
Stuttgart, Germany

ISSN 1439-7358   ISSN 2197-7100   (electronic)
Lecture Notes in Computational Science and Engineering
ISBN 978-3-319-80309-8   ISBN 978-3-319-28262-6   (eBook)
DOI 10.1007/978-3-319-28262-6

Mathematics Subject Classification (2010): 65D99, 65M12, 65N99, 65Y20, 65N12, 62H99

Printed on acid-free paper

Springer International Publishing AG Switzerland is part of Springer Science+Business Media
(www.springer.com)

Preface

Sparse grids are a popular approach for the numerical treatment of high-dimensional problems. Where classical numerical discretization schemes fail in more than three or four dimensions, sparse grids, in their different flavors, are frequently the method of choice, be it spatially adaptive in the hierarchical basis or via the dimensionally adaptive combination technique.

The third Workshop on Sparse Grids and Applications (SGA2014), which took place at the University of Stuttgart from September 1 to 5 in 2014, demonstrated once again the importance of this numerical discretization scheme. Organized by Hans-Joachim Bungartz, Jochen Garcke, Michael Griebel, Markus Hegland, Dirk Pflüger, and Clayton Webster, almost 60 participants from 8 different countries have presented and discussed the current state of the art of sparse grids and their applications. Thirty-eight talks covered their numerical analysis as well as efficient data structures and new forms of adaptivity and a range of applications from clustering and model order reduction to uncertainty quantification settings and optimization. As a novelty, the topic high-performance computing covered several talks, targeting exascale computing and related tasks. Besides data structures and communication patterns with excellent parallel scalability, fault tolerance was introduced to the SGA series, the hierarchical approach providing novel approaches to the treatment of hardware failures without checkpoint restart. This volume of *LNCSE* collects selected contributions from attendees of the workshop.

We thank the SimTech Cluster of Excellence and the Informatik Forum Stuttgart (infos e.V.) for their financial support. Furthermore, we thank Mario Heene, Fabian Franzelin, and David Pfander for their assistance with the local organization.

Bonn, Germany   Jochen Garcke
Fraunhofer SCAI
Sankt Augustin, Germany
Stuttgart, Germany   Dirk Pflüger
October 2015

Contents

Adaptive Sparse Grid Model Order Reduction for Fast Bayesian Estimation and Inversion

Peng Chen and Christoph Schwab

Abstract We present new sparse-grid based algorithms for fast Bayesian estimation and inversion of parametric operator equations. We propose Reduced Basis (RB) acceleration of numerical integration based on Smolyak sparse grid quadrature. To tackle the curse-of-dimensionality in high-dimensional Bayesian inversion, we exploit sparsity of the parametric forward solution map as well as of the Bayesian posterior density with respect to the random parameters. We employ an dimension adaptive Sparse Grid method (aSG) for both, offline-training the reduced basis as well as for deterministic quadrature of the conditional expectations which arise in Bayesian estimates. For the forward problem with nonaffine dependence on the random variables, we perform further affine approximation based on the Empirical Interpolation Method (EIM) proposed in [1]. A novel combined algorithm to adaptively refine the sparse grid used for quadrature approximation of the Bayesian estimates, of the reduced basis approximation and to compress the parametric forward solutions by empirical interpolation is proposed. The theoretically predicted computational efficiency which is independent of the number of active parameters is demonstrated in numerical experiments for a model, nonaffine-parametric, stationary, elliptic diffusion problem, in two spacial and in parameter space dimensions up to 1024.

1 Introduction

Bayesian estimation, ie. the "most likely" prediction of responses of ordinary and partial differential differential equations (ODEs and PDEs, for short), subject to uncertain input data, given noisy observation data, is a key topic in computational statistics and in computational science. We refer to [16, 34] and the references there for a survey.

P. Chen (✉) • C. Schwab
HG G 56.1, Seminar for Applied Mathematics, ETH Zürich, Rämistrasse 101, CH-8092 Zürich, Switzerland
e-mail: peng.chen@sam.math.ethz.ch; christoph.schwab@sam.math.ethz.ch

© Springer International Publishing Switzerland 2016
J. Garcke, D. Pflüger (eds.), *Sparse Grids and Applications – Stuttgart 2014*,
Lecture Notes in Computational Science and Engineering 109,
DOI 10.1007/978-3-319-28262-6_1

As a rule, Monte-Carlo (MC) based sampling methods are used. Then, the methods are computationally intensive due to the slow convergence of MC methods (implying a rather large number of samples) and due to the high cost of forward solves per "sample" (which, in PDE models, amounts to one numerical PDE solve per sample). Distributed uncertain inputs such as, for example, uncertain diffusion coefficients, require forward solves of *infinite-dimensional, parametrized PDEs*.

Recent mathematical results [12–14, 20, 22, 23] indicate that the parameter-to-solution maps of these parametric operator equations exhibit *sparsity* in the sense that their n-widths are small, independent of the number of parameters which are activated in the approximation. This has lead to the proposal of *sparse, dimension-adaptive interpolation schemes* for the exploitation of this sparsity in the solution map, see, e.g. [11], and to deterministic *sparse, adaptive quadrature methods in Bayesian inversion [31]*. Small n-widths of parametric PDE solution maps can, in principle, be exploited by greedy approximation strategies, according to recent mathematical results [2, 35]. This observation was used in recent works by one of the authors [3–6, 8] to accelerate forward solves of parametric PDEs.

Similar accelerations are possible for the widely used Markov-chain Monte Carlo (MCMC) methods. Specific accerelations via sparse, generalized polynomial chaos (gPC)-based surrogates of the parametric forward maps are analyzed in [24].

Accelerations of forward solutions via *model order reduction (MOR for short) approaches* [15, 17, 27] have also been proposed. In the present note, we consider acceleration via *reduced basis methods* (RB) relying on adaptive sparse grid (aSG) collocation for both interpolation and integration. At each sparse grid node generated in the parameter space U by the adaptive algorithm, we evaluate the RB surrogate density of the Bayesian posterior or of some other Quantity of Interest (QoI for short). If the surrogate is not determined to sufficient accuracy, as certified by some reliable and efficient goal-oriented error estimator (which is an essential, and novel, element of the presently proposed algorithm), then a full forward problem is to be solved numerically to evaluate the posterior density. The RB approximation is refined by the full ("HiFi") solution at this grid node. The efficient online evaluation of the RB surrogate and the error estimator introduced in [9] depends on an affine-parametric structure of the underlying PDEs that enables efficient offline-online decomposition. For more general nonaffine problems such as those considered in the present paper, we propose in the present paper to compute their affine approximations by empirical interpolation methods [1, 7] in combination with aSG. To this end, a combined adaptive sparse grid, empirical interpolation and reduced basis methods (aSG-EIM-RB) are developed to reliably and efficiently accelerate the Bayesian estimation and inversion. In this paper, we present the first detailed description of the corresponding algorithm, and the first numerical experiments performed with the proposed numerical inversion strategy. Dimension-independent convergence rates are demonstrated with parameter space dimensions up to 1024.

This paper is organized as follows: the formulation of parametric Bayesian inversion in function space is presented in Sect. 2. The admissible forward maps are countably-parametric operator equations. Section 3 presents a constructive algorithm based on adaptive sparse grid interpolation for gPC approximation of the parametric solution, of the posterior density and of the related QoI. Section 4 is devoted to the development of reduced basis acceleration for the Bayesian inversion, where a combined aSG-EIM-RB algorithm is presented. Numerical experiments in demonstrating the dimension-independent convergence rates and computational efficiency of the proposed algorithm are provided in Sect. 5, followed by conclusions in Sect. 6.

2 Bayesian Inversion

We review the mathematical setting of Bayesian inversion of partial differential equations with distributed uncertainty, in the mathematical framework of [16].

By *uncertainty parametrization* together with a Bayesian prior on the (generally infinite-dimensional) parameter space, the problem of Bayesian estimation is converted to a problem of quadrature of the parametric, deterministic posterior density.

Sparsity of the parametric forward solution map and of the Bayesian posterior density as well as its integration with respect to the infinite-dimensional parameter sequence will be presented. Dimension-independent convergence rates of sparsity-exploiting Smolyak quadratures are stated and verified in numerical experiments.

2.1 Formulation of Bayesian Inversion

We consider Bayesian inversion problems that consist of: given observation data subject to additive, centered gaussian observation noise of some system output, and given a prior distribution of the uncertain, distributed system input, Bayes' theorem yields the posterior distribution and an estimate of the "most likely" system response for some quantity of interest (QoI), that depends on the system state [16]. More precisely, we consider a system with state variable q belonging to a separable Hilbert space $\mathcal{X}$ (with (anti-)dual $\mathcal{X}'$), which is determined through the system forward map $G : X \to \mathcal{X}$ by an uncertain system input variable u that takes values in a separable Banach space X. The observation data $\delta \in Y = \mathbb{R}^K$, $K \in \mathbb{N}$, is assumed to correspond to a true system response $G(u)$ observed through K sensors $O(\cdot) = (o_1, \ldots, o_K) \in (\mathcal{X}')^K$, whose readings we assume to be corrupted by additive Gaussian observation noise $\eta \sim \mathcal{N}(0, \Gamma)$ with symmetric positive definite correlation matrix $\Gamma \in \mathbb{R}^{K \times K}$, i.e.

$$\delta = O(G(u)) + \eta . \tag{1}$$

We assume that a prior probability distribution π_0 defined on a Banach space X of uncertain input data is prescribed for the uncertain system input u with $\pi_0(X) = 1$.

Under appropriate continuity conditions on the uncertainty-to-observation map $\mathcal{G} := O \circ G(\cdot) : X \to Y$, Bayes' rule guarantees the existence of a posterior distribution π^δ that is absolutely continuous with density $\Theta : X \to \mathbb{R}$ with respect to the prior distribution. Various concrete, sufficient conditions for absolute posterior continuity are provided in the surveys [16, 34]. We parametrize the uncertain input u taking values in the separable Banach space X by postulating a countable basis $(\phi_j)_{j \in \mathbb{J}} \in X$, where $\mathbb{J} = \{1, \ldots, J\}$ with $J \in \mathbb{N}$ or $\mathbb{J} = \mathbb{N}$. We assume that the uncertain input u can be represented by these bases: there is an *affine parametrization*, i.e. $u = u(\mathbf{y})$ with the parameter (after possible rescaling and shifting) $\mathbf{y} = (y_j)_{j \in \mathbb{J}} \in U$, being $U = [-1, 1]^{\mathbb{J}}$ a reference parameter space. Under appropriate re-scaling of the basis ϕ_j, we may reparametrize the distributed, uncertain input data u as

$$u(\mathbf{y}) = \bar{u} + \sum_{j \in \mathbb{J}} y_j \phi_j, \quad \mathbf{y} \in U, \tag{2}$$

where $\bar{u} \in X$ is the "nominal" value of the uncertain data u and $u - \bar{u}$ entails possible fluctuations of u through $\mathbf{y} \in U$. An example of the expression (2) is Karhunen–Loève expansion of a random field u with mean field $\bar{u}$ and (rescaled) eigenfunctions $(\phi_j)_{j \in \mathbb{J}}$. In practice, the parametrization may also be a nonlinear transformation of an affine input, i.e. the parameters can not be separated from the bases. For instance in describing a positive permeability field κ, we may assume that $\kappa = e^u$ with u defined in (2), so that κ is positive at each parameter value $\mathbf{y} \in U$ but nonaffine with respect to $\mathbf{y}$. An example for a "non-affine" uncertainty parameterization is a Karhunen–Loève expansion of $\log(\kappa)$, which typically arises in log-gaussian models for u, in which case the Bayesian prior is a Gaussian measure on X. Under the parametrization, for prescribing a prior distribution of the uncertain input data u we only need to prescribe a prior measure for the parameters, i.e.

$$\pi_0 = \bigotimes_{j \in \mathbb{J}} \frac{1}{2} \lambda^1 \text{ or } d\pi_0(\mathbf{y}) = \bigotimes_{j \in \mathbb{J}} \frac{1}{2} dy_j, \tag{3}$$

where λ^1 denotes the Lebesgue measure on $[-1, 1]$. By Bayes' theorem, there exists a posterior measure which is absolutely continuous with respect to the prior. For the corresponding Radon–Nikodym derivative holds

$$\frac{d\pi^\delta}{d\pi_0}(\mathbf{y}) = \frac{1}{Z} \Theta(\mathbf{y}), \tag{4}$$

where the (rescaled) posterior density Θ is given by

$$\Theta(\mathbf{y}) = \exp\left(-\frac{1}{2}(\delta - O(G(u(\mathbf{y}))))^\top \Gamma^{-1} (\delta - O(G(u(\mathbf{y}))))\right), \tag{5}$$

and the renormalization constant Z is defined as

$$Z := \mathbb{E}^{\pi_0}[\Theta] = \int_U \Theta(\mathbf{y}) d\pi_0(\mathbf{y}) . \tag{6}$$

Under the posterior distribution, we can evaluate some quantity of interest (QoI) Ψ that depends on the system input u, e.g. the system input itself $\Psi(u) = u$ or the system response $\Psi(u) = G(u)$, as well as some statistics of the QoI, e.g. the expectation

$$\mathbb{E}^{\pi^\delta}[\Psi(u)] = \mathbb{E}^{\pi_0}[\Psi(u)\Theta] = \int_U \Psi(u(\mathbf{y}))\Theta(\mathbf{y}) d\pi_0(\mathbf{y}) . \tag{7}$$

2.2 Parametric Operator Equations

Under the above parametrization of the system input, we consider a class of parametric operator equations for the modelling of the system, which read as: for any parameter value $\mathbf{y} \in U$, find the solution $q(\mathbf{y}) \in X$ such that

$$A(\mathbf{y})q(\mathbf{y}) = f(\mathbf{y}) \quad \text{in } \mathcal{Y}' , \tag{8}$$

where $\mathcal{Y}$ is a separable Hilbert space with anti-dual $\mathcal{Y}'$, $A(\mathbf{y})$ is a parametric operator and $f(\mathbf{y})$ is a parametric right hand side, both depending on the parameter $\mathbf{y}$ through the uncertain system input $u(\mathbf{y})$. In particular, we consider linear systems modelled by countably-parametric, linear operator families $A(\mathbf{y}) \in \mathcal{L}(X, \mathcal{Y}')$. We associate the parametric operator $A(\mathbf{y})$ and $f(\mathbf{y})$ with sesquilinear and antilinear forms, respectively, in the Hilbert spaces X and $\mathcal{Y}$ over $\mathbb{C}$ as

$$a(\mathbf{y}; w, v) =_{\mathcal{Y}} \langle v, A(\mathbf{y})w \rangle_{\mathcal{Y}'} \text{ and } f(\mathbf{y}; v) :=_{\mathcal{Y}} \langle v, f(\mathbf{y}) \rangle_{\mathcal{Y}'} \quad \forall w \in X, v \in \mathcal{Y} . \tag{9}$$

The weak formulation of the parametric operator equation (8) reads: for any parameter value $\mathbf{y} \in U$, find the solution $q(\mathbf{y}) \in X$ such that

$$a(\mathbf{y}; q(\mathbf{y}), v) = f(\mathbf{y}; v) \quad \forall v \in \mathcal{Y} . \tag{10}$$

For the well-posedness of problem (10) and for the approximation of its solution, of the corresponding Bayesian posterior density, and of the QoI in Bayesian prediction, we make the following assumptions.

Assumption 1

A1 For $\epsilon > 0$ and $0 < p < 1$, there exists a positive sequence $(b_j)_{j \in \mathbb{J}} \in \ell^p(\mathbb{J})$ such that for any sequence $\rho := (\rho_j)_{j \in \mathbb{J}}$ with $\rho_j > 1$ for all $j \in \mathbb{J}$ and

$$\sum_{j \in \mathbb{J}} (\rho_j - 1) b_j \leq \epsilon, \tag{11}$$

the parametric maps a and f in (10) admit holomorphic extensions to certain cylindrical sets $O_\rho = \otimes_{j\in\mathbb{J}} O_{\rho_j}$, where $O_{\rho_j} \subset \mathbb{C}$ is an open set containing the Bernstein ellipse $\mathcal{E}_{\rho_j}$ with foci at ± 1 and with semi axes of length $(\rho_j + \rho_j^{-1})/2 > 1$ and $(\rho_j - \rho_j^{-1})/2 > 0$.

A2　　*There exist constants $0 < \beta < \gamma < \infty$ and $\theta > 0$ such that these extensions satisfy for all $\mathbf{z} \in O_\rho$ the uniform continuity conditions*

$$\sup_{v\in\mathcal{Y}} \frac{f(\mathbf{z};v)}{||v||_{\mathcal{Y}}} \le \theta \ and \ \sup_{w\in\mathcal{X}}\sup_{v\in\mathcal{Y}} \frac{a(\mathbf{z};w,v)}{||w||_{\mathcal{X}}||v||_{\mathcal{Y}}} \le \gamma \tag{12}$$

and the uniform inf-sup conditions

$$\inf_{0\ne w\in\mathcal{X}}\sup_{0\ne v\in\mathcal{Y}} \frac{|a(\mathbf{z};w,v)|}{||w||_{\mathcal{X}}||v||_{\mathcal{Y}}} \ge \beta \ and \ \inf_{0\ne v\in\mathcal{Y}}\sup_{0\ne w\in\mathcal{X}} \frac{|a(\mathbf{z};w,v)|}{||w||_{\mathcal{X}}||v||_{\mathcal{Y}}} \ge \beta . \tag{13}$$

We point out that the abstract assumptions **A1** and **A2** are valid for a host of countably-parametric problems, which are not necessarily of the specific form (10). We mention only elliptic and parabolic problems in uncertain domains (see, e.g. [12]) and nonlinear, parametric initial values ODEs (see, e.g., [20]). The following approximations results are key to dimension-robust convergence rate of the model order reduction methods; we refer to [12, 33] and [31] for proofs.

Theorem 1 *Under Assumption 1, there exists a positive constant $C < \infty$ depending on $\theta, \gamma, \beta, p, \epsilon$ and ρ, such that the operator equation (8) admits a unique uniformly bounded solution satisfying a generalized polynomial chaos expansion (gPC)*

$$\sup_{\mathbf{z}\in O_\rho} ||q(\mathbf{z})||_{\mathcal{X}} \le C \ and \ the \ gpc \ expansion \ q(\mathbf{y}) = \sum_{v\in\mathcal{F}} q_v P_v(\mathbf{y}) \tag{14}$$

holds. Here $P_v(\mathbf{y}) := \prod_{j\in\mathbb{J}} P_{v_j}(y_j)$, with P_n denoting the univariate Legendre polynomial of degree n for the interval $[-1, 1]$ normalized such that $||P_n||_{L^\infty([-1,1])} = 1$. In (14), $\mathcal{F}$ denotes the countable set of all finitely supported sequences $v \in \mathbb{N}_0^{\mathbb{J}}$, and the convergence is unconditional and in the supremum norm over $\mathbf{y} \in U$.

Moreover, there exists a nested sequence $\{\Lambda_M\}_{M\ge 1}$ of downward closed index sets $\Lambda_M \subset \mathcal{F}$ ("dc set", for short)[1] with at most M indices such that the dimension-independent convergence rate holds

$$\sup_{\mathbf{y}\in U} ||q(\mathbf{y}) - \sum_{v\in\Lambda_M} q_v P_v(\mathbf{y})||_{\mathcal{X}} \le C_q M^{-s}, \quad s = \frac{1}{p} - 1 . \tag{15}$$

---

[1] A subset $\Lambda \subset \mathcal{F}$ is a dc set if for every $v \in \Lambda_M$ also $\mu \in \Lambda_M$ for any $\mu \preceq v$ ($\mu_j \le v_j$ for all $j \in \mathbb{J}$)

Here the constant C_q neither depends on M nor on the number of active coordinates, ie. $\max\{\#\{j \in \mathbb{N} : \nu_j \neq 0\} : \nu \in \Lambda_M\}$. The same convergence rate (15) also holds for the approximation of the posterior density $\Theta(\mathbf{y})$ as well as for the QoI $\Psi(\mathbf{y})$.

3 Adaptive Sparse Grid Approximation

Theorem 1 in the last section guarantees the existence of sparse generalized polynomial approximations of the forward solution map and of the posterior density which approximate these quantities with dimension-independent convergence rate. We exploit this sparsity in two ways: first, in the choice of sparse parameter samples during the offline-training phase of model order reductions, and, as already proposed in [32], for adaptive, Smolyak-based numerical integration for the evaluation of the Bayesian estimate.

Both are based on constructive algorithms for the computation of such sparse polynomial approximations, To this end, we present an adaptive sparse grid algorithm for both interpolation and integration, based on the approaches in [9, 12, 18], see also [28, 29, 36] for anisotropic and, in certain cases, quasi-optimal sparse grid interpolation in the high-dimensional parameter space U.

3.1 Adaptive Univariate Approximation

In the univariate case $U = [-1, 1]$, given a set of interpolation nodes $-1 \leq y_1 < \cdots < y_m \leq 1$, we define the interpolation operator $\mathcal{I} : C(U; \mathcal{Z}) \to \mathcal{P}_{m-1}(U) \otimes \mathcal{Z}$ as

$$\mathcal{I}g(y) = \sum_{k=1}^{m} g(y_k)\ell_k(y), \tag{16}$$

where the function $g \in C(U; \mathcal{Z})$, representing e.g. the parametric forward solution map q with $\mathcal{Z} = \mathcal{X}$ or the posterior density Θ with $\mathcal{Z} = \mathbb{R}$; $\ell_k(y)$, $1 \leq k \leq m$, are the associated Lagrange polynomials in $\mathcal{P}_{m-1}(U)$, the space of polynomials of degree at most $m - 1$. To define the sparse collocation, as usual the interpolation operator defined in (16) is recast as telescopic sum, ie.,

$$\mathcal{I}_L g(y) = \sum_{l=1}^{L} \mathcal{D}^l g(y), \tag{17}$$

where L represents the level of interpolation grid; $\mathcal{D}^l := \mathcal{I}_l - \mathcal{I}_{l-1}$ with $\mathcal{I}_0 g \equiv 0$. Let Ξ^l denote the set of all interpolation nodes in the grid of level l, such that the grid is nested, i.e. $\Xi^l \subset \Xi^{l+1}$, $l = 0, \ldots, L - 1$, with $\Xi^0 = \emptyset$ and $\Xi^L = \{y_1, \ldots, y_m\}$. As

$I_{l-1}g(y) = g(y)$ for any $y \in \Xi^{l-1}$, we have $I_{l-1} = I_l \circ I_{l-1}$ and, with the notation $\Xi_{\mathcal{D}}^l = \Xi^l \setminus \Xi^{l-1}$, the interpolation operator (17) can be written in the form

$$I_L g(y) = \sum_{l=1}^{L} \sum_{y_k^l \in \Xi_{\mathcal{D}}^l} (I_l - I_l \circ I_{l-1})g(y) = \sum_{l=1}^{L} \sum_{y_k^l \in \Xi_{\mathcal{D}}^l} \underbrace{(g(y_k^l) - I_{l-1}g(y_k^l))}_{s_k^l} \ell_k^l(y),$$

(18)

where s_k^l represents the interpolation error of $I_{l-1}g$ evaluated at the node $y_k^l \in \Xi_{\mathcal{D}}^l$, $k = 1, \ldots, |\Xi_{\mathcal{D}}^l|$, so that we can use it as a posteriori error estimator for adaptive construction of the interpolation (18). More precisely, we start from the root level $L = 1$ with the root interpolation node $y = 0$, whenever the interpolation error estimator

$$\mathcal{E}_i := \max_{y_k^L \in \Xi_{\mathcal{D}}^L} |s_k^L|$$

(19)

is larger than a given tolerance, we refine the interpolation to the next level $L + 1$ by taking new interpolation node, for instance one Leja node

$$y_1^{L+1} = \underset{y \in U}{\operatorname{argmax}} \prod_{l=1}^{L} |y - y^l|,$$

(20)

or Clenshaw–Curtis nodes

$$y_k^{L+1} = \cos\left(\frac{k}{2^{L-1}}\pi\right), \quad k = 0, 1 \text{ for } L = 1; k = 1, 3, \ldots, 2^{L-1} - 1 \text{ for } L \geq 2.$$

(21)

Based on the adaptive interpolation, an associated quadrature formula is given by

$$\mathbb{E}[g] \approx \mathbb{E}[I_L g] = \sum_{l=1}^{L} \sum_{y_k^l \in \Xi_{\mathcal{D}}^l} s_k^l w_k^l, \quad \text{being } w_k^l = \mathbb{E}[\ell_k^l],$$

(22)

for which the integration error estimator can be taken as

$$\mathcal{E}_e := \left| \sum_{y_k^L \in \Xi_{\mathcal{D}}^L} s_k^L w_k^L \right|.$$

(23)

3.2 Adaptive Sparse Grid Approximation

In multiple dimensions $\mathbf{y} \in U = [-1, 1]^J$, we construct an adaptive sparse grid (aSG) interpolation by tensorizing the univariate interpolation formula (17)

$$S_{\Lambda_M} g(\mathbf{y}) = \sum_{\nu \in \Lambda_M} \left(\mathcal{D}_1^{\nu_1} \otimes \cdots \otimes \mathcal{D}_J^{\nu_J} \right) g(\mathbf{y}) , \tag{24}$$

where Λ_M is a downward closed index set defined in Theorem 1. As $\Lambda_1 \subset \cdots \subset \Lambda_M$ and the interpolation nodes are nested, the aSG formula (24) can be rewritten as

$$S_{\Lambda_M} g(\mathbf{y}) = \sum_{m=1}^{M} \sum_{\mathbf{y}_\mathbf{k}^{\nu^m} \in \Xi_{\mathcal{D}}^{\nu^m}} \underbrace{\left(g(\mathbf{y}_\mathbf{k}^{\nu^m}) - S_{\Lambda_{m-1}} g(\mathbf{y}_\mathbf{k}^{\nu^m}) \right)}_{s_\mathbf{k}^{\nu^m}} \ell_\mathbf{k}^{\nu^m}(\mathbf{y}) , \tag{25}$$

where $\Xi_{\mathcal{D}}^{\nu^m}$ is the set of added nodes corresponding to the index $\nu^m = (\nu_1^m, \ldots, \nu_J^m) = \Lambda_m \backslash \Lambda_{m-1}$; $\ell_\mathbf{k}^{\nu^m}(\mathbf{y}) = \ell_{k_1}^{\nu_1}(y_1) \otimes \cdots \otimes \ell_{k_J}^{\nu_J}(y_J)$, is the multidimensional Lagrange polynomial; $s_\mathbf{k}^{\nu^m}$ denotes the interpolation error of $S_{\Lambda_{m-1}} g$ evaluated at $\mathbf{y}_\mathbf{k}^{\nu^m}$, which can be used as an interpolation error estimator for the construction of the aSG.

More explicitly, we start from the initial index $\nu = \mathbf{1} = (1, \ldots, 1)$, thus $\Lambda_1 = \{\mathbf{1}\}$, with root node $\mathbf{y} = \mathbf{0} = (0, \ldots, 0)$. We then look for the maximal *active index set* Λ_M^a such that $\Lambda_M \cup \{\nu\}$ remains downward closed for any $\nu \in \Lambda_M^a$, e.g. for $\Lambda_M = \{\mathbf{1}\}$ when $M = 1$, we have $\Lambda_M^a = \{\mathbf{1} + \mathbf{e}_j, j = 1, \ldots, J\}$, being $\mathbf{e}_j = (0, \ldots, j, \ldots, 0)$ whose j-th entry is one and all other entries are zeros. For each $\nu \in \Lambda_M^a$, we evaluate the errors of the interpolation $S_{\Lambda_M} g$ at the nodes $\Xi_{\mathcal{D}}^{\nu}$, and enrich the index set $\Lambda_{M+1} = \Lambda_M \cup \{\nu^{M+1}\}$ with the new index

$$\nu^{M+1} := \operatorname*{argmax}_{\nu \in \Lambda_M^a} \max_{\mathbf{y}_\mathbf{k} \in \Xi_{\mathcal{D}}^{\nu}} \frac{1}{|\Xi_{\mathcal{D}}^{\nu}|} |s_\mathbf{k}^{\nu}| , \tag{26}$$

where the error is balanced by the work measured in the number of new nodes $|\Xi_{\mathcal{D}}^{\nu}|$. An adaptive sparse grid quadrature can be constructed similar to (24) as

$$\mathbb{E}[g] \approx \mathbb{E}[S_{\Lambda_M} g] = \sum_{m=1}^{M} \sum_{\mathbf{y}_\mathbf{k}^{\nu^m} \in \Xi_{\mathcal{D}}^{\nu^m}} s_\mathbf{k}^{\nu^m} w_\mathbf{k}^{\nu^m} , \quad \text{being } w_\mathbf{k}^{\nu^m} = \mathbb{E}[\ell_\mathbf{k}^{\nu^m}] , \tag{27}$$

for which can enrich the index set with the new index

$$\nu^{M+1} := \operatorname*{argmax}_{\nu \in \Lambda_M^a} \frac{1}{|\Xi_{\mathcal{D}}^{\nu}|} \left| \sum_{\mathbf{y}_\mathbf{k}^{\nu} \in \Xi_{\mathcal{D}}^{\nu}} s_\mathbf{k}^{\nu} w_\mathbf{k}^{\nu} \right| . \tag{28}$$

To terminate the aSG algorithm for either interpolation or quadrature, we monitor
the following error estimators compared to some prescribed tolerances, respectively:

$$\mathcal{E}_i := \max_{\nu \in \Lambda_M^a} \max_{\mathbf{y}_\mathbf{k}^\nu \in \Xi_{\mathcal{D}}^\nu} |s_\mathbf{k}^\nu| \quad \text{and} \quad \mathcal{E}_e := \left| \sum_{\nu \in \Lambda_M^a} \sum_{\mathbf{y}_\mathbf{k}^\nu \in \Xi_{\mathcal{D}}^\nu} s_\mathbf{k}^\nu w_\mathbf{k}^\nu \right|. \tag{29}$$

The following convergence results can be obtained for the aSG interpolation and
integration errors based on that for gPC approximation in Theorem 1, see [12, 31].

Theorem 2 *Under Assumption 1, there exists a downward closed set Λ_M such that
the interpolation error*

$$\sup_{\mathbf{y} \in U} \|q(\mathbf{y}) - S_{\Lambda_M} q(\mathbf{y})\|_X \leq C_i M^{-s}, \quad s = \frac{1}{p} - 1 , \tag{30}$$

*where C_i is independent of M. Analogously, there exists a nested sequence of dc sets
Λ_M of cardinality not exceeding M and a constant $C_e > 0$ which is independent
of parameter dimension activated by interpolation in $\mathbb{P}_{\Lambda_M}$ such that the integration
error*

$$\|\mathbb{E}^{\pi_0}[q] - \mathbb{E}^{\pi_0}[S_{\Lambda_M} q]\|_X \leq C_e M^{-s}, \quad s = \frac{1}{p} - 1 , \tag{31}$$

*where C_e is independent of M. The same convergence rate holds also for the aSG
interpolation and integration errors of the posterior density Θ and the QoI Ψ.*

Remark 1 When $p \leq 2/3$ in Assumption 1 **A1**, the aSG integration error can
converge faster the (dimension-independent) rate $M^{-1/2}$, which is the convergence
rate of Monte Carlo integration (in L^2 norm, however). This dimension-independent,
possibly higher convergence rate renders aSG integration preferable.

4   Model Order Reduction

The evaluation of the posterior density Θ, the renormalization constant Z, the
QoI Ψ as well as its statistics, e.g. $\mathbb{E}^{\pi^\delta}[\Psi]$, requires the solutions of the forward
parametric equation (10) at many interpolation or integration nodes $\mathbf{y} \in U$ by
the aSG algorithms. This section is devoted to the development of model order
reduction techniques, in particular the *reduced basis method* (see e.g. [2, 4, 30, 35]
for a general introduction and the convergence analysis) combined (for nonaffine-
parametric problems) with the empirical interpolation method (denoted by EIM
for short. We refer to [1, 7, 19, 25] for the basic ideas and error analysis of the
EIM), to effectively reduce the computational cost for the forward solutions. We
extend these ideas and of the numerical evaluation of the Bayesian posterior density

and related QoIs. A crucial aspect of this work is the extension of the RB and of EIM techniques to high-dimensional parameter spaces such as U, which has been developed in recent years in [5, 8, 21].

4.1 High-Fidelity Petrov-Galerkin Approximation

For the solution of the forward parametric problem (10) at any given parameter $\mathbf{y}$, we introduce the finite-dimensional trial space $\mathcal{X}_h$ and test space $\mathcal{Y}_h$, with $\dim(\mathcal{X}_h) = \dim(\mathcal{Y}_h) = \mathcal{N}$, $\mathcal{N} \in \mathbb{N}$. Here h denotes a discretization parameter, such as the mesh width of finite element discretization or the reciprocal of polynomial degree for spectral discretization. The Petrov–Galerkin (PG), high-fidelity (HiFi for short) approximation of problem (10) reads: for any $\mathbf{y} \in U$, find $q_h(\mathbf{y}) \in \mathcal{X}_h$ such that

$$a(\mathbf{y}; q_h(\mathbf{y}), v_h) = f(\mathbf{y}; v_h) \quad \forall v_h \in \mathcal{Y}_h \tag{32}$$

We proceed under the hypothesis that Assumption 1 holds also in the finite-dimensional spaces $\mathcal{X}_h$ and $\mathcal{Y}_h$, in particular the inf-sup condition (13) is satisfied with constant $\beta_h > 0$ uniformly w.r. to $\mathbf{y}$. The parametric Bayesian posterior density $\Theta(\mathbf{y})$ in (5) can then be approximated by

$$\Theta_h(\mathbf{y}) = \exp\left(-\frac{1}{2}(\delta - O_h(q_h(\mathbf{y})))^\top \Gamma^{-1}(\delta - O_h(q_h(\mathbf{y})))\right), \tag{33}$$

where O_h represents the finite-dimensional approximation of the observation functional O. Similarly, the QoI Ψ can be approximated by the corresponding quantity Ψ_h. Under Assumption 1 in $\mathcal{X}_h$ and $\mathcal{Y}_h$, the well-posedness and gPC as well as the aSG approximation properties in Theorem 1 and Theorem 2 hold with the same convergence rates.

4.2 Reduced Basis Approximation

In order to compute an approximation subject to a prescribed error tolerances for the quantities q, Θ and Ψ, the dimension $\mathcal{N}$ of the finite-dimensional spaces used in the PG approximation problem (32) is, typically, large. Thus, the numerical solution of the HiFi problem (32) is generally expensive, rendering the aSG approximation that requires one solution at each of many interpolation/integration nodes computationally unfeasible in many cases. To reduce it, we propose a model order reduction technique based on reduced basis (RB) approximations constructed by a greedy algorithm with goal-oriented a-posteriori error estimation and Offline-Online decomposition.

4.2.1 Reduced Basis Construction

Analogous to the HiFi -PG approximation, we look for a RB trial space $X_N \subset X_h$ and a RB test space $\mathcal{Y}_N \subset \mathcal{Y}_h$ with $\dim(X_N) = \dim(\mathcal{Y}_N) = N$, $N \in \mathbb{N}$, $N << \mathcal{N}$.

Then we approximate the forward solution map by solving a PG-RB problem: for any $\mathbf{y} \in U$, find $q_N(\mathbf{y}) \in X_N$ such that

$$a(\mathbf{y}; q_N(\mathbf{y}), v_N) = f(\mathbf{y}; v_N) \quad \forall v_N \in \mathcal{Y}_N . \tag{34}$$

For accurate and efficient approximation of the solution manifold $\mathcal{M}_h = \{q_h(\mathbf{y}), \mathbf{y} \in U\}$, RB takes the HiFi solutions $q_h(\mathbf{y})$ at N carefully chosen parameter values $\mathbf{y} = \mathbf{y}^n$, $1 \le n \le N$, called *snapshots*, as the basis functions of the trial space, i.e.

$$X_N = \text{span}\{q_h(\mathbf{y}^n), 1 \le n \le N\} . \tag{35}$$

In order to select "most representative snapshots" for the approximation of the posterior density Θ (or the QoI Ψ, which can be approximated in the same way),

$$\Theta_N(\mathbf{y}) = \exp\left(-\frac{1}{2}(\delta - O_h(q_N(\mathbf{y})))^\top \Gamma^{-1}(\delta - O_h(q_N(\mathbf{y})))\right) , \tag{36}$$

which is nonlinear with respect to the solution q_N, we propose a greedy algorithm based on a goal-oriented a-posteriori error estimator $\triangle_N^\Theta(\mathbf{y})$ for the RB approximation error of the posterior density, $|\Theta_h(\mathbf{y}) - \Theta_N(\mathbf{y})|$ for any $\mathbf{y} \in U$. We start with the first parameter value $\mathbf{y}^1$, e.g. the center of U or a random sample, and construct the initial RB trial space as $X_1 = \text{span}\{q_h(\mathbf{y}^1)\}$. Then, for $N = 1, 2, \ldots$, we pick the next parameter value by

$$\mathbf{y}^{N+1} := \underset{\mathbf{y} \in U}{\text{argmax}} \, \triangle_N^\Theta(\mathbf{y}) , \tag{37}$$

and enrich the RB space as $X_{N+1} = X_N \oplus \text{span}\{q_h(\mathbf{y}^{N+1})\}$. In practice, instead of solving a high-dimensional optimization problem (37), we can replace the parameter domain U by a suitable training set Ξ_{train}, e.g. the sparse grid nodes. The basis functions for the test space $\mathcal{Y}_N$ are chosen such that the PG-RB approximation is stable. In the case that the bilinear form $a(\mathbf{y}; \cdot, \cdot)$ is coercive in $X_h \times \mathcal{Y}_h$ for $\mathcal{Y}_h = X_h$, the choice $\mathcal{Y}_N = X_N$ satisfies the stability condition for the PG-RB approximation. For noncoercive problems, we construct the RB test space through the *supremizer operator* $T^\mathbf{y} : X_h \to \mathcal{Y}_h$ defined as

$$(T^\mathbf{y} w_h, v_h)_\mathcal{Y} = a(\mathbf{y}; w_h, v_h) \quad \forall v_h \in \mathcal{Y}_h . \tag{38}$$

Then $T^y w_h \in \mathcal{Y}_h$ is the supremizer for the element $w_h \in \mathcal{X}_h$ with respect to the functional $a(\mathbf{y}; w_h, \cdot) : \mathcal{Y}_h \to \mathbb{R}$, i.e.

$$T^y w_h = \operatorname*{argsup}_{v_h \in \mathcal{Y}_h} \frac{a(\mathbf{y}; w_h, v_h)}{||w_h||_{\mathcal{X}} ||v_h||_{\mathcal{Y}}} . \tag{39}$$

For any $\mathbf{y} \in U$, the $\mathbf{y}$-dependent RB test space $\mathcal{Y}_N^{\mathbf{y}}$ is defined as

$$\mathcal{Y}_N^{\mathbf{y}} := \operatorname{span}\{T^y q_h(\mathbf{y}^n), 1 \le n \le N\} . \tag{40}$$

It can be shown (see [9]) that

$$\beta_N(\mathbf{y}) := \inf_{0 \neq w_N \in \mathcal{X}_N} \sup_{v_N \in \mathcal{Y}_N^{\mathbf{y}}} \frac{|a(\mathbf{y}; w_N, v_N)|}{||w_N||_{\mathcal{X}} ||v_N||_{\mathcal{Y}}} \ge \beta_h > 0 , \tag{41}$$

ie., the PG-RB approximation problem (34) is uniformly well-posed w.r.to $\mathbf{y} \in U$.

4.2.2 A-Posteriori Error Estimator

The goal-oriented a-posteriori error estimator $\triangle_N^\Theta$ plays a key role in constructing the RB spaces, which should be reliable and efficient, i.e. there exist two constants $0 < c_\triangle \le C_\triangle < \infty$ such that

$$c_\triangle |\Theta_h(\mathbf{y}) - \Theta_N(\mathbf{y})| \le \triangle_N^\Theta(\mathbf{y}) \le C_\triangle |\Theta_h(\mathbf{y}) - \Theta_N(\mathbf{y})| . \tag{42}$$

As we can view the function $\Theta_h : U \to \mathbb{R}$ as a functional $\Theta_h(\cdot) : \mathcal{X}_h \to \mathbb{R}$ through $\Theta(\mathbf{y}) = \Theta_h(q_h(\mathbf{y}))$, following the derivation in [9], smooth dependence of the posterior on the parameters in the forward map implies a formal Taylor expansion of $\Theta_h(q_h(\mathbf{y}))$ about $q_N(\mathbf{y})$:

$$\Theta_h(q_h(\mathbf{y})) = \Theta_h(q_N(\mathbf{y})) + \frac{\partial \Theta_h}{\partial q_h}\Big|_{q_N(\mathbf{y})} (q_h(\mathbf{y}) - q_N(\mathbf{y})) + O(||q_h(\mathbf{y}) - q_N(\mathbf{y})||_{\mathcal{X}}^2) , \tag{43}$$

where the second term of the right hand side is the Fréchet derivative of Θ_h at $q_N(\mathbf{y})$ with respect to q_h, evaluated at the error $e_N^h(\mathbf{y}) = q_h(\mathbf{y}) - q_N(\mathbf{y})$. As the first term $\Theta_h(q_N(\mathbf{y})) = \Theta_N(\mathbf{y})$, as long as the last term is dominated by the second term, we can define the error estimator for $|\Theta_h(\mathbf{y}) - \Theta_N(\mathbf{y})|$ as the second term in (43), i.e.

$$\triangle_{N,h}^\Theta(\mathbf{y}) := \frac{\partial \Theta_h}{\partial q_h}\Big|_{q_N(\mathbf{y})} (e_N^h(\mathbf{y})) . \tag{44}$$

In order to evaluate $\triangle_{N,h}^{\Theta}(\mathbf{y})$ more efficiently, we propose a dual HiFi PG approximation [9, 26]: for any $\mathbf{y} \in U$, find the dual solution $\varphi_h(\mathbf{y}) \in \mathcal{Y}_h$ such that

$$a(\mathbf{y}; w_h, \varphi_h(\mathbf{y})) = \left. \frac{\partial \Theta_h}{\partial q_h} \right|_{q_N(\mathbf{y})} (w_h) \quad \forall w_h \in X_h . \tag{45}$$

Then, with the definition of the residual for the primal HiFi problem (32) evaluated at the RB solution of (34), i.e.

$$r(\mathbf{y}; v_h) := f(\mathbf{y}; v_h) - a(\mathbf{y}; q_N(\mathbf{y}), v_n) \quad \forall v_h \in \mathcal{Y}_h , \tag{46}$$

we obtain, as the primal HiFi equation (32) holds for $\varphi_h \in \mathcal{Y}_h$,

$$r(\mathbf{y}; \varphi_h(\mathbf{y})) = f(\mathbf{y}; \varphi_h(\mathbf{y})) - a(\mathbf{y}; q_N(\mathbf{y}), \varphi_h(\mathbf{y})) = a(\mathbf{y}; e_N^h(\mathbf{y}), \varphi_h(\mathbf{y})) , \tag{47}$$

which, together with definition (44) and (45), imply

$$\triangle_{N,h}^{\Theta}(\mathbf{y}) = r(\mathbf{y}; \varphi_h(\mathbf{y})) . \tag{48}$$

As it is computationally expensive to obtain the solution $\varphi_h(\mathbf{y})$, we propose to use RB approximation for the HiFi -PG approximation of the dual problem (45) following the same development as for the primal HiFi problem in the last section. With the dual RB solution $\varphi_N(\mathbf{y})$ (where number N of degrees of freedom of the dual problem could be different from N which was used in the RB-PG approximation of the primal problem), we define the a-posteriori error estimator for the error $|\Theta_h(\mathbf{y}) - \Theta_N(\mathbf{y})|$ as

$$\triangle_N^{\Theta}(\mathbf{y}) = r(\mathbf{y}; \varphi_N(\mathbf{y})) , \tag{49}$$

whose difference from $\triangle_h^{\Theta}(\mathbf{y})$ can be bounded by

$$|\triangle_h^{\Theta}(\mathbf{y}) - \triangle_N^{\Theta}(\mathbf{y})| = r(\mathbf{y}; \varepsilon_N^h(\mathbf{y})) = a(\mathbf{y}; e_N^h(\mathbf{y}), \varepsilon_N^h(\mathbf{y})) \le \gamma \|e_N^h(\mathbf{y})\|_X \|\varepsilon_N^h(\mathbf{y})\|_Y , \tag{50}$$

where $\varepsilon_N^h(\mathbf{y}) = \varphi_h(\mathbf{y}) - \varphi_N(\mathbf{y})$ and γ represents the continuity constant of the bilinear form a. In general, the primal and dual RB errors $e_N^h(\mathbf{y})$ and $\varepsilon_N^h(\mathbf{y})$ tend to zero so that, asymptotically, (50) and the second order term in (43) are both dominated by the first order term of (43), we can expect to obtain a reliable and efficient, computable a-posteriori error estimator $\triangle_N^{\Theta}(\mathbf{y})$ for the error $|\Theta_h(\mathbf{y}) - \Theta_N(\mathbf{y})|$, with the corresponding constants $c_\triangle$ and $C_\triangle$ in (42) close to one uniformly w.r. to $\mathbf{y}$.

4.2.3 Offline-Online Computation

To this end, we make a crucial assumption that the HiFi PG discretization of the parametric problem, (32) is affine, i.e. $\forall w_h \in X_h, v_h \in \mathcal{Y}_h$, the bilinear and linear forms can be written as

$$a(\mathbf{y}; w_h, v_h) = \sum_{m=1}^{M_a} \lambda_m^a(\mathbf{y}) a_m(w_h, v_h) \quad \text{and} \quad f(\mathbf{y}; v_h) = \sum_{m=1}^{M_f} \lambda_m^f(\mathbf{y}) f_m(v_h) . \tag{51}$$

For instance, for a diffusion problem with affine-parametric diffusion coefficient (2), we have $\lambda_m^a(\mathbf{y}) = y_m$ and $a_m(w_h, v_h) = (\phi_m \nabla w_h, \nabla v_h)$, $1 \le m \le M_a = J$. We defer the discussion of linearization in parameter space, ie., the approximation of the non-affine parametric problem by an affine parametric model in (51).

For the sake of algebraic stability of the PG-RB approximation (34), we compute the orthonormal bases $(w_N^n)_{n=1}^N$ of X_N obtained by Gram–Schmidt orthonormalization algorithm for the bases $(q_h(\mathbf{y}^n))_{n=1}^N$. Then the RB solution of problem (34) at any $\mathbf{y} \in U$ can be represented by

$$q_N(\mathbf{y}) = \sum_{n=1}^N q_N^n(\mathbf{y}) w_N^n , \tag{52}$$

where $\mathbf{q}_N(\mathbf{y}) = (\mathbf{q}_N^1(\mathbf{y}), \dots, \mathbf{q}_N^N(\mathbf{y}))^\top \in \mathbb{R}^N$, denoting the coefficient of $q_N(\mathbf{y})$. In the coercive case where $\mathcal{Y}_N = X_N$ with basis $v_N^n = w_N^n$, $1 \le n \le N$, the algebraic system of the PG-RB problem (34) becomes

$$\left(\sum_{m=1}^{M_a} \lambda_m^a(\mathbf{y}) A_m \right) \mathbf{q}_N(\mathbf{y}) = \sum_{m=1}^{M_f} \lambda_m^f(\mathbf{y}) \mathbf{f}_m , \tag{53}$$

where the RB matrix A_m, $1 \le m \le M_a$, and the RB vector $\mathbf{f}_m$, $1 \le m \le M_f$, are given respectively by

$$(A_m)_{n',n} = a_m(w_N^n, v_N^{n'}) \quad \text{and} \quad (\mathbf{f}_m)_n = f_m(v_N^n) \quad n, n' = 1, \dots, N , \tag{54}$$

which do not depend on the parameter $\mathbf{y} \in U$ and can therefore be assembled and stored once and for all in the Offline stage. Given any $\mathbf{y} \in U$, the algebraic system (53) can be assembled and solved Online with $O(M_a N^2 + M_f N)$ and $O(N^3)$ operations, respectively, which do not depend on the number $\mathcal{N}$ of high-fidelity degrees of freedom. In the noncoercive case, for any $\mathbf{y} \in U$, the test basis v_N^n, $1 \le n \le N$, is given by

$$v_N^n = T^{\mathbf{y}} w_N^n = \sum_{m=1}^{M_a} \lambda_m^a(\mathbf{y}) T_m w_N^n , \tag{55}$$

where $T_m w_N^n$, $1 \le m \le M_a$, $1 \le n \le N$, is the solution of

$$(T_m w_N^n, v_h)_{\mathcal{Y}} = a_m(w_N^n, v_h) \quad \forall v_h \in \mathcal{Y}_h , \tag{56}$$

which does not depend on $\mathbf{y} \in U$ and which can be computed and stored once and for all during the Offline stage. The corresponding algebraic system of the PG-RB problem (34) is given by

$$\left(\sum_{m}^{M_a} \sum_{m'}^{M_a} \lambda_m^a(\mathbf{y}) \lambda_{m'}^a(\mathbf{y}) A_{m,m'} \right) \mathbf{q}_N(\mathbf{y}) = \sum_{m=1}^{M_f} \lambda_m^f(\mathbf{y}) \mathbf{f}_m , \tag{57}$$

where the (densely populated) RB matrix $A_{m,m'}$, $1 \le m, m' \le M_a$, is given by

$$(A_{m,m'})_{n',n} = a_m(w_N^n, T_{m'} w_N^{n'}) \quad 1 \le n, n' \le N . \tag{58}$$

This matrix does not depend on $\mathbf{y}$ and can be computed and stored once and for all during the Offline stage. Given any $\mathbf{y} \in U$, the algebraic system (57) can be assembled and solved Online in $O(M_a^2 N^2 + M_f N)$ and $O(N^3)$ operations.

The dual RB solution $\varphi_N(\mathbf{y})$ can be computed by the same Offline-Online procedure. The a-posteriori error estimator (49) takes the explicit form

$$\Delta_N^\Theta(\mathbf{y}) = \sum_{m=1}^{M_f} \sum_{n=1}^{N} \lambda_m^f(\mathbf{y}) f_m(v_N^{n,du}) \varphi_N^n(\mathbf{y})$$

$$\tag{59}$$

$$- \sum_{m=1}^{M_a} \sum_{n=1}^{N} \sum_{n'=1}^{N} \lambda_m^a(\mathbf{y}) q_N^n(\mathbf{y}) a_m(w_N^n, v_N^{n,du}) \varphi_N^{n'}(\mathbf{y}) ,$$

where $\varphi_N^{n'}(\mathbf{y})$ is the coefficient of the dual RB solution $\varphi_N(\mathbf{y})$ on the trial RB basis $v_N^{n',du} \in \mathcal{Y}_N^{du}$, $1 \le n' \le N$. As $f_m(v_N^{n,du})$, $1 \le m \le M_f$ and $a_m(w_N^n, v_N^{n,du})$, $1 \le m \le M_a$, $1 \le n, n' \le N$, are independent of $\mathbf{y}$, they can be computed and stored once during the Offline stage and the error estimator (59) can be assembled during the Online stage for any given $\mathbf{y} \in U$ with $O(M_f N + M_a N^2)$ operations.

Finally, the RB posterior density $\Theta_N(\mathbf{y})$ can be computed by

$$\Theta_N(\mathbf{y}) = \exp\left(-\frac{1}{2} \left(\delta - O_K^N \mathbf{q}_N(\mathbf{y}) \right)^\top \Gamma^{-1} \left(\delta - O_K^N \mathbf{q}_N(\mathbf{y}) \right) \right) , \tag{60}$$

where the observation matrix $O_K^N \in \mathbb{R}^{K \times N}$ with elements $(O_K^N)_{k,n} = o_k(w_N^n)$, $1 \le k \le K$, $1 \le n \le N$, is computed and stored for once during Offline stage and $\Theta_N(\mathbf{y})$ is assembled for any $\mathbf{y} \in U$ during the Online stage in $O(NK^2)$ operations.

As the error estimator $\Delta_N^\Theta(\mathbf{y})$ is an approximation of the second term in the Taylor expansion (43) for $\Theta_h(\mathbf{y})$, we correct the RB approximation $\Theta_N(\mathbf{y})$ by

$$\Theta_N^\Delta(\mathbf{y}) = \Theta_N(\mathbf{y}) + \Delta_N^\Theta(\mathbf{y}) , \tag{61}$$

which is generally more accurate than $\Theta_N(\mathbf{y})$.

Theorem 3 ([9]) *Under Assumption 1, the RB error for the posterior density satisfies*

$$\sup_{y \in U} |\Theta_h(\mathbf{y}) - \Theta_N^\Delta(\mathbf{y})| \leq C_\Theta^\Delta N^{-2s}, \quad s = \frac{1}{p} - 1, \tag{62}$$

where the constant C_Θ^Δ is independent of the number of RB bases N and the active dimension J. The same convergence rate holds for RB approximation of the QoI Ψ.

4.3 Empirical Interpolation Method (EIM)

As the computational reduction due to the $\mathcal{N}$-independent Online RB evaluation crucially depends on the assumption (51), which is however not necessarily valid in practice: we mention only diffusion problems with lognormal diffusion coefficient given by $\kappa = e^u$. We outline the Empirical Interpolation Method (EIM) for affine-parametric approximation of problems with nonaffine parameter dependence. More precisely, suppose $\mathcal{X}_h$ is defined in the domain $D \subset \mathbb{R}^d$, $d \in \mathbb{N}$, with the finite set of discretization nodes $D_h \in D$, we seek to approximate an arbitrary, non-affine function $g : D_h \times U \to \mathbb{R}$ in the bilinear and linear forms by

$$g(x, \mathbf{y}) \approx \mathcal{J}_M[g](x, \mathbf{y}) = \sum_{m=1}^{M} \lambda_m(\mathbf{y}) g_m(x), \tag{63}$$

which results in an approximation of the problem (32) with affine representation (51). For instance, when g is the diffusion coefficient of a diffusion problem, we obtain (51) with $\lambda_m^a(\mathbf{y}) = \lambda_m(\mathbf{y})$ and $a_m(w_h, v_h) = (g_m \nabla w_h, \nabla v_h)$, $1 \leq m \leq M_a = M$.

One choice for the approximation (63) is by the aSG interpolation based on some structured interpolation nodes, e.g. Leja nodes or Clenshaw-Curtis nodes, presented in Sect. 3. As the work for each Online RB evaluation is proportional to the number M of affine terms, it is important to keep M as small as possible. To this end, we propose an adaptive construction of a sparse interpolation set by the following greedy algorithm. We start by searching for the first parameter value $\mathbf{y}^1 \in U$ and the first discretization node $x^1 \in D_h$ such that

$$\mathbf{y}^1 = \operatorname*{argsup}_{\mathbf{y} \in U} \max_{x \in D_h} |g(x, \mathbf{y})| \text{ and } x^1 = \operatorname*{argmax}_{x \in D_h} |g(x, \mathbf{y}^1)|. \tag{64}$$

The first basis g_1 is taken as $g_1(x) = g(x, \mathbf{y}^1)/g(x^1, \mathbf{y}^1)$, $x \in D_h$. We define the EIM node set $S_1 = \{x^1\}$. For $M = 1, 2, \ldots$, for any $\mathbf{y} \in U$, the coefficient $\lambda_m(\mathbf{y})$, $1 \leq m \leq M$, of the interpolation (63) is obtained by Lagrange interpolation at the selected discretization nodes, i.e.

$$\sum_{m=1}^{M} \lambda_m(\mathbf{y}) g_m(x) = g(x, \mathbf{y}) \quad \forall x \in S_M . \tag{65}$$

Then we define the *empirical interpolation residual* as

$$r_{M+1}(x, \mathbf{y}) = g(x, \mathbf{y}) - \sum_{m=1}^{M} \lambda_m(\mathbf{y}) g_m(x) . \tag{66}$$

The next parameter sample $\mathbf{y}^{M+1}$ and discretization node x^{M+1} are chosen as

$$\mathbf{y}^{M+1} = \operatorname*{argsup}_{\mathbf{y} \in U} \max_{x \in D_h} |r_{M+1}(\mathbf{y})| \quad \text{and} \quad x^{M+1} = \operatorname*{argmax}_{x \in D_h} |r_{M+1}(x, \mathbf{y}^{M+1})| . \tag{67}$$

We define $\mathcal{E}_{EIM}(\mathbf{y}^{M+1}) := |r_{M+1}(x^{M+1}, \mathbf{y}^{M+1})|$ and $S_{M+1} := S_M \cup \{x^{M+1}\}$ and choose the next basis function g_{M+1} according to

$$g_{M+1}(x) = \frac{r_{M+1}(x, \mathbf{y}^{M+1})}{r_{M+1}(x^{M+1}, \mathbf{y}^{M+1})} \quad x \in D_h . \tag{68}$$

We remark that in practice the parameter domain U is replaced with a finite training set Ξ_{train} to avoid solving a continuous, high-dimensional maximization problem (67). Details and error bounds is available in, e.g. [7, 10, 25].

4.4 Adaptive aSG-EIM-RB Algorithm

In this section, we propose an adaptive algorithm for the evaluations of the posterior density Θ as well as its expectation Z for Bayesian inversion with nonaffine forward map by incorporation of approximations of aSG, EIM and RB in order to reduce the total computational cost. The same algorithm applies for the evaluation of the QoI Ψ and its statistical moments as well. The basic idea is that at each step of the construction of aSG with new interpolation or integration nodes, we refine the EIM approximation of the nonaffine parametric function and refine the RB approximation of Θ when their approximation errors are larger than prescribed tolerances at the new nodes. In the end, instead of solving a large number of HiFi problems for the evaluation of $\Theta_h(\mathbf{y})$ at all aSG nodes, we approximate $\Theta_h(\mathbf{y})$ by the $\Theta_N(\mathbf{y})$ resp. by $\Theta_N^{\Delta}(\mathbf{y})$, which only requires inexpensive RB solutions. The main procedure of simultaneous aSG-EIM-RB construction and evaluation is provided in Algorithm 1.

In the adaptive refinement of EIM interpolation, we may replace the set of discretization nodes D_h in (67), which depends on the HiFi degree of freedom N, by (i) a smaller number of randomly selected discretization nodes in $D_h \setminus S_M$; or

---

Algorithm 1: Adaptive aSG-EIM-RB Algorithm

---

1. Specify the tolerances ε_{aSG}, ε_{EIM} and ε_{RB} and the maximum numbers of nodes M_{aSG}^{max}, M_{EIM}^{max} for aSG, EIM and bases N_{max} for RB approximations, respectively, set $\mathcal{E}_{aSG} = 2\varepsilon_{aSG}$;
2. Initialize the aSG, EIM and RB approximation with $M_{aSG} = M_{EIM} = N = 1$:

   a. solve the primal and dual HiFi problems (32) and (45) at the root node $\mathbf{y}^1 = \mathbf{0} \in U$;
   b. initialize the index set $\Lambda_1 = \{1\}$, and construct the aSG approximation, either the interpolation as $\mathcal{S}_{\Lambda_1}\Theta_h(\mathbf{y}) = \Theta_h(\mathbf{y}^1)$ or the integration as $\mathbb{E}[\mathcal{S}_{\Lambda_1}\Theta_h] = \Theta_h(\mathbf{y}^1)$;
   c. set the first EIM basis as $\mathcal{J}_1[g](\mathbf{y}) = g(\mathbf{y}^1)$, set $x^1 \in \text{argmax}_{x \in D_h} |g(x, \mathbf{y}^1)|$;
   d. construct the first RB primal trial space $X_1 = \text{span}\{q_h(\mathbf{y}^1)\}$ and dual trial space $\mathcal{Y}_1^{du} = \text{span}\{\varphi_h(\mathbf{y}^1)\}$, compute and store all quantities in Offline stage.

3. **While** $M_{aSG} < M_{aSG}^{max}$ and $\mathcal{E}_{aSG} > \varepsilon_{aSG}$

   a. compute the active index set $\Lambda_{M_{aSG}}^a$ for the aSG approximation;
   b. **For each** $\boldsymbol{\nu} \in \Lambda_{M_{aSG}}^a$

      i. compute the set of added nodes $\Xi_{\mathcal{D}}^{\nu}$ associated to $\boldsymbol{\nu}$;
      ii. **For each** $\mathbf{y} \in \Xi_{\mathcal{D}}^{\nu}$

         A. compute EIM interpolation of g at $\mathbf{y}$ and the interpolation error $\mathcal{E}_{EIM}(\mathbf{y})$;
         B. **If** $M_{EIM} < M_{EIM}^{max}$ and $\mathcal{E}_{EIM}(\mathbf{y}) > \varepsilon_{EIM}$

            • refine the EIM interpolation with the new basis $g(\mathbf{y})$, select $x^{M_{EIM}+1}$;
            • set $M_{EIM} = M_{EIM} + 1$;

            EndIf

         C. compute the RB solution and $\Theta_N^{\triangle}(\mathbf{y})$ and the error estimator $\mathcal{E}_{RB}(\mathbf{y}) = \Delta_N^{\Theta}(\mathbf{y})$;
         D. **If** $N < N_{max}$ and $\mathcal{E}_{RB}(\mathbf{y}) > \varepsilon_{RB}$

            • enrich the RB trial spaces X_N with $q_h(\mathbf{y})$ and $\mathcal{Y}_N^{du}$ with $\varphi_h(\mathbf{y})$;
            • compute and save the all Offline quantities;
            • set $N = N + 1$;

            EndIf

      EndFor

   EndFor

   c. compute the aSG error estimator $\mathcal{E}_{aSG}$ as one of (29) with the RB approximation $\Theta_N^{\triangle}$;
   d. enrich $\Lambda_{M_{aSG}}$ by $\boldsymbol{\nu}^{M_{aSG}+1}$ according to (26) for interpolation or (28) for integration;
   e. set $M_{aSG} = M_{aSG} + 1$;

   EndWhile

(ii) the last s (e.g. $s = 1, 2, \ldots$) selected nodes $\{x^M, x^{M-1}, x^{M-s+1}\}$ and use the first $M - s$ EIM bases to evaluate the error estimator $\mathcal{E}_{EIM}(\mathbf{y})$.

Remark 2 In practice, the specification of the tolerances ε_{aSG}, ε_{EIM} and ε_{RB} depends on the given problem and the user specified accuracy. A general (heuristic) guideline for their selection is that the EIM error be smaller than or equal to the RB error, which should be in turn smaller than or equal to the aSG error. Therefore, we propose to set $\varepsilon_{aSG} > \varepsilon_{RB} > \varepsilon_{EIM}$ (as was done in all numerical experiments reported below). As the number of aSG nodes is considerably larger than the number of

EIM and RB bases, we set $M_{aSG}^{max} > M_{EIM}^{max}$ and $M_{aSG}^{max} > N_{max}$. Concrete examples for the specification of these parameters will be given in the ensuing numerical experiments.

5 Numerical Experiments

We consider a linear diffusion problem in the physical domain $D = (0, 1)^2$: for $\mathbf{y} \in U$, find the solution $q(\mathbf{y}) \in H_0^1(D)$ such that

$$- div(\kappa(\mathbf{y}) \nabla q(\mathbf{y})) = f, \tag{69}$$

where we set $f = 1$ and prescribe homogeneous Dirichlet boundary conditions $q(\mathbf{y}) = 0$ on ∂D; the parametric diffusion coefficient $\kappa(\mathbf{y})$ in (69) is given by

$$\kappa(\mathbf{y}) = e^{u(\mathbf{y})} \text{ with } u(\mathbf{y}) = 1 + \sum_{j=1}^{J} y_j \frac{1}{j^3} \sin((j_1 + 1)\pi x_1) \sin((j_2 + 1)\pi x_2), \tag{70}$$

where $j_1, j_2 = 1, \ldots, \sqrt{J}$ such that $j = j_1 + j_2 \sqrt{J}$ for a square J; $x = (x_1, x_2) \in D$. Note that $u(\mathbf{y})$ is nonaffine with respect to $\mathbf{y}$. We perform interpolation for an affine decomposition of $\kappa(\mathbf{y})$ by applying both aSG and EIM. We first investigate the convergence of the aSG interpolation error with respect to the number of dimensions J. For simplicity, we only consider the interpolation for the function $\kappa(\mathbf{y})$ at a sample node $x = (0.3, 0.6)$ (interpolation at any other node (or set of nodes) or the worst case scenario measured in $L^\infty(D)$-norm can be performed in the same way, but with much more computational cost for the latter case). We test the cases of $J = 16, 64, 256$, and 1024, and construct the aSG (by Algorithm 1 where we replace EIM-RB construction and evaluation, i.e. procedure ii. in Algorithm 1, by a HiFi solution and density estimation at each $\mathbf{y} \in \Xi_D^\nu$) using Clenshaw–Curtis nodes defined in (21) with the maximum number of interpolation nodes set to 10^5. Figure 1 displays the convergence of the interpolation error estimator defined in (29) with respect to the number of interpolation nodes. We observe that the convergence rate converges to the one close to M^{-2} when the number active parameters increases from 16 to 1024, which is in agreement with the theoretical prediction of the error convergence rate in Theorem 2 for high-(infinite-)dimensional sparse interpolation.

In the numerical convergence study of the empirical interpolation error, we consider the $J = 64$ dimensional case for uniform, triangular meshes with mesh widths $h = 1/16, 1/32, 1/64$, and $1/128$. The tolerance is chosen as 10^{-8} and the same 1000 random samples as the training samples for the construction of EIM are selected. $M = 161, 179, 179$, and 179 EIM bases are constructed for $h = 1/16, 1/32, 1/64$ and $1/128$, respectively. This shows that at a given level of accuracy, the number of EIM bases is independent of HiFi mesh width, provided it is sufficiently fine. We use $h = 1/32$, i.e. with Finite Element nodes

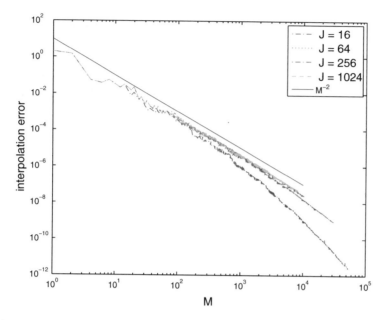

Fig. 1 Decay of the aSG interpolation error with respect to the number of interpolation nodes M for $\kappa(\mathbf{y})$ in $J = 16, 64, 256$ and 1024 parameter dimensions at a sample node $x = (0.3, 0.6)$

$x = (i_1/32, i_2/32), i_1, i_2 = 0, \ldots, 32$, and 1000 random training samples to evaluate the convergence of EIM error with respect to the number of dimensions $J = 16, 64, 256$, and 1024, which is shown in Fig. 2. We observe that as J increases, the convergence rate tends to M^{-2}, as could be expected from the results in the affine-parametric setting in [9]. However, as the number of EIM bases increases beyond the dimension J of the set of active parameters, the convergence for EIM error exceeds the rate M^{-2} and becomes much faster (in fact, exponential) than the aSG error that still converges with a rate close to M^{-2}. This is further demonstrated in the case $J = 64$ in Fig. 3, where the aSG is constructed for interpolation only at the sample node $x = (0.3, 0.6)$ and at the Finite Element nodes $x = (i_1/32, i_2/32)$, $i_1, i_2 = 0, \ldots, 32$, respectively. The EIM bases are constructed with all previously computed aSG nodes (5×10^4) as training samples.

From Fig. 3 we see that the worst aSG interpolation error over all mesh nodes (where a single sparse grid is adaptively constructed using the largest error indicator over all the mesh nodes) decays at a lower rate (with rate about $M^{-1.2}$) than the theoretical prediction M^{-2} in Theorem 2 and that of aSG at only one sample node. This indicates that the aSG constructed to minimize the maximal interpolation error over all mesh nodes can produce approximations which do not converge at the rate afforded by the N-approximation results.

We also see that, in order to achieve the same interpolation accuracy, a much smaller number of EIM bases is needed compared to that of aSG nodes. For example, only 50 EIM bases are needed in order to achieve the same accuracy

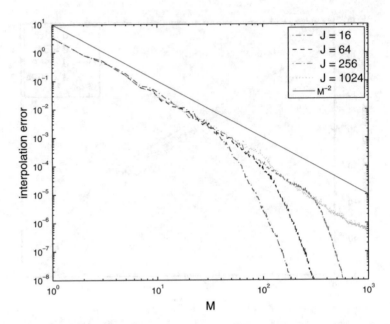

Fig. 2 Decay of the EIM interpolation error with respect to the number of interpolation nodes M for $\kappa(\mathbf{y})$ in $J = 16, 64, 256$ and 1024 dimensions uniformly at all Finite Element nodes

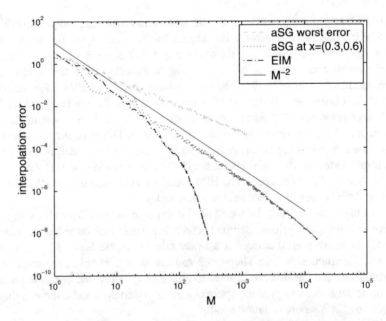

Fig. 3 Decay of interpolation error with respect to the number of interpolation nodes M for $\kappa(\mathbf{y})$ in $J = 64$ dimensions by aSG at the sample node $x = (0.3, 0.6)$ (aSG at x=(0.3,0.6)) and at the Finite Element nodes (aSG worst error), and by EIM at the Finite Element nodes

3.7×10^{-4} as for the worst case scenario aSG that requires 1248 interpolation nodes, while 289 EIM bases are needed to attain the interpolation accuracy 4.5×10^{-9}, for which about 1.3×10^7 interpolation nodes are expected (according to the estimated convergence rate $M^{-1.2}$) for the worst case scenario aSG, even only 15748 nodes are needed for aSG interpolation at a single mesh sample point $x = (0.3, 0.6)$. Therefore, in the affine approximation of the nonaffine function $\kappa(\mathbf{y})$ for this example with $J = 64$ parameters, EIM is much more efficient than aSG. For the higher dimensional case, e.g. for $J = 1024$, the same conclusion can be drawn as the worst aSG interpolation error converges at a lower rate (about $M^{-1.2}$) than EIM, which converges at a rate of about M^{-2} when the number of EIM bases is smaller than J and much faster than M^{-2} when the number of EIM bases becomes larger than J.

To study the convergence of the RB errors and the error estimator as well as its effectivity for the approximation of the posterior density Θ in different dimensions, here in particular $J = 16, 64, 256$ and 1024, we first construct the EIM approximation of the nonaffine random field using 1000 random samples with tolerance 10^{-8} (selected so small that EIM interpolation error is dominated by the RB error). We next construct the RB approximation for the posterior density using the same 1000 samples with tolerance 10^{-8}. Then, the RB approximation errors of the posterior density, defined as $e_N = |\Theta_h(\mathbf{y}_{max}) - \Theta_N(\mathbf{y}_{max})|$, where $\mathbf{y}_{max} = \text{argmax}_{\mathbf{y} \in \Xi_{test}} |\Theta_h(\mathbf{y}) - \Theta_N(\mathbf{y})|$, $e_N^\Delta = |\Theta_h(\mathbf{y}_{max}) - \Theta_N^\Delta(\mathbf{y}_{max})|$, and the RB error estimator $\triangle_N^\Theta(\mathbf{y}_{max})$ defined in (49), are computed in a test set Ξ_{test} with 100 random samples that are independent of the 1000 training samples. Figure 4 displays the convergence of the RB errors and the error estimator with respect to the number of RB bases in different dimensions. We can see that the RB error e_N can hardly be distinguished from the error estimator $\triangle_N^\Delta$, which implies that the error estimator is very effective. As parameter space dimension J increases, the approximation error becomes larger. The corrected density Θ_N^Δ is more accurate than Θ_N in all cases, especially when N is relatively small. In fact, a convergence rate N^{-2} can be observed for e_N compared to N^{-4} for e_N^Δ when N is small. When N and J become larger, both errors converge with *a dimension-independent, asymptotic convergence rate N^{-4}*, which is in complete agreement with Theorem 3. We remark that the convergence rate depends on the sparsity of the problem. For problems whose parametric solution families exhibit less sparsity (e.g. when $2/3 < p < 1$), the convergence rate may be inferior than $N^{-1/2}$. In this case sampling (ie., Monte-Carlo) methods might be preferable.

In the last experiment, we consider the influence of the tolerance for RB training to the accuracy of the RB approximation of the posterior density Θ and its integration Z using the aSG-EIM-RB Algorithm 1, where we set the maximum number of the sparse grid nodes as 10^4 and 2×10^3 for the interpolation of Θ and the integration of Z, respectively, and set the tolerance for RB training as 10^{-4}, 10^{-6} and 10^{-8} for the construction of aSG interpolation, and 10^{-5}, 10^{-7} and 10^{-9} for the construction of aSG integration. Figure 5 shows the convergence rates of the aSG interpolation and integration error estimators defined in (29), which are

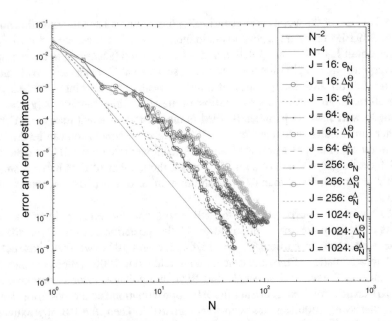

Fig. 4 Decay of the RB approximation errors $e_N(\mathbf{y}_{max})$, $e_N^\Delta(\mathbf{y}_{max})$, and the RB error estimator $\Delta_N^\Theta(\mathbf{y}_{max})$ with respect to the number of RB bases N in $J = 16, 64, 256$ and 1024 dimensions. The training set consists of 1000 random samples for the construction of RB approximation with EIM and RB tolerances set as 10^{-8}. The test set Ξ_{test} consists of another 100 random samples

close to M^{-2} (the same as theoretical prediction in Theorem 2) and M^{-3} (faster than the theoretical prediction M^{-2}). Figure 5 also displays the number of RB and its approximation accuracy with different tolerances. We see that in order to achieve the same approximation accuracy for pointwise evaluation of the posterior density Θ, the number of RB required is considerably smaller than the number of aSG nodes, e.g. 74 RB bases compared to 3476 aSG nodes. This entails the need to solve a smaller number of high-fidelity problems by RB. For the approximation of Z, for which we need a combination of aSG for integration and RB for pointwise evaluation, 96 RB bases are constructed out of 2×10^3 aSG nodes, which preserves the same integration accuracy as aSG with 584 nodes. Note that in this test, we set the tolerance of EIM as 10^{-9} for the interpolation of Θ and as 10^{-10} for the integration of Z, both of which are negligible compared to the accuracy/tolerance of RB and aSG. When the tolerances for the EIM were selected smaller, the number of EIM bases, whose cost of construction depends linearly on $\mathcal{N}$, are relatively large. In order to balance the errors from aSG, EIM and RB to reach a prescribed numerical accuracy at minimum computational cost, an algorithm will be presented in [10].

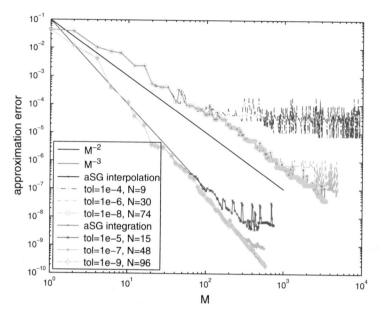

Fig. 5 Decay of aSG interpolation and integration errors with respect to the number of aSG nodes in $J = 64$ dimensions; RB is trained adaptively at the aSG nodes with different tolerances. The lines of aSG interpolation and aSG integration are corresponding to using Algorithm 1 without EIM-RB but only HiFi solution. Note also that the line of aSG interpolation is overlapped with that of tol=1e-8, $N = 74$. The error of aSG integration practically coincides with the the error produced by tol=1e-9, $N = 96$

6 Conclusion

We investigated acceleration of computational Bayesian inversion for PDEs with distributed parameter uncertainty. Upon reformulation, forward models which are given in terms of PDEs with random input data take the form of countably-parametric, deterministic operator equations. Sparsity of the parameter-to-solution maps is exploited computationally by the reduced basis approach. Sparse grids enter the proposed numerical methods in several ways: first, sparse dimension-adaptive quadratures are used to evaluate conditional expectations in Bayesian estimates and second, sparse grids are used in the offline stage of the reduced basis algorithms (in particular, the empirical interpolation method) to "train" the greedy algorithms and to facilitate the greedy searches over the high-dimensional parameter spaces. For a model diffusion problem, we present detailed numerical experiments of the proposed algorithms, indicating their essentially dimension-independent performance and convergence rates which are only limited by the sparsity in the data-to-solution map.

In the present work, we considered only a model problem with uniform Bayesian prior on the parameter space U. The proposed approach is, however, also applicable directly to priors with separable densities w.r. to uniform priors. Generalizations to nonseparable prior densities will be presented elsewhere. The present methods and results can also be extended to nonlinear parametric problems with non-affine parameter dependence, in particular problems with uncertain domain of definition [10].

Acknowledgements This work is supported in part by the European Research Council (ERC) under AdG427277.

References

1. M. Barrault, Y. Maday, N. Nguyen, A. Patera, An empirical interpolation method: application to efficient reduced-basis discretization of partial differential equations. Comptes Rendus Mathematique, Analyse Numérique **339**(9), 667–672 (2004)
2. P. Binev, A. Cohen, W. Dahmen, R. DeVore, G. Petrova, P. Wojtaszczyk, Convergence rates for greedy algorithms in reduced basis methods. SIAM J. Math. Anal. **43**(3), 1457–1472 (2011)
3. P. Chen, A. Quarteroni, Accurate and efficient evaluation of failure probability for partial differential equations with random input data. Comput. Methods Appl. Mech. Eng. **267**(0), 233–260 (2013)
4. P. Chen, A. Quarteroni, Weighted reduced basis method for stochastic optimal control problems with elliptic PDE constraints. SIAM/ASA J. Uncertain. Quantif. **2**(1), 364–396 (2014)
5. P. Chen, A. Quarteroni, A new algorithm for high-dimensional uncertainty quantification based on dimension-adaptive sparse grid approximation and reduced basis methods. J. Comput. Phys. **298**, 176–193 (2015)
6. P. Chen, A. Quarteroni, G. Rozza, Comparison of reduced basis and stochastic collocation methods for elliptic problems. J. Sci. Comput. **59**, 187–216 (2014)
7. P. Chen, A. Quarteroni, G. Rozza, A weighted empirical interpolation method: a priori convergence analysis and applications. ESAIM: Math. Model. Numer. Anal. **48**, 943–953, 7 (2014)
8. P. Chen, A. Quarteroni, G. Rozza, Reduced order methods for uncertainty quantification problems. ETH Zurich, SAM Report 03, Submitted, 2015
9. P. Chen, C. Schwab, Sparse grid, reduced basis Bayesian inversion. Comput. Methods Appl. Mech. Eng. 297, 84–115 (2015)
10. P. Chen, C. Schwab, Sparse grid, reduced basis Bayesian inversion: nonaffine-parametric nonlinear equations. ETH Zurich, SAM Report 21, Submitted, 2015
11. A. Chkifa, A. Cohen, R. DeVore, C. Schwab, Adaptive algorithms for sparse polynomial approximation of parametric and stochastic elliptic PDEs. M2AN Math. Mod. Num. Anal. **47**(1), 253–280 (2013)
12. A. Chkifa, A. Cohen, C. Schwab, Breaking the curse of dimensionality in sparse polynomial approximation of parametric pdes. Journal de Mathématiques Pures et Appliquées. **103**(2), 400–428 (2014)
13. A. Cohen, R. DeVore, Kolmogorov widths under holomorphic mappings (2014). arXiv:1502.06795
14. A. Cohen, R. DeVore, C. Schwab, Analytic regularity and polynomial approximation of parametric and stochastic elliptic PDE's. Anal. Appl. **9**(01), 11–47 (2011)
15. T. Cui, Y. Marzouk, K. Willcox, Data-driven model reduction for the bayesian solution of inverse problems. Int. J. Numer. Methods Eng. **102**(5), 966–990 (2015)

16. M. Dashti, A. Stuart, The Bayesian approach to inverse problems (2016). arXiv:1302.6989, to appear in Springer Handbook of Uncertainty Quantification, Editor: Ghanem et al.
17. D. Galbally, K. Fidkowski, K. Willcox, O. Ghattas, Non-linear model reduction for uncertainty quantification in large-scale inverse problems. Int. J. Numer. Methods Eng. **81**(12), 1581–1608 (2010)
18. T. Gerstner, M. Griebel, Dimension–adaptive tensor–product quadrature. Computing **71**(1), 65–87 (2003)
19. M. Grepl, Y. Maday, N. Nguyen, A. Patera, Efficient reduced-basis treatment of nonaffine and nonlinear partial differential equations. ESAIM: Math. Model. Numer. Anal. **41**(03), 575–605 (2007)
20. M. Hansen, C. Schwab, Sparse adaptive approximation of high dimensional parametric initial value problems. Vietnam J. Math. **41**(2), 181–215 (2013)
21. J. S. Hesthaven, B. Stamm, S. Zhang, Efficient greedy algorithms for high-dimensional parameter spaces with applications to empirical interpolation and reduced basis methods. ESAIM Math. Model. Numer. Anal. **48**(1), 259–283 (2014)
22. V. Hoang, C. Schwab, Analytic regularity and polynomial approximation of stochastic, parametric elliptic multiscale pdes. Anal. Appl. (Singap.) **11**(1), 1350001 (2013)
23. V. Hoang, C. Schwab, Sparse tensor galerkin discretizations for parametric and random parabolic pdes – analytic regularity and gpc approximation. SIAM J. Math. Anal. **45**(5), 3050–3083 (2013)
24. V. Hoang, C. Schwab, A. Stuart, Complexity analysis of accelerated mcmc methods for bayesian inversion. Inverse Probl. **29**(8), 085010 (2013)
25. Y. Maday, N. Nguyen, A. Patera, G. Pau, A general, multipurpose interpolation procedure: the magic points. Commun. Pure Appl. Anal. **8**(1), 383–404 (2009)
26. Y. Maday, A. Patera, D. Rovas, A blackbox reduced-basis output bound method for noncoercive linear problems. Stud. Math. Appl. **31**, 533–569 (2002)
27. N. Nguyen, G. Rozza, D. Huynh, A. Patera, Reduced basis approximation and a posteriori error estimation for parametrized parabolic PDEs; application to real-time Bayesian parameter estimation. *Biegler, Biros, Ghattas, Heinkenschloss, Keyes, Mallick, Tenorio, van Bloemen Waanders, and Willcox, editors, Computational Methods for Large Scale Inverse Problems and Uncertainty Quantification* (John Wiley, Hoboken, 2009)
28. F. Nobile, L. Tamellini, R. Tempone, Convergence of quasi-optimal sparse grid approximation of Hilbert-valued functions: application to random elliptic PDEs. EPFL MATHICSE report 12, 2014
29. F. Nobile, R. Tempone, C. Webster, An anisotropic sparse grid stochastic collocation method for partial differential equations with random input data. SIAM J. Numer. Anal. **46**(5), 2411–2442 (2008)
30. G. Rozza, D. Huynh, A. Patera, Reduced basis approximation and a posteriori error estimation for affinely parametrized elliptic coercive partial differential equations. Arch. Comput. Methods Eng. **15**(3), 229–275 (2008)
31. C. Schillings, C. Schwab, Sparse, adaptive Smolyak quadratures for Bayesian inverse problems. Inverse Probl. **29**(6), 065011 (2013)
32. C. Schillings, C. Schwab, Sparsity in Bayesian inversion of parametric operator equations. Inverse Probl. **30**(6), 065007 (2014)
33. C. Schwab, A. Stuart, Sparse deterministic approximation of bayesian inverse problems. Inverse Probl. **28**(4), 045003 (2012)
34. A. Stuart, Inverse problems: a Bayesian perspective. Acta Numer. **19**(1), 451–559 (2010)
35. G. Turinici, C. Prud'Homme, A. Patera, Y. Maday, A. Buffa, A priori convergence of the greedy algorithm for the parametrized reduced basis method. ESAIM: Math. Model. Numer. Anal. 46(3):595 (2012)
36. D. Xiu, *Numerical Methods for Stochastic Computations: A Spectral Method Approach* (Princeton University Press, Princeton, 2010)

From Data to Uncertainty: An Efficient Integrated Data-Driven Sparse Grid Approach to Propagate Uncertainty

Fabian Franzelin and Dirk Pflüger

Abstract We present a novel data-driven approach to propagate uncertainty. It consists of a highly efficient integrated adaptive sparse grid approach. We remove the gap between the subjective assumptions of the input's uncertainty and the unknown real distribution by applying sparse grid density estimation on given measurements. We link the estimation to the adaptive sparse grid collocation method for the propagation of uncertainty. This integrated approach gives us two main advantages: First, the linkage of the density estimation and the stochastic collocation method is straightforward as they use the same fundamental principles. Second, we can efficiently estimate moments for the quantity of interest without any additional approximation errors. This includes the challenging task of solving higher-dimensional integrals. We applied this new approach to a complex subsurface flow problem and showed that it can compete with state-of-the-art methods. Our sparse grid approach excels by efficiency, accuracy and flexibility and thus can be applied in many fields from financial to environmental sciences.

1 Introduction

There are different types of uncertainty [33] that influence the outcome of large systems that support risk assessment, planning, decision making, validation, etc. Uncertainties can enter the system due to missing knowledge about the physical domain, think of subsurface flow simulations, or there are inherent stochastic processes driving the system, such as Brownian motion. The quantification of the influence of such stochastic components on some quantity of interest is the task of forward propagation of uncertainty in the field of uncertainty quantification (UQ). This is challenging since the statistical characteristics of the uncertainties can be unknown or don't have an analytic representation. Furthermore, the systems or models we are interested in can be arbitrarily complex (highly nonlinear, discontinuous, etc.).

F. Franzelin (✉) • D. Pflüger
Department of Simulation Software Engineering, IPVS, University of Stuttgart, Stuttgart, Germany
e-mail: fabian.franzelin@ipvs.uni-stuttgart.de; dirk.pflueger@ipvs.uni-stuttgart.de

© Springer International Publishing Switzerland 2016
J. Garcke, D. Pflüger (eds.), *Sparse Grids and Applications – Stuttgart 2014*,
Lecture Notes in Computational Science and Engineering 109,
DOI 10.1007/978-3-319-28262-6_2

Therefore, we need efficient and reliable algorithms and software that can make expensive statistical analysis feasible.

In this paper we want to focus on data-driven quantification of uncertainties in numerical simulations. There are two main problems we face in that context: First, the uncertainty of the quantity of interest depends strongly on the uncertainty in the input. Therefore, one needs objective measures to get an unbiased representation of the uncertainty of the input. Second, to quantify the uncertainties of some quantity of interest, we need to evaluate the model, which can be very costly and involves to run a whole simulation. However, the accuracy of the quantities we compute should be very high, which means in general that we need to evaluate the model often, and the main challenge is to balance costs and accuracy.

The first problem of obtaining objective measures for the input's uncertainties has been assessed in various articles in the past. One idea is to use data and estimate the stochastic properties using density estimation techniques. The authors of [7] used kernel density estimators, and [1] proposed to use kernel moment matching methods, for example. A comparison between data-driven approaches and approaches based supervised estimation by experts can be found in [21]. However, often the combination of expert knowledge and data is essential if the reliability in the data is low. A very popular approach to combine them is Bayesian inference [35, 39].

The incorporation of data or estimated densities into a UQ forward problem depends on the method that is required by the application to propagate the uncertainty. For non-intrusive methods, for example, there has been done a lot of work in the field of polynomial chaos expansion (PCE) [37, 38]. The generalized PCE, however, is defined for analytic, independent marginal distributions. It has therefore been extended to the arbitrary PCE [34] that supports also dependent marginal distributions [19] and data [23]. However, global polynomials are not always the best choice to propagate uncertainty [6]. Stochastic collocation methods [2, 36] became popular in the last years, especially due to sparse grids [15, 20]. They are used to obtain a surrogate model of the expensive model function. They can overcome the curse of dimensionality to some extent [3], can handle large gradients [8] or even discontinuities in the response functions [13, 29].

In this paper we present a new approach to incorporate data into the UQ forward propagation pipeline. We propose an integrated data-driven sparse grid method, where we estimate the unknown density of the input using the sparse grid density estimation (SGDE) [24, 26] method and propagate the uncertainty using sparse grid collocation (SGC) with adaptively refined grids. The SGDE method has been widely used for Data Mining problems and can be applied for either small or large data sets. It is highly efficient with respect to learning, evaluating and sampling. It interacts seamlessly with SGC since both are based on the same fundamental principles, which can be exploited to reduce the numerical errors in the forward propagation pipeline.

This paper is structured as follows: First, we give a formal definition of a data-driven UQ forward-propagation problem in Sect. 2. In Sect. 3 we describe the methods we use for density estimation and for uncertainty propagation. Then we compare the performance of our approach with other techniques in Sect. 4. We present a lower-dimensional analytic example and a higher-dimensional subsurface flow problem in porous media. In Sect. 5 we summarize the paper and give an outlook to future work.

2 Problem Formulation

We define (Ω, Σ, P) being a complete probability space with Ω being the set of outcomes, $\Sigma \subset 2^{\Omega}$ the σ-algebra of events and $P: \Sigma \rightarrow [0, 1]$ a probability measure. Let $\boldsymbol{\xi} = (\xi_1, \ldots, \xi_d) \in \Omega$ be a random sample and let the probability law of $\boldsymbol{\xi}$ be completely defined by the probability density function $f : \Omega \rightarrow \mathbb{R}^+$ with $\int_{\Omega} f(\boldsymbol{\xi}) d\boldsymbol{\xi} = 1$. Consider a model $\mathcal{M}$ defined on a bounded physical domain $\mathbf{x} \in D \subset \mathbb{R}^{d_s}$ with $1 \leq d_s \leq 3$, a temporal domain $t \in T \subset \mathbb{R}$ and the probability space (Ω, Σ, P) describing the uncertainty in the model inputs as

$$u(\mathbf{x}, t, \boldsymbol{\xi}) = \mathcal{M}(\mathbf{x}, t, \boldsymbol{\xi}): D \times T \times \Omega \rightarrow \mathbb{R}^{d_r} , \tag{1}$$

with $0 < d_r \in \mathbb{N}$. We restrict ourselves without loss of generality to scalar quantities of u and define an operator Q, which extracts the quantity we are interested in, i.e.

$$Q[u(\mathbf{x}, t, \boldsymbol{\xi})]: \mathbb{R}^{d_r} \rightarrow \mathbb{R}. \tag{2}$$

The outcome of u becomes uncertain due to its uncertain inputs $\boldsymbol{\xi}$. This uncertainty is what we want to quantify. The probability law of $\boldsymbol{\xi}$, of course, influences heavily the probability law of u. Therefore, in data-driven UQ one assumes to have a set of samples $\mathcal{D} := \{\boldsymbol{\xi}^{(k)}\}_{k=1}^n$, with $\boldsymbol{\xi}^{(k)} = (\xi_1^{(k)}, \ldots, \xi_d^{(k)}) \in \Omega$, which are drawn from the unknown probability density f. $\mathcal{D}$ is an objective measure describing the uncertainty we want to propagate. A schematic representation of the data-driven UQ pipeline is given in Fig. 1.

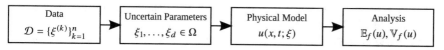

Fig. 1 Data-driven UQ forward pipeline. The data set $\mathcal{D}$ describes the stochastic characteristics of the uncertain parameters $\boldsymbol{\xi}$ for some physical model u. The underlying probability density function f is unknown. The stochastic analysis of the uncertain outcome of u depends strongly on f

3   Methodology

In this section we introduce the methods we use to propagate uncertainty. We formally introduce stochastic collocation and the concept of sparse grids based on an interpolation problem. We present the sparse grid density estimation method and describe how it can be used to estimate efficiently moments of some quantity of interest.

3.1   Stochastic Collocation

In stochastic collocation we search for a function g that approximates the unknown model function u. We solve N deterministic problems of u at a set of collocation points $\Xi_N := \{\xi^{(k)}\}_{k=1}^N \subset \Omega$ and impose

$$g^{(i)}(\xi^{(k)}) := Q[u(x_i, t_i, \xi^{(k)})], \quad \forall \xi^{(k)} \in \Xi_N . \tag{3}$$

at a selected point in space $x_i \in D$ and time $t_i \in T$. This is, of course, nothing else than an interpolation problem. A common choice for $g^{(i)}$ is to use a sum of ansatz functions on some mesh with either global [17] or local support [8, 12, 15]. The expensive stochastic analysis is then done on the cheap surrogate $g^{(i)}$.

For simplicity in the notation we omit in the following the index i on g and focus on the approximation in Ω. Without loss of generality, we assume furthermore that there exists a bijective transformation from Ω to the unit hypercube and assume the collocation nodes $\xi^{(k)}$ in the following to stem from $[0, 1]^d$.

3.2   Sparse Grids

We introduce here the most important properties of sparse grids in the context of interpolation problems. The general idea of sparse grids is based on a hierarchical definition of a one-dimensional basis. This means that the basis is inherently incremental. We exploit this property in higher-dimensional settings to overcome the curse of dimensionality to some extent. For details and further reading we recommend [3, 25]. For adaptive sparse grids and efficient implementations of sparse grid methods we refer to [27], for suitable refinement criteria for UQ problems you may read [8, 15, 16].

Suppose we are searching for an interpolant g of an unknown multivariate function $u(\xi) \in \mathbb{R}$ on the unit hypercube, i.e. $\xi = (\xi_1, \xi_2, \ldots, \xi_d) \in [0, 1]^d$. For g we restrict ourselves to the space of piecewise d-linear functions V_ℓ with ℓ being the maximum discretization level in each dimension.

Let $\mathbf{l} = \{l_1, \ldots, l_d\}$ and $\mathbf{i} = \{i_1, \ldots, i_d\}$ with $l_k > 0$ and $i_k > 0$ be multi-indices. We define a nested index set

$$\mathcal{I}_{\mathbf{l}} := \{(\mathbf{l}, \mathbf{i}) : 1 \le i_k < 2^{l_k}, i_k \text{ odd}, k = 1, \ldots, d\} \tag{4}$$

of level-index vectors defining the grid points

$$\boldsymbol{\xi}_{\mathbf{l},\mathbf{i}} := (2^{-l_1} i_1, \ldots, 2^{-l_d} i_d) . \tag{5}$$

For each grid point we use the general one-dimensional reference hat function $\psi(\xi) := \max(1 - |\xi|, 0)$ to obtain the linear one-dimensional hierarchical hat functions $\psi_{l,i}(\xi)$ centered at the grid points by scaling and translation according to level l and index i as $\psi_{l,i}(\xi) := \psi(2^l \xi - i)$, see Fig. 2 (left). We obtain the higher-dimensional basis via a tensor-product approach,

$$\psi_{\mathbf{l},\mathbf{i}}(\boldsymbol{\xi}) := \prod_{k=1}^{d} \psi_{l_k,i_k}(\xi_k) . \tag{6}$$

Note that the level-index vectors $(\mathbf{l}, \mathbf{i}) \in \mathcal{I}_{\mathbf{l}}$ define a unique set of hierarchical increment spaces $W_{\mathbf{l}} := \text{span}(\{\psi_{\mathbf{l},\mathbf{i}} : (\mathbf{l}, \mathbf{i}) \in \mathcal{I}_{\mathbf{l}}\})$, which are shown in the center of Fig. 2. All increment spaces up to $|\mathbf{l}|_\infty = \max_i l_i \le \ell$ span the space of piecewise d-linear functions on a full grid.

Now we take advantage of the hierarchical definition of the basis and reduce the number of grid points by choosing just those spaces $W_{\mathbf{l}}$ that contribute most to our approximation. An optimal choice is possible a priori if u is sufficiently smooth, i.e. if u is a function of the mixed Sobolev space H_{mix}^2 where the mixed, weak derivatives

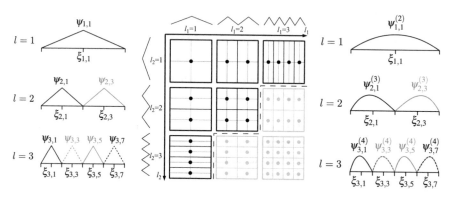

Fig. 2 One-dimensional piecewise linear basis functions up to level 3 (*left*), polynomial ones (*right*), and the tableau of hierarchical increments $W_{\mathbf{l}}$ up to level 3 in two dimensions (*center*)

up to order 2 are bounded. We define the sparse grid space $V_\ell^{(1)}$ as

$$V_\ell^{(1)} := \bigoplus_{|\mathbf{l}|_1 \le \ell+d-1} W_\mathbf{l} \,, \tag{7}$$

where we select just those subspaces that fulfill $|\mathbf{l}|_1 = \sum_{k=1}^d l_k \le \ell + d - 1$, see the upper triangle in the center of Fig. 2. We define a sparse grid function $g_{I_1} \in V_\ell^{(1)}$ as

$$g_{I_1}(\xi) = \sum_{(\mathbf{l},\mathbf{i}) \in I_1, |\mathbf{l}|_1 \le \ell+d-1} v_{\mathbf{l},\mathbf{i}} \psi_{\mathbf{l},\mathbf{i}}(\xi) \,, \tag{8}$$

where $v_{\mathbf{l},\mathbf{i}}$ are the so-called hierarchical coefficients. Note that we omit the index $\mathbf{l}$ when we refer to g_I being a sparse grid function defined on an adaptively refined grid.

The sparse grid space $V_\ell^{(1)}$ has one main advantage over the full tensor space V_ℓ: The number of grid points is reduced significantly from $O((2^\ell)^d)$ for a full grid to $O(2^\ell \ell^{d-1})$ while the interpolation accuracy is of order $O((2^{-\ell})^2 \ell^{d-1})$, which is just slightly worse than the accuracy of a full grid $O((2^{-\ell})^2)$ [3].

If we can impose a higher smoothness for u in a sense that all the weak mixed derivatives up to order $p + 1$ are bounded, it makes sense to employ a higher-order piecewise polynomial basis $\psi_{\mathbf{l},\mathbf{i}}^{(p)}$ with maximum degree $1 \le p \in \mathbb{N}$ in each dimension. Note that these polynomials are defined locally, see Fig. 2 (right). Therefore, we don't suffer Runge's phenomenon even though we use equidistant grid spacing in each dimension. For details about the construction of the basis we refer to [3]. The number of grid points, of course, is the same as for the piecewise linear case. However, the interpolation accuracy is now of order $O((2^{-\ell})^{p+1} \ell^{d-1})$.

3.3 Sparse Grid Density Estimation Method

The sparse grid density estimation (SGDE) method is based on a variational problem presented in [11], first mentioned in the context of sparse grids in [9] and first developed in [26].

We want to estimate some unknown but existing probability density function f from which a set of observations/samples are available, $\mathcal{D} := \{\xi^{(k)}\}_{k=1}^n \subset \Omega$. The SGDE method can be interpreted as a histogram approach with piecewise linear ansatz functions. We search for a sparse grid function $\hat{f}_\mathcal{K} \in V_\ell^{(1)}$ with $|\mathcal{K}| = M$ grid points that minimizes the following functional [9]

$$R(\hat{f}_\mathcal{K}) = \left\| \hat{f}_\mathcal{K} \right\|_{L_2}^2 - \frac{1}{n} \sum_{\xi^{(k)} \in \mathcal{D}} \hat{f}_\mathcal{K}(\xi^{(k)}) + \lambda \left\| S\hat{f}_\mathcal{K} \right\|^2 \,, \tag{9}$$

where S is some regularization operator and $0 \le \lambda \in \mathbb{R}$ a regularization parameter.

In the unit domain the second term is the discrete version of the first one given a set of observations $\mathcal{D}$. $\mathcal{D}$ is a realization of f, so the second term implicitly has larger weights where the probability is larger. This is done explicitly in the first term by the multiplication of the pay-off function with the density $\hat{f}_{\mathcal{K}}$.

By minimizing $R(\hat{f}_{\mathcal{K}})$ we therefore search for a piecewise continuous density $\hat{f}_{\mathcal{K}}$ for which the first two terms are equal. Note that the first term is equal to the definition of the expectation value where the pay off function is the probability density function itself.

From the point of view of histograms, we can say that the sparse grid discretization defines the (overlapping) buckets. The first term in R collects the density mass in all the buckets while the second term does the same for the observations available for each hierarchical ansatz function. The penalty term balances fidelity in the data and smoothness of $\hat{f}_{\mathcal{K}}$ via the regularization parameter λ and the regularization operator S.

Solving $R(\hat{f}_{\mathcal{K}})$ leads to a system of linear equations [11]. The system matrix is the mass matrix of $\hat{f}_{\mathcal{K}}$, which depends only on the number of grid points and is therefore independent of the number of samples n. We obtain the regularization parameter via cross validation: we split $\mathcal{D}$ in a training and a test set, solve the optimization problem on the training set and compute the L_2-norm of the residual of the system of linear equations applied on the test set. For details see [26]. The estimated density function has unit integrand if we choose $S = \nabla$ [26].

Positivity, however, is not guaranteed with this approach. For the numerical examples in this paper we forced $\hat{f}_{\mathcal{K}}$ to be positive by employing a local full grid search on $\hat{f}_{\mathcal{K}}$. For piecewise linear ansatz functions there exists a simple algorithm, see Algorithm 1. A sparse grid function can locally be negative if the coefficient of an arbitrary level-index vector $(\mathbf{l}, \mathbf{i})$ is negative. If this is the case, then, due to monotony of $\hat{f}_{\mathcal{K}}$ between grid points, it is sufficient to apply a full grid search on the support of $(\mathbf{l}, \mathbf{i})$. We add grid points whenever its function value is negative and

---

Algorithm 1: Forcing the sparse grid density to be positive everywhere

---

Data: training sample set $\mathcal{D}$ and sparse grid $\mathcal{I}$
Result: positive sparse grid function $(\mathcal{I}, \mathbf{v})$ with unit integrand
done $\leftarrow$ *False*;
while *not done* **do**
    $\mathbf{v} \leftarrow doSparseGridDensityEstimation(\mathcal{D}, \mathcal{I})$;
    newGridPoints $\leftarrow$ list();
    for $(\mathbf{l}, \mathbf{i}) \in \{(\mathbf{l}, \mathbf{i}) \in \mathcal{I} : v_{\mathbf{l},\mathbf{i}} < 0\}$ **do**
        negativeGridPoints $\leftarrow$ *findNegativeFullGridPointsLocally*$(\mathcal{I}, \mathbf{v}, \mathbf{l}, \mathbf{i})$;
        newGridPoints $\leftarrow$ *append*(newGridPoints, negativeGridPoints);
    if *newGridPoints is not empty* **then**
        $\mathcal{I} \leftarrow addGridPoints(\mathcal{I}, newGridPoints)$;
    else
        done $\leftarrow$ *True*
return $\mathbf{v}, \mathcal{I}$;

---

obtain the hierarchical coefficients for these grid points by learning the density on
the extended grid. We repeat this process until we don't find any negative function
value. This algorithm is, of course, just feasible if the number of local full grid points
to be examined is small. Note that for a piecewise polynomial basis the algorithm
doesn't work because the maximum of each ansatz function is not at a grid point.

3.4 Sparse Grid Collocation

The sparse grid collocation method (SGC) is based on the sparse grid discretization
scheme (see Sect. 3.2) of the stochastic input space. The level-index vectors $(\mathbf{l}, \mathbf{i}) \in$
$\mathcal{I}$ of some sparse grid with $|\mathcal{I}| = N$ define our set of collocation nodes as

$$\Xi_N := \{\boldsymbol{\xi}_{\mathbf{l},\mathbf{i}}\}_{(\mathbf{l},\mathbf{i})\in\mathcal{I}} \tag{10}$$

with $\boldsymbol{\xi}_{\mathbf{l},\mathbf{i}}$ being the grid points, see Eq. (5). We evaluate u at every collocation node
of Ξ_N and solve the interpolation problem

$$g_I(\boldsymbol{\xi}_{\mathbf{l},\mathbf{i}}) := Q[u(\cdot, \boldsymbol{\xi}_{\mathbf{l},\mathbf{i}})], \forall \boldsymbol{\xi}_{\mathbf{l},\mathbf{i}} \in \Xi_N , \tag{11}$$

by a basis transformation from the nodal basis to the hierarchical basis. Efficient
transformation algorithms for both linear and polynomial bases can be found in
[3, 27]. We can furthermore employ adaptive refinement to consider local features
in u. Suitable refinement criteria can be found in [8, 12, 13, 15, 28].

3.5 Moment Estimation

Let $\mu = \mathbb{E}_f(u)$ and $\sigma^2 = \mathbb{V}_f(u)$ be the unknown stochastic quantities of u for the
true density f we want to estimate, g_I be a sparse grid surrogate model for u and
therefore $\mu \approx \mathbb{E}_f(g_I)$ and $\sigma^2 \approx \mathbb{V}_f(g_I)$. Let $\hat{f}$ be an estimated density for the
unknown probability density f obtained by a set of samples $\mathcal{D} := \{\boldsymbol{\xi}^{(k)}\}_{k=1}^n$ drawn
from f.

To estimate the expectation value and the variance we need to solve integrals,
which can be higher-dimensional, depending on the correlations of $\boldsymbol{\xi}$ and the density
estimation technique we use. An easy method that can be applied to any estimated
density we can sample or evaluate is vanilla Monte Carlo quadrature. We generate a
new set of samples $\hat{\mathcal{D}} = \{\boldsymbol{\xi}^{(k)}\}_{k=1}^{\hat{n}}$ drawn from $\hat{f}$ with $\hat{n} \gg n$. We can now substitute
$\hat{f}$ by the discrete density

$$\hat{f}_\delta(\xi) = \frac{1}{\hat{n}} \sum_{\xi^{(k)}\in\hat{\mathcal{D}}} \delta(\xi - \boldsymbol{\xi}^{(k)}) , \tag{12}$$

with δ being the Dirac delta function and estimate the expectation value as

$$\mathbb{E}_{\hat{f}}(g_I) = \int_\Omega g_I(\boldsymbol{\xi})\hat{f}(\boldsymbol{\xi})\mathrm{d}\boldsymbol{\xi} \approx \int_\Omega g_I(\boldsymbol{\xi})\hat{f}_\delta(\boldsymbol{\xi})\mathrm{d}\boldsymbol{\xi} = \frac{1}{\hat{n}} \sum_{\boldsymbol{\xi}^{(k)} \in \hat{\mathcal{D}}} g_I(\boldsymbol{\xi}^{(k)}) . \tag{13}$$

We obtain the result for the sample variance using the same approach for the numerically stable two-pass algorithm [4]

$$\mathbb{V}_{\hat{f}}(g_I) \approx \frac{1}{\hat{n}-1} \sum_{\boldsymbol{\xi}^{(k)} \in \hat{\mathcal{D}}} (g_I(\boldsymbol{\xi}^{(k)}) - \mathbb{E}_{\hat{f}}(g_I))^2 . \tag{14}$$

Due to the substitution of $\hat{f}$ by $\hat{f}_\delta$ we can solve the higher-dimensional integrals easily but we introduce a new numerical error, which converges slowly with respect to $\hat{n}$.

However, this substitution is not necessary if the estimated density $\hat{f}$ is a sparse grid function $\hat{f}_\mathcal{K}$. Due to the tensor-product approach we can decompose the higher-dimensional integral into one-dimensional ones and solve them separately without a numerical error larger than the machine precision ϵ. Let us additionally define some arbitrary order on the collocation points so that we can iterate over them in a predefined order. The expectation value of g_I with respect to $\hat{f}_\mathcal{K}$ can be computed as

$$\mathbb{E}_{\hat{f}_\mathcal{K}}(g_I) = \int_\Omega g_I(\boldsymbol{\xi})\hat{f}_\mathcal{K}(\boldsymbol{\xi})\mathrm{d}\boldsymbol{\xi}$$

$$= \int_\Omega \sum_{(\mathbf{l},\mathbf{i}) \in I} v_{\mathbf{l},\mathbf{i}} \psi_{\mathbf{l},\mathbf{i}}^{(p)}(\boldsymbol{\xi}) \sum_{(\mathbf{k},\mathbf{j}) \in \mathcal{K}} w_{\mathbf{k},\mathbf{j}} \varphi_{\mathbf{k},\mathbf{j}}^{(q)}(\boldsymbol{\xi})\mathrm{d}\boldsymbol{\xi}$$

$$= \sum_{(\mathbf{l},\mathbf{i}) \in I} v_{\mathbf{l},\mathbf{i}} \sum_{(\mathbf{k},\mathbf{j}) \in \mathcal{K}} w_{\mathbf{k},\mathbf{j}} \int_\Omega \psi_{\mathbf{l},\mathbf{i}}^{(p)}(\boldsymbol{\xi})\varphi_{\mathbf{k},\mathbf{j}}^{(q)}(\boldsymbol{\xi})\mathrm{d}\boldsymbol{\xi}$$

$$= \sum_{(\mathbf{l},\mathbf{i}) \in I} v_{\mathbf{l},\mathbf{i}} \sum_{(\mathbf{k},\mathbf{j}) \in \mathcal{K}} w_{\mathbf{k},\mathbf{j}} \underbrace{\int_{\Omega_1} \psi_{l_1,i_1}^{(p)}\varphi_{k_1,j_1}^{(q)}\,\mathrm{d}\xi_1 \cdot \ldots \cdot \int_{\Omega_d} \psi_{l_d,i_d}^{(p)}\varphi_{k_d,j_d}^{(q)}\,\mathrm{d}\xi_d}_{=:A_{(\mathbf{l},\mathbf{i}),(\mathbf{k},\mathbf{j})}}$$

$$= \mathbf{v}^T A \mathbf{w} . \tag{15}$$

The same holds for the variance for which we use Steiners translation theorem

$$\mathbb{V}_{\hat{f}_\mathcal{K}}(g_I) = \mathbb{E}_{\hat{f}_\mathcal{K}}(g_I^2) - \mathbb{E}_{\hat{f}_\mathcal{K}}(g_I)^2 \tag{16}$$

and compute

$$
\begin{aligned}
\mathbb{E}_{\hat{f}_{\mathcal{K}}}(g_I^2) &= \int_\Omega g_I^2(\boldsymbol{\xi}) \hat{f}_{\mathcal{K}}(\boldsymbol{\xi}) d\boldsymbol{\xi} \\
&= \int_\Omega \left[\sum_{(\mathbf{l},\mathbf{i}) \in I} v_{\mathbf{l},\mathbf{i}} \psi_{\mathbf{l},\mathbf{i}}^{(p)}(\boldsymbol{\xi}) \right] \left[\sum_{(\tilde{\mathbf{l}},\tilde{\mathbf{i}}) \in I} v_{\tilde{\mathbf{l}},\tilde{\mathbf{i}}} \psi_{\tilde{\mathbf{l}},\tilde{\mathbf{i}}}^{(p)}(\boldsymbol{\xi}) \right] \hat{f}_{\mathcal{K}}(\boldsymbol{\xi}) d\boldsymbol{\xi} \\
&= \sum_{(\mathbf{l},\mathbf{i}) \in I} v_{\mathbf{l},\mathbf{i}} \sum_{(\tilde{\mathbf{l}},\tilde{\mathbf{i}}) \in I} v_{\tilde{\mathbf{l}},\tilde{\mathbf{i}}} \underbrace{\int_\Omega \psi_{\mathbf{l},\mathbf{i}}^{(p)}(\boldsymbol{\xi}) \psi_{\tilde{\mathbf{l}},\tilde{\mathbf{i}}}^{(p)}(\boldsymbol{\xi}) \hat{f}_{\mathcal{K}}(\boldsymbol{\xi}) d\boldsymbol{\xi}}_{=:B_{(\mathbf{l},\mathbf{i}),(\tilde{\mathbf{l}},\tilde{\mathbf{i}})}} \\
&= \mathbf{v}^T B \mathbf{v},
\end{aligned}
\tag{17}
$$

where the matrix entries $B_{(\mathbf{l},\mathbf{i}),(\tilde{\mathbf{l}},\tilde{\mathbf{i}})}$ are

$$
\begin{aligned}
B_{(\mathbf{l},\mathbf{i}),(\tilde{\mathbf{l}},\tilde{\mathbf{i}})} &= \sum_{(\mathbf{k},\mathbf{j}) \in \mathcal{K}} w_{\mathbf{k},\mathbf{j}} \int_\Omega \psi_{\mathbf{l},\mathbf{i}}^{(p)}(\boldsymbol{\xi}) \psi_{\tilde{\mathbf{l}},\tilde{\mathbf{i}}}^{(p)}(\boldsymbol{\xi}) \varphi_{\mathbf{k},\mathbf{j}}^{(q)}(\boldsymbol{\xi}) d\boldsymbol{\xi} \\
&= \sum_{(\mathbf{k},\mathbf{j}) \in \mathcal{K}} w_{\mathbf{k},\mathbf{j}} \underbrace{\int_{\Omega_1} \psi_{l_1,i_1}^{(p)} \psi_{\tilde{l}_1,\tilde{i}_1}^{(p)} \varphi_{k_1,j_1}^{(q)} d\boldsymbol{\xi}_1 \cdot \ldots \cdot \int_{\Omega_d} \psi_{l_d,i_d}^{(p)} \psi_{\tilde{l}_d,\tilde{i}_d}^{(p)} \varphi_{k_d,j_d}^{(q)} d\boldsymbol{\xi}_d}_{=:b_{(\mathbf{k},\mathbf{j})}} \\
&= \mathbf{w}^T \mathbf{b}.
\end{aligned}
\tag{18}
$$

The runtime $O(N \cdot M)$ for a naive implementation for the expectation value is determined by the matrix vector product. For the variance it holds $O(N^2 \cdot M)$ accordingly. In the inner loop we need to compute the scalar products of the one-dimensional basis functions, which can be done a-priori using Gauss-Legendre quadrature for the corresponding polynomial degree. However, we can reduce the quadratic dependency of the runtime on the number of sparse grid points to just a linear dependency by employing the UpDown-scheme. In the UpDown-scheme we exploit the tree-structure of the grid and apply the uni-directional principle to compute the inner-products [27]. This reduces the quadratic run time for the expectation value to be just linear in the number of grid points, i.e. $O(N + M)$.

3.6 The Sparse Grid Data-Driven UQ Forward Pipeline

Here we want to discuss the numerical properties of the sparse grid based data-driven UQ pipeline for forward problems (see Fig. 3). It consists of four steps: (1) The data set $\mathcal{D}$ is a randomly chosen set of n samples drawn from f. The quality of the set, how good it represents the moments of f, can vary significantly depending on it's size. Furthermore, it makes the estimated density $\hat{f}_{\mathcal{K}}$ (2) to be a random variable.

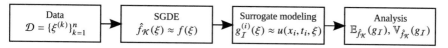

Fig. 3 Data-driven sparse grid UQ forward pipeline

Peherstorfer et al. [26] showed in this context consistency for the SGDE method, i.e.

$$P\left(\left\|f - \hat{f}_{\mathcal{K}}\right\|_{L_2}^2 = 0\right) = 1 \,, \tag{19}$$

for $|\mathcal{K}| \to \infty$ and $n \to \infty$. The accuracy of the sparse grid surrogate model g_I (3) depends on the smoothness of u. Bungartz and Griebel [3] showed that with a piecewise polynomial basis of degree p for regular sparse grids it holds

$$\|u - g_I\|_{L_2} \in O((2^{-\ell})^{p+1} \cdot \ell^{d-1}) \,. \tag{20}$$

For very specific adaptively refined sparse grids we refer to the results in [15] for a convergence proof. The estimated moments of u in the last step become random variables due to $\hat{f}_{\mathcal{K}}$ being a random variable. Therefore, when we talk about convergence of the moments of u we need to consider the mean accuracy with respect to density estimations based on different realizations of f. We define the mean relative error for the expectation value as

$$\mathbb{E}||(\mu - \mathbb{E}_{\hat{f}_{\mathcal{K}}}(g_I))/\mu|| \,, \tag{21}$$

and for the variance as

$$\mathbb{E}||(\sigma^2 - \mathbb{V}_{\hat{f}_{\mathcal{K}}}(g_I))/\sigma^2|| \,, \tag{22}$$

where μ and σ^2 are the true solutions. The quadrature step itself adds numerical errors in the order of the machine precision $O(\epsilon)$ per operation since it consists just of computing scalar products of one-dimensional polynomials.

4 Numerical Examples

In this section we discuss an analytic example, consisting of independent marginal Beta-distributions, and a three-dimensional subsurface flow problem with borehole data.

4.1 Analytic Example

In preparation of the subsurface problem we consider a two-dimensional analytic
scenario with ξ_1, ξ_2 being two independent Beta-distributed random variables,

$$\xi_1 \sim \mathcal{B}(\alpha_1, \beta_1); \xi_2 \sim \mathcal{B}(\alpha_2, \beta_2) , \tag{23}$$

with shape parameters $\alpha_1 = 5$, $\beta_1 = 4$, $\alpha_2 = 3$ and $\beta_2 = 2$ and sample space
$\Omega_1 = \Omega_2 = [0, 1]$. The corresponding density functions are defined as

$$f_k(\xi_k) = c_k \xi_k^{\alpha_k - 1} (1 - \xi_k)^{\beta_k - 1}, \ k \in \{1, 2\} , \tag{24}$$

where $c_k = \Gamma(\alpha_k + \beta_k)/(\Gamma(\alpha_k)\Gamma(\beta_k))$ with Γ being the gamma function. Let
$f(\xi_1, \xi_2) = f_1(\xi_1)f_2(\xi_2)$ be the joint probability density function defined on $\Omega = \Omega_1 \times \Omega_2$. One realization $\mathcal{D}$ of f of size n was created in two steps: first, we generated
n uniformly distributed samples in each dimension. Second, we applied the inverse
cumulative distribution function of f_k and obtained samples from the beta space Ω.
As a model function u we use a simple parabola with $u(\partial \Omega) = 0$ and $u(0.5, 0.5) = 1$

$$u(\xi) = \prod_{k=1}^{2} 4\xi_k (1 - \xi_k) . \tag{25}$$

This model function has two main advantages: First, we can compute analytic
solutions for the expectation value μ and the variance σ^2 of u as

$$\mu = \frac{c_1 c_2}{4725} \approx 0.71111 \tag{26}$$

$$\sigma_2 = \frac{c_1^3 c_2^3}{75{,}014{,}100{,}000} - \frac{2c_1^2 c_2^2}{22{,}325{,}625} + \frac{4c_1 c_2}{24{,}255} \approx 0.04843 , \tag{27}$$

Second, u can be approximated perfectly with a sparse grid function g_I of level
1 with a piecewise quadratic basis. This means that the numerical error in the
surrogate model vanishes. The only two errors remaining are the approximation
error estimating the density $\hat{f}$ and the quadrature error in moment estimation if we
use Monte Carlo. The second one can be minimized by increasing the sample size;
the first one, however, is limited to the amount of information there is about f, which
is encoded in the data $\mathcal{D}$ we use for density estimation. Due to the randomness in
the data we measure the error in the expectation value and the variance according
to Eqs. (21) and (22). We compare these two errors for (1) vanilla Monte Carlo
($k = 200$ realizations of f), omitting the density estimation step, (2) a kernel-density
estimator using the libagf library [18] ($k = 20$ realizations), (3) density trees (dtrees)
[30] ($k = 50$ realizations) and (4) SGDE using the SG++ library [27] ($k = 20$
realizations). For the SGDE approach we used regular sparse grids with different
levels, estimated the density according to [26] and chose the best approximation

Table 1 KL-divergence (KL) and cross entropy (L) for different density estimation methods and sizes for the training data sets $\mathcal{D}$. The test data set $\mathcal{T}$ to compute the measure had size $m = 10^4$

# samples	libagf		dtrees		SGDE	
	KL	L	KL	L	KL	L
50	0.2655	−0.7046	2.829	1.85	0.2157	−0.7491
75	0.2387	−0.7314	1.837	0.858	0.1838	−0.781
100	0.213	−0.7571	1.424	0.446	0.1533	−0.8115
500	0.1081	−0.8655	0.4294	−0.5488	0.1157	−0.849
1000	0.07851	−0.8951	0.2744	−0.7027	0.06948	−0.8953
5000	0.03964	−0.9356	0.1185	−0.8598	0.02778	−0.937
10,000	0.03001	−0.9352	0.09217	−0.8847	0.02014	−0.9446

with respect to the minimal cross entropy L for a test set $\mathcal{T}$ of size $m = 10^4$.

$$L_{\mathcal{T}}(\hat{f}) = -\frac{1}{m} \sum_{\xi^{(k)} \in \mathcal{T}} \log_2(\hat{f}(\xi^{(k)})) . \tag{28}$$

The test set $\mathcal{T}$ is generated analogous to $\mathcal{D}$ with a different seed. This measure is known to minimize the Kullback-Leibler-divergence (KL) and is therefore a suitable criterion [32].

Table 1 shows the KL-divergence and the cross entropy for the different density estimation methods and different sizes of training sets. We can see three main aspects: First, as expected, the cross entropy minimizes the KL-divergence. Second, the cross entropy decreases monotonically with the sample size for all density estimation methods. This means that the density estimation methods are able to capture the increasing information they get from the larger sample sets. Third, while SGDE and libagf have very similar results for all number of samples, the density trees have a poor performance especially for smaller sample sizes. The KL-divergence of the density trees is 10 times larger for 50–100 training samples than the ones of SGDE and libagf. This is a significant drawback for applications where the real costs lie in obtaining the samples. Think of boreholes that need to be drilled into the ground to obtain information about the physical domain of subsurface flow problems. If we look at the convergence of the expected error in the expectation value, see Fig. 4 (left), SGDE performs almost one order of magnitude better than libagf. Both methods converge with $n^{-1/2}$, which is basically the Monte Carlo convergence for the quadrature problem. The convergent phase for density trees starts later at a size of 1000 samples. We see basically the same picture if we look at the variance, see Fig. 4 (right). SGDE performs best compared to the other density estimation methods. However, it seems that the density estimation methods can not outperform Monte Carlo. The reason is that density estimation is based on Monte Carlo estimates for the moments of the distribution, see Eq. (9). It would pay off if there could be gained additional information from the data by extrapolation or regularization. There is no extrapolation in this case due to the definition of the

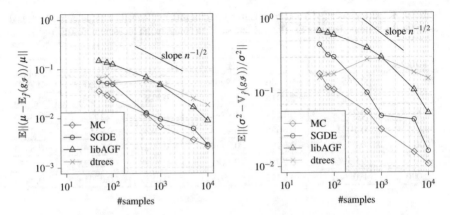

Fig. 4 Decay of the average error in the expectation value (*left*) and the variance (*right*) for Monte Carlo (MC), the sparse grid density estimation (SGDE) method, the kernel density estimator (libagf), and the density trees (dtrees)

joint probability density function f, which is zero at the boundaries. Neither does regularization, since there is no noise in the data and we measured the mean error over several realizations of f and made it therefore independent of single realizations where regularization could pay off. However, if the number of samples is limited, as it is in the CO_2 benchmark problem, the data-driven forward propagation approach with density estimation will reduce the error compared to Monte Carlo.

4.2 Multivariate Stochastic Application

In this application we simulate carbon sequestration based on the CO_2 benchmark model defined by [5]. We will not introduce here the modeling of this highly non-linear multiphase flow in a porous media problem but rather refer to [5] and focus on the stochastic part. The basic setting however is the following: We inject CO_2 through an injection well into a subterranean aquiferous reservoir. The CO_2 starts spreading according to the geological characteristics of the reservoir until it reaches a leaky well where the CO_2 rises up again to a shallower aquifer. A schematic view on the problem is shown in Fig. 5. The CO_2 leakage at the leaky well is the quantity we are interested in. It depends on the plume development in the aquifer and the pressure built up in the aquifer due to injection. While we can control the injection pressure, we cannot control the geological properties of the reservoir like porosity, permeability, etc. This is where the uncertainty comes in and for which we use data to describe it.

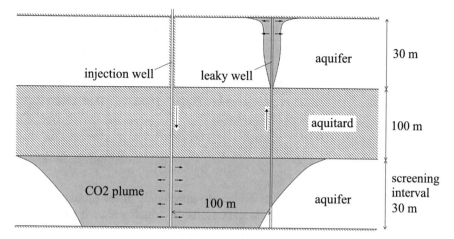

Fig. 5 Cross-section through the subterranean reservoir [22]

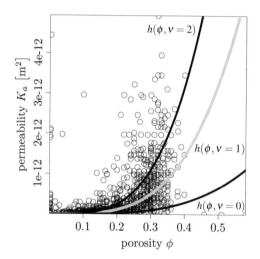

Fig. 6 Raw data for porosity ϕ and permeability K_a from the U.S. National Petroleum Council Public Database including 1270 data points. The *curves* show the upper and the lower bound of the transformed and truncated sample space with respect to the variation v

4.2.1 Stochastic Formulation

The CO_2 benchmark model has three stochastic parameters with respect to the geological properties of the reservoir and the leaky well: (1) the reservoir porosity ϕ, (2) the reservoir permeability K_a, and (3) the leaky well permeability K_L. For the description of the reservoir, i.e. K_a and ϕ, we use a raw data set from the U.S. National Petroleum Council Public Database including 1270 reservoirs, shown in Fig. 6, see also [14]. The data set is assumed to be one realization of the unknown

Table 2 Marginal densities and ranges used in the *analytic* approach to describe the uncertainty of the CO_2 benchmark model. They are fitted to the decorrelated samples according to [22]

Uncertain parameter	Probability density function	Range Ω_k
Porosity ϕ	$f_1(\phi) = \frac{1}{\phi\sigma\sqrt{2\pi}}\exp\left(-\frac{(\ln(\phi)-\mu)^2}{2\sigma^2}\right)$ with $\mu = -27.6310, \sigma = 0.3579$	$[0.0896, 0.2511]$
Permeability K_a	$f_2(K_L) = \frac{1}{K_L\sigma\sqrt{2\pi}}\exp\left(-\frac{(\ln(K_L)-\mu)^2}{2\sigma^2}\right)$ with $\mu = -1.8971, \sigma = 0.2$	$[3.88, 25.8] \times 10^{-13}$
Variation v	$f_3(v) = \frac{\Gamma(\alpha+\beta)}{\Gamma(\alpha)\Gamma(\beta)}v^{\alpha-1}(1-v)^{\beta-1}$ with $\alpha = 3, \beta = 3$	$[0, 2]$

probability density function f that defines the uncertainty of the problem. For the leaky well permeability there is no data available, which makes it necessary to make further assumptions. We make here the same assumptions as in Oladyshkin et al. in [22], see permeability in Table 2. Furthermore, Oladyshkin et al. presented in the same paper an integrative approach to quantify the stochastic outcome of the CO_2 benchmark model using analytic densities based on the data at hand, see Table 2. They defined a sample space that is different to the parameter space of the simulation. They substituted $K_a \in \Omega_{K_a}$ by a variation parameter $v \in \Omega_v$ and encoded the correlation in a transformation function from the new sample space to the parameter space, i.e.

$$h: \Omega_\phi \times \Omega_v \to \Omega_{K_a}$$
$$\phi, v \mapsto c_1\phi^{c_2}[1 + c_3 v] \,, \tag{29}$$

with parameters $c_1 = 4.0929 \times 10^{-11}, c_2 = 3.6555, c_3 = -2$. In the new sample space $\Omega := \Omega_\phi \times \Omega_v \times \Omega_{K_L}$ all variables are decorrelated. Therefore we can define analytic independent marginal densities and can propagate the resulting uncertainty directly through the model function using polynomial chaos expansion, for example. We refer to this approach as the *analytic* approach. To apply stochastic collocation to this problem, we truncated the infinite sample space Ω such that in directions of ϕ and K_a we collect 99.99% of their mass around the corresponding mean, see Table 2. We denote the samples that lie within this truncated space as $\mathcal{D} := \{(K_a^{(k)}, \phi^{(k)})\}_{k=1}^{413}$, see Fig. 7. We use $\mathcal{D}$ for density estimation to obtain objective measures of the input's uncertainty.

4.2.2 Results

In this section we want to illustrate the ability of the integrated sparse grid approach to predict accurate expectation values and variances for the CO_2 benchmark model with the given input data set. We assume that a surrogate model using adaptive sparse grid collocation is available and approximates the unknown model function so well that the expectation value and the variance can be estimated accurately with

Fig. 7 The plot shows the transformed raw data points $h^{-1}(\phi, K_a)$ from Fig. 6 that lie within the parameter ranges given in Table 2. They are additionaly linearly scaled to $[0, 1]^2$. The remaining number of samples is 413

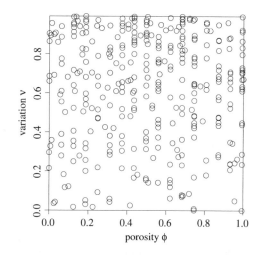

Table 3 Mean cross entropy (L) for different density estimation methods and for 10 randomly chosen training data sets of size 363 and test data sets of size 50

# training	# test	analytic (L)	libagf (L)	dtrees (L)	SGDE (L)
363	50	0.7279	0.00278	−0.1314	−0.3042

respect to the analytic approach ($N = 114$, refinement according to the *variance surplus refinement*, see [8] for details). We compare these results of the (1) analytic approach with (2) SGDE, (3) kernel-densities, (4) density trees, and (5) Monte Carlo with bootstrapping on the available data $\mathcal{D}$.

The results of the density estimations with respect to the cross entropy for a test set $\mathcal{T}$ with $m = 50$ are listed in Table 3. The analytic approach has by far the largest cross entropy, which suggests that it doesn't capture the underlying data as good as the others do. The SGDE method performs best as it has the lowest cross entropy. Note that some samples we use for training are located close to the boundary of the domain (see Fig. 7). This affects the SGDE method since we need to consider the boundary in the discretization of the sample space. For this problem we used trapezoidal boundary sparse grids [27] where each inner grid point has two ancestors in each dimension that lie on the corresponding boundary of the domain.

If we look at the results of the estimation of the expectation value, see Fig. 8 (left), and the variance, see Fig. 8 (right), we obtain surprising results. Let us assume that the results of the Monte Carlo quadrature approach using bootstrapping on the available data is our ground truth. By this we say indirectly that we have large confidence in the available data. Compared to this ground truth, the analytic approach overestimates the expectation value and underestimates the variance significantly. We call this difference the "subjective gap", which has been introduced by expert knowledge. The other density estimation methods lie in between these approaches. The density trees match the expectation value of the data almost exactly. The SGDE method overestimates the expectation value slightly, the kernel density

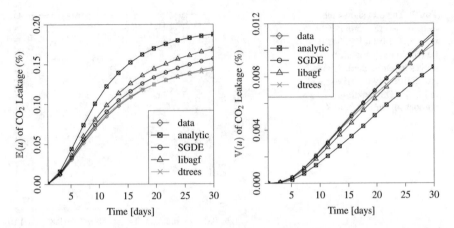

Fig. 8 Expectation value (*left*) and variance (*right*) over simulation time for the bootstrapping method with the raw data (data), the analytic approach as in [22] (analytic), the sparse grid density estimation (SGDE) method, the kernel density estimator (libagf), and the density trees (dtrees)

does as well. But if we look at the variance then the SGDE method gives the best results compared to the data, while all others underestimate it.

The overestimation of the expectation value for SGDE with respect to the data can be explained by extrapolation: We used a sparse grid with trapezoidal boundary for SGDE because some samples are located close to the boundary of the domain. Furthermore, we impose smoothness on the unknown density. These two facts let the SGDE method extrapolate towards the boundary resulting in a larger density than there is in the data. This is not even wrong since the boundaries of our transformed and truncated domain Ω are located in the middle of the parametric domain in which the raw data lies, see Fig. 6 (left). This leads then to the higher expectation values we see in Fig. 8 (left). In fact, if we use a sparse grid without boundary points for density estimation we match the expectation value of the data as well as the density trees do. However, the cross entropy for this sparse grid density is larger compared to the others ($L = 0.2632$) indicating a worse estimation. And indeed, with this estimation we overestimate now the variance significantly.

Due to these arguments, we question the ground truth, which in this application was based on a very limited data set. Of course, this makes the comparison of different methods difficult. But, since the balancing between fidelity in the data and smoothness of the density function is defined clearly for the SGDE approach, we consider it a reliable and robust approach in the context of UQ.

5 Conclusions

In this paper we presented a new integrated sparse grid approach for data-driven uncertainty quantification forward problems. It has two main advantages over common approaches for such problems: First, it is an unsupervised approach that relies on the data at hand. It is not influenced by expert knowledge. Second, the integrated sparse grid approach allows a seamless interaction with the stochastic collocation method with respect to adaptive refinement and quadrature.

Furthermore, the numerical experiments showed that the SGDE method gives good approximations of the unknown probability density function already for small sample sets. It did better than the very popular kernel density. Even newer approaches such as density trees showed worse results compared to SGDE.

However, the SGDE method has drawbacks when it comes to statistical applications for which we need to assure unit integrand and positivity. We presented one way to overcome these problems by suitable discretization and using appropriate regularization. This approach however is limited in terms of the problem's dimensionality since it implies a local full grid search. There is ongoing work in this field, see [10], to overcome these problems.

In this paper we focused on small sample sets. When it comes to large data sets one can speed up this process significantly using fast algorithms such as SGDE. For example, when one uses Markov chain Monte Carlo to obtain a discrete posterior density in an inverse UQ setting. Such densities are often correlated and pose problems to established forward propagation methods such as the generalized polynomial chaos expansion. The Rosenblatt transformation [31] and the inverse Rosenblatt transformation can play an important role in this context. They can be computed very efficiently without additional numeric errors using sparse grids.

Acknowledgements The authors acknowledge the German Research Foundation (DFG) for its financial support of the project within the Cluster of Excellence in Simulation Technology at the University of Stuttgart.

References

1. A. Alwan, N. Aluru, Improved statistical models for limited datasets in uncertainty quantification using stochastic collocation. J. Comput. Phys. **255**, 521–539 (2013)
2. I. Babuška, F. Nobile, R. Tempone, A stochastic collocation method for elliptic partial differential equations with random input data. SIAM J. Numer. Anal. **45**(3), 1005–1034 (2007)
3. H.-J. Bungartz, M. Griebel, Sparse grids. Acta Numer. **13**, 147–269 (2004)
4. T.F. Chan, G.H. Golub, R.J. LeVeque, Algorithms for computing the sample variance: analysis and recommendations. Am. Stat. **37**(3), 242–247 (1983)
5. H. Class, A. Ebigbo, R. Helmig, H. Dahle, J. Nordbotten, M. Celia, P. Audigane, M. Darcis, J. Ennis-King, Y. Fan, B. Flemisch, S. Gasda, M. Jin, S. Krug, D. Labregere, A. Naderi Beni, R. Pawar, A. Sbai, S. Thomas, L. Trenty, L. Wei, A benchmark study on problems related to CO_2 storage in geologic formations. Comput. Geosci. **13**(4), 409–434 (2009)

6. M. Eldred, J. Burkardt, Comparison of non-intrusive polynomial chaos and stochastic collocation methods for uncertainty quantification, in *Aerospace Sciences Meetings* (American Institute of Aeronautics and Astronautics, Reston, 2009)

7. H.C. Elman, C.W. Miller, Stochastic collocation with kernel density estimation. Comput. Methods Appl. Mech. Eng. **245–246**(0), 36–46 (2012)

8. F. Franzelin, P. Diehl, D. Pflüger, Non-intrusive uncertainty quantification with sparse grids for multivariate peridynamic simulations, in *Meshfree Methods for Partial Differential Equations VII*, ed. by M. Griebel, M.A. Schweitzer. Volume 100 of Lecture Notes in Computational Science and Engineering (Springer, Cham, 2015), pp. 115–143

9. J. Garcke, Maschinelles Lernen durch Funktionsrekonstruktion mit verallgemeinerten dünnen Gittern, Ph.D. thesis, Universität Bonn, Institut für Numerische Simulation, 2004

10. M. Griebel, M. Hegland. A finite element method for density estimation with gaussian process priors. SIAM J. Numer. Anal. **47**(6), 4759–4792 (2010)

11. M. Hegland, G. Hooker, S. Roberts, Finite element thin plate splines in density estimation. ANZIAM J. **42**, 712–734 (2000)

12. J. Jakeman, S. Roberts, Local and dimension adaptive stochastic collocation for uncertainty quantification, in *Sparse Grids and Applications*, ed. by J. Garcke, M. Griebel. Volume 88 of Lecture Notes in Computational Science and Engineering (Springer, Berlin/Heidelberg, 2013), pp. 181–203

13. J.D. Jakeman, R. Archibald, D. Xiu, Characterization of discontinuities in high-dimensional stochastic problems on adaptive sparse grids. J. Comput. Phys. **230**(10), 3977–3997 (2011)

14. A. Kopp, H. Class, H. Helmig, Investigations on CO_2 storage capacity in saline aquifers – part 1: dimensional analysis of flow processes and reservoir characteristics. Int. J. Greenh. Gas Control **3**, 263–276 (2009)

15. X. Ma, N. Zabaras, An adaptive hierarchical sparse grid collocation algorithm for the solution of stochastic differential equations. J. Comput. Phys. **228**(8), 3084–3113 (2009)

16. X. Ma, N. Zabaras, An adaptive high-dimensional stochastic model representation technique for the solution of stochastic partial differential equations. J. Comput. Phys. **229**(10), 3884–3915 (2010)

17. O.P.L. Maître, O.M. Knio, *Spectral Methods for Uncertainty Quantification: With Applications to Computational Fluid Dynamics*. Scientific Computation (Springer, Dordrecht/New York, 2010)

18. P. Mills, Efficient statistical classification of satellite measurements. Int. J. Remote Sens. **32**(21), 6109–6132 (2011)

19. M. Navarro, J. Witteveen, J. Blom, Polynomial chaos expansion for general multivariate distributions with correlated variables, Technical report, Centrum Wiskunde & Informatica, June 2014

20. F. Nobile, R. Tempone, C.G. Webster, A sparse grid stochastic collocation method for partial differential equations with random input data. SIAM J. Numer. Anal. **46**(5), 2309–2345 (2008)

21. S. Oladyshkin, H. Class, R. Helmig, W. Nowak, A concept for data-driven uncertainty quantification and its application to carbon dioxide storage in geological formations. Adv. Water Resour. **34**(11), 1508–1518 (2011)

22. S. Oladyshkin, H. Class, R. Helmig, W. Nowak, An integrative approach to robust design and probabilistic risk assessment for CO_2 storage in geological formations. Comput. Geosci. **15**(3), 565–577 (2011)

23. S. Oladyshkin, W. Nowak, Data-driven uncertainty quantification using the arbitrary polynomial chaos expansion. Reliab. Eng. Syst. Saf. **106**(0), 179–190 (2012)

24. B. Peherstorfer, Model order reduction of parametrized systems with sparse grid learning techniques. Ph.D. thesis, Technical University of Munich, Aug 2013

25. B. Peherstorfer, C. Kowitz, D. Pflüger, H.-J. Bungartz, Selected recent applications of sparse grids. Numer. Math. Theory Methods Appl. **8**(01), 47–77, Feb 2015

26. B. Peherstorfer, D. Pflüger, H.-J. Bungartz, Density estimation with adaptive sparse grids for large data sets, in *Proceedings of the 2014 SIAM International Conference on Data Mining*, Philadelphia, 2014, pp. 443–451

27. D. Pflüger, *Spatially Adaptive Sparse Grids for High-Dimensional Problems* (Verlag Dr. Hut, Munich, 2010)
28. D. Pflüger, Spatially adaptive refinement, in *Sparse Grids and Applications*, ed. by J. Garcke, M. Griebel. Lecture Notes in Computational Science and Engineering (Springer, Berlin/Heidelberg, 2012), pp. 243–262
29. D. Pflüger, B. Peherstorfer, H.-J. Bungartz, Spatially adaptive sparse grids for high-dimensional data-driven problems. J. Complex. **26**(5), 508–522 (2010). Published online Apr 2010
30. P. Ram, A.G. Gray, Density estimation trees, in *Proceedings of the 17th ACM SIGKDD International Conference on Knowledge Discovery and Data Mining*, San Diego, 21–24 Aug 2011, pp. 627–635
31. M. Rosenblatt, Remarks on a multivariate transformation. Ann. Math. Stat. **23**(1952), 470–472 (1952)
32. B. Silverman, *Density Estimation for Statistics and Data Analysis*, 1st edn. (Chapman & Hall, London/New York, 1986)
33. W. Walker, P. Harremoes, J. Rotmans, J. van der Sluijs, M. van Asselt, P. Janssen, M.K. von Krauss, Defining uncertainty: a conceptual basis for uncertainty management in model-based decision support. Integr. Assess. **4**(1), 5–17 (2005)
34. X. Wan, E. Karniadakis, Multi-element generalized polynomial chaos for arbitrary probability measures. SIAM J. Sci. Comput. **28**(3), 901–928 (2006)
35. M. Xiang, N. Zabaras, An efficient Bayesian inference approach to inverse problems based on an adaptive sparse grid collocation method. Inverse Probl. **25**(3), 035013 (2009)
36. D. Xiu, J.S. Hesthaven, High-order collocation methods for differential equations with random inputs. SIAM J. Sci. Comput. **27**(3), 1118–1139 (2005)
37. D. Xiu, G. Karniadakis, The Wiener–askey polynomial chaos for stochastic differential equations. SIAM J. Sci. Comput. **24**(2), 619–644 (2002)
38. D. Xiu, G.E. Karniadakis, Modeling uncertainty in flow simulations via generalized polynomial chaos. J. Comput. Phys. **187**(1), 137–167 (2003)
39. G. Zhang, D. Lu, M. Ye, M.D. Gunzburger, C.G. Webster, An adaptive sparse-grid high-order stochastic collocation method for Bayesian inference in groundwater reactive transport modeling. Water Resour. Res. **49**(10), 6871–6892 (2013)

Combination Technique Based Second Moment Analysis for Elliptic PDEs on Random Domains

Helmut Harbrecht and Michael Peters

Abstract In this article, we propose the sparse grid combination technique for the second moment analysis of elliptic partial differential equations on random domains. By employing shape sensitivity analysis, we linearize the influence of the random domain perturbation on the solution. We derive deterministic partial differential equations to approximate the random solution's mean and its covariance with leading order in the amplitude of the random domain perturbation. The partial differential equation for the covariance is a tensor product Dirichlet problem which can efficiently be determined by Galerkin's method in the sparse tensor product space. We show that this Galerkin approximation coincides with the solution derived from the combination technique if the detail spaces in the related multiscale hierarchy are constructed with respect to Galerkin projections. This means that the combination technique does not impose an additional error in our construction. Numerical experiments quantify and qualify the proposed method.

1 Introduction

Various problems in science and engineering can be formulated as boundary value problems for an unknown function. In general, the numerical simulation is well understood provided that the input parameters are known exactly. In many applications, however, the input parameters are not known exactly. Especially, the treatment of uncertainties in the computational domain has become of growing interest, see e.g. [5, 21, 38, 41]. In this article, we consider the elliptic diffusion equation

$$- \operatorname{div} \big(\alpha \nabla u(\omega) \big) = f \text{ in } D(\omega), \quad u(\omega) = 0 \text{ on } \partial D(\omega) \tag{1}$$

as a model problem, where the underlying domain $D(\omega) \subset \mathbb{R}^d$ or respectively its boundary $\partial D(\omega)$ are random. For example, one might think of tolerances in the

H. Harbrecht (✉) • M. Peters
Departement Mathematik und Informatik, Universität Basel, Spiegelgasse 1, 4051 Basel, Schweiz

e-mail: helmut.harbrecht@unibas.ch; michael.peters@unibas.ch

© Springer International Publishing Switzerland 2016
J. Garcke, D. Pflüger (eds.), *Sparse Grids and Applications – Stuttgart 2014*,
Lecture Notes in Computational Science and Engineering 109,
DOI 10.1007/978-3-319-28262-6_3

shape of products fabricated by line production or shapes which stem from inverse problems, like for example from tomography. Of course, besides a scalar diffusion coefficient $\alpha(\mathbf{x})$, one can also consider a diffusion matrix $\mathbf{A}(\mathbf{x})$, cf. [18]. Even so, the emphasis of our considerations will be laid on the case $\alpha(\mathbf{x}) \equiv 1$, that is the Poisson equation.

Except for the fictitious domain approach considered in [5], one might essentially distinguish two approaches: the *domain mapping method*, cf. [6, 23, 27, 38, 41], and the *perturbation method*. They result from a description of the random domain either in Lagrangian coordinates or in Eulerian coordinates, see e.g. [37]. The latter approach will be dealt with in this article.

The perturbation method starts with a prescribed perturbation field

$$\mathbf{V}(\omega) \colon D_{\text{ref}} \to \mathbb{R}^d$$

for a fixed reference domain D_{ref} and uses a *shape Taylor expansion* with respect to this perturbation field to represent the solution to the model problem, see e.g. [19, 21]. In fact, as we will see later on, it is sufficient to know the perturbation field in a vicinity of ∂D_{ref}, i.e.

$$\mathbf{V}(\omega) \colon \partial D_{\text{ref}} \to \mathbb{R}^d.$$

This is a remarkable feature since it might in practice be much easier to obtain measurements from the outside of a work-piece to estimate the perturbation field $\mathbf{V}(\omega)$ rather than from its interior.

The starting point for our considerations will be the knowledge of an appropriate description of the the random field $\mathbf{V}(\omega)$. To that end, we assume that the random vector field is described in terms of its mean

$$\mathbb{E}[\mathbf{V}] \colon D_{\text{ref}} \to \mathbb{R}^d, \quad \mathbb{E}[\mathbf{V}](\mathbf{x}) = \big[\mathbb{E}[V_1](\mathbf{x}), \dots, \mathbb{E}[V_d](\mathbf{x})\big]^{\mathsf{T}}$$

and its (matrix valued) covariance function

$$\text{Cov}[\mathbf{V}] \colon D_{\text{ref}} \times D_{\text{ref}} \to \mathbb{R}^{d \times d}, \quad \text{Cov}[\mathbf{V}](\mathbf{x}, \mathbf{y}) = \begin{bmatrix} \text{Cov}_{1,1}(\mathbf{x}, \mathbf{y}) & \cdots & \text{Cov}_{1,d}(\mathbf{x}, \mathbf{y}) \\ \vdots & & \vdots \\ \text{Cov}_{d,1}(\mathbf{x}, \mathbf{y}) & \cdots & \text{Cov}_{d,d}(\mathbf{x}, \mathbf{y}) \end{bmatrix}.$$

For the perturbation method, this representation of the random vector field is already sufficient. Having the mean and the covariance of the random vector field at hand, we aim at approximating the corresponding statistics of the unknown random solution.

Making use of sensitivity analysis, we linearize the solution's nonlinear dependence on the random vector field $\mathbf{V}(\omega)$. Based on this, we derive deterministic equations, which compute, to leading order, the mean field and the covariance. In particular, the covariance solves a tensor product boundary value problem on the product domain $D_{\text{ref}} \times D_{\text{ref}}$. This linearization technique has already been applied to random diffusion coefficients or even to elliptic equations on random domains in [18, 20, 21]. In difference to these previous works, we do not explicitly use wavelets [21, 33, 34] or multilevel frames [18, 20] for the discretization in a sparse tensor product space. Instead, we define the complement spaces which enter the sparse tensor product construction by Galerkin projections. The Galerkin discretization leads then to a system of linear equations which decouples into subproblems with respect to full tensor product spaces of small size. These subproblems can be solved by standard multilevel finite element methods. In our particular realization, we need only the access to the stiffness matrix, the BPX preconditioner, cf. [3], and the sparse grid interpolant, cf. [4], of the covariance function of the random vector field under consideration. In this sense, our approach can be considered to be weakly intrusive. The resulting representation of the covariance is known as the *combination technique* [14]. Nevertheless, in difference to [14, 28, 32, 42], this representation is a consequence of the Galerkin method in the sparse tensor product space and is not an additional approximation step.

The rest of this article is structured as follows. In Sect. 2, we introduce the underlying framework. Here, we define random vector fields and the related Lebesgue-Bochner spaces. Moreover, we briefly refer to the Karhunen-Loève expansion of random vector fields. Section 3 is devoted to shape sensitivity analysis. Especially, the shape Taylor expansion is introduced here. In Sect. 4, we apply the shape Taylor expansion to our model problem and derive deterministic equations for the mean and the covariance. Section 5 deals with the approximation of tensor product Dirichlet problems. In Sect. 6, we present in detail the sparse grid combination technique for the solution of tensor product Dirichlet problems. The efficient implementation of the proposed method is non-trivial. Therefore, we believe it is justified to dedicate Sect. 7 to this topic. Finally, in Sect. 8 we present our numerical results.

Throughout this article, in order to avoid the repeated use of generic but unspecified constants, by $C \lesssim D$ we mean that C can be bounded by a multiple of D, independently of parameters which C and D may depend on. Obviously, $C \gtrsim D$ is defined as $D \lesssim C$, and $C \eqsim D$ as $C \lesssim D$ and $C \gtrsim D$.

2 Preliminaries

The natural environment for the consideration of random vector fields are the so called *Lebesgue-Bochner spaces*. These spaces quantify the integrability of Banach space valued functions and have originally been introduced in [1]. In this section, we shall provide some facts and results on Lebesgue-Bochner spaces. For more details on this topic, we refer to [24].

Let $(\Omega, \mathcal{F}, \mathbb{P})$ denote a complete and separable probability space with σ-algebra $\mathcal{F}$ and probability measure $\mathbb{P}$. Here, complete means that $\mathcal{F}$ contains all $\mathbb{P}$-null sets.

The separability is e.g. obtained if $\mathcal{F}$ is countably generated, cf. [16, Theorem 40.B]. Furthermore, let $D_{\mathrm{ref}} \subset \mathbb{R}^d$ denote a sufficiently smooth domain.

Definition 1 For $p \geq 0$, the *Lebesgue-Bochner space* $L_{\mathbb{P}}^p(\Omega; L^2(D_{\mathrm{ref}}; \mathbb{R}^d))$ consists of all equivalence classes of strongly $\mathbb{P}$-measurable maps $u: \Omega \to L^2(D_{\mathrm{ref}}; \mathbb{R}^d)$ with finite norm

$$\|u\|_{L_{\mathbb{P}}^p(\Omega; L^2(D_{\mathrm{ref}}; \mathbb{R}^d))} := \begin{cases} \left(\int_{\Omega} \|u(\omega, \cdot)\|_{L^2(D_{\mathrm{ref}}; \mathbb{R}^d)}^p \, d\mathbb{P} \right)^{1/p}, & p < \infty \\ \operatorname*{ess\,sup}_{\omega \in \Omega} \|u(\omega, \cdot)\|_{L^2(D_{\mathrm{ref}}; \mathbb{R}^d)}, & p = \infty. \end{cases} \tag{2}$$

Two functions $u, v: \Omega \to L^2(D_{\mathrm{ref}}; \mathbb{R}^d)$ are identified if they coincide $\mathbb{P}$-almost everywhere, i.e. if $\mathbb{P}(\{u \neq v\}) = 0$. Moreover, the space $L^2(D_{\mathrm{ref}}; \mathbb{R}^d)$ is equipped with the inner product

$$(\mathbf{u}, \mathbf{v})_{L^2(D_{\mathrm{ref}}; \mathbb{R}^d)} := \int_{D_{\mathrm{ref}}} \langle \mathbf{u}, \mathbf{v} \rangle \, d\mathbf{x} \quad \text{for all } \mathbf{u}, \mathbf{v} \in L^2(D_{\mathrm{ref}}; \mathbb{R}^d),$$

where $\langle \cdot, \cdot \rangle$ denotes the canonical inner product in $\mathbb{R}^d$.

In the definition, the term *strongly $\mathbb{P}$-measurable* refers to functions which are measurable in the classical sense and in addition essentially separable valued. The second condition is automatically met for functions $u: \Omega \to L^2(D_{\mathrm{ref}}; \mathbb{R}^d)$ which are measurable in the classical sense.

The spaces $L_{\mathbb{P}}^p(\Omega; L^2(D_{\mathrm{ref}}; \mathbb{R}^d))$ are complete for all $p \in [1, \infty]$ with respect to the norm defined in (2) and thus Banach spaces, see e.g. [24] for a proof of this statement. For $p = 1$, the space $L_{\mathbb{P}}^1(\Omega; L^2(D_{\mathrm{ref}}; \mathbb{R}^d))$ coincides with the space of *Bochner integrable* functions, cf. [9, Theorem 2.4]. It is moreover well known that $L_{\mathbb{P}}^2(\Omega)$ is separable if $(\Omega, \mathcal{F}, \mathbb{P})$ is separable, cf. [16, Exercise 43.(1)]. Hence, for $p = 2$, the Bochner space $L_{\mathbb{P}}^2(\Omega; L^2(D_{\mathrm{ref}}; \mathbb{R}^d))$ is a separable Hilbert space equipped with the inner product

$$(u, v)_{L_{\mathbb{P}}^2(\Omega; L^2(D_{\mathrm{ref}}; \mathbb{R}^d))} := \int_{\Omega} \left(u(\omega, \cdot), v(\omega, \cdot) \right)_{L^2(D_{\mathrm{ref}}; \mathbb{R}^d)} \, d\mathbb{P}.$$

In particular, it holds $L_{\mathbb{P}}^2(\Omega; L^2(D_{\mathrm{ref}}; \mathbb{R}^d)) \cong L_{\mathbb{P}}^2(\Omega) \otimes L^2(D_{\mathrm{ref}}; \mathbb{R}^d)$.

We summarize some important facts about the Bochner integral from [24].

Theorem 1

(a) *The Bochner integral*

$$\int_{\Omega} \cdot \, d\mathbb{P}: L_{\mathbb{P}}^1(\Omega; L^2(D_{\mathrm{ref}}; \mathbb{R}^d)) \to L^2(D_{\mathrm{ref}}; \mathbb{R}^d)$$

is a linear map.

(b) *For* $u \in L^1_{\mathbb{P}}\big(\Omega; L^2(D_{\text{ref}}; \mathbb{R}^d)\big)$ *it holds*

$$\left\| \int_A u(\omega, \cdot) \, d\mathbb{P} \right\|_{L^2(D_{\text{ref}}; \mathbb{R}^d)} \leq \int_A \|u(\omega, \cdot)\|_{L^2(D_{\text{ref}}; \mathbb{R}^d)} \, d\mathbb{P}$$

for all $A \in \mathcal{F}$.

(c) *Let* $\{u_n\}_n$ *be a sequence of Bochner integrable functions with* $\lim_{n \to \infty} u_n = u$ *in* $\mathbb{P}$-measure *and* g *a Lebesgue integrable function on* Ω *such that* $\|u_n\|_{L^2(D_{\text{ref}}; \mathbb{R}^d)} \leq g$ $\mathbb{P}$-almost everywhere. *Then,* u *is Bochner integrable and*

$$\lim_{n \to \infty} \int_A u_n \, d\mathbb{P} = \int_A u \, d\mathbb{P}$$

for all $A \in \mathcal{F}$. *Moreover, it holds*

$$\lim_{n \to \infty} \int_\Omega \|u_n - u\|_{L^2(D_{\text{ref}}; \mathbb{R}^d)} \, d\mathbb{P} = 0.$$

(d) *Let* $T: U \to \mathscr{B}$ *be a closed linear operator for some Banach space* $\mathscr{B}$ *and* $U \subseteq L^2(D_{\text{ref}}; \mathbb{R}^d)$. *If* u *and* Tu *are Bochner integrable, then*

$$T\left(\int_A u \, d\mathbb{P} \right) = \int_A Tu \, d\mathbb{P}$$

for all $A \in \mathcal{F}$.

Let the random vector field $\mathbf{V} \in L^2_{\mathbb{P}}\big(\Omega; L^2(D_{\text{ref}}; \mathbb{R}^d)\big)$ be represented according to

$$\mathbf{V}(\omega, \mathbf{x}) = [V_1(\omega, \mathbf{x}), \dots, V_d(\omega, \mathbf{x})]^{\mathsf{T}}.$$

Then, we can define the *mean* of $\mathbf{V}$ in terms of the Bochner integral

$$\mathbb{E}[\mathbf{V}](\mathbf{x}) := \int_\Omega \mathbf{V}(\omega, \mathbf{x}) \, d\mathbb{P} \in L^2(D_{\text{ref}}; \mathbb{R}^d).$$

Especially, it holds $\mathbb{E}[V_i](\mathbf{x}) = \int_\Omega V_i(\omega, \mathbf{x}) \, d\mathbb{P}$. With respect to the *centered* random field

$$\mathbf{V}_0 = \mathbf{V} - \mathbb{E}[\mathbf{V}],$$

we introduce the (matrix valued) *covariance function* of $\mathbf{V}$ according to

$$\text{Cov}[\mathbf{V}](\mathbf{x}, \mathbf{y}) = [\text{Cov}_{i,j}(\mathbf{x}, \mathbf{y})]_{i,j=1}^d$$

with

$$\text{Cov}_{i,j}(\mathbf{x}, \mathbf{y}) = \mathbb{E}\big[V_{0,i}(\omega, \mathbf{x})V_{0,j}(\omega, \mathbf{y})\big]. \tag{3}$$

The boundedness of $\text{Cov}_{i,j}(\mathbf{x}, \mathbf{y})$ in $L^2(D_{\text{ref}} \times D_{\text{ref}})$ follows from the Cauchy-Schwarz inequality and the application of Fubini's theorem. Since $\text{Cov}_{i,j}(\mathbf{x}, \mathbf{y}) \in L^2(D_{\text{ref}} \times D_{\text{ref}})$ holds, we conclude $\text{Cov}[\mathbf{V}](\mathbf{x}, \mathbf{y}) \in L^2(D_{\text{ref}} \times D_{\text{ref}}; \mathbb{R}^{d \times d})$.

In order to make the random vector field $\mathbf{V}(\omega, \mathbf{x}) \in L^2_{\mathbb{P}}\big(\Omega; L^2(D_{\text{ref}}; \mathbb{R}^d)\big)$ feasible for numerical computations, e.g. for a (quasi-) Monte Carlo method, we shall introduce its *Karhunen-Loève expansion*, cf. [26]. Since we may identify $L^2_{\mathbb{P}}\big(\Omega; L^2(D_{\text{ref}}; \mathbb{R}^d)\big) \cong L^2_{\mathbb{P}}(\Omega) \otimes L^2(D_{\text{ref}}; \mathbb{R}^d)$, one can show that $\mathbf{V}_0(\omega, \mathbf{x})$ exhibits the orthogonal decomposition

$$\mathbf{V}_0 = \sum_{i \in I} \sigma_i X_i \otimes \boldsymbol{\varphi}_i,$$

where $\{\boldsymbol{\varphi}_i\}_{i \in I} \subset L^2(D_{\text{ref}}; \mathbb{R}^d)$ and $\{X_i\}_{i \in I} \subset L^2_{\mathbb{P}}(\Omega)$ are orthonormal families. With respect to the canonical map

$$L^2_{\mathbb{P}}(\Omega) \otimes L^2(D_{\text{ref}}; \mathbb{R}^d) \to L^2_{\mathbb{P}}\big(\Omega; L^2(D_{\text{ref}}; \mathbb{R}^d)\big), \quad X \otimes \varphi \mapsto X(\omega)\varphi(\mathbf{x}),$$

we end up with the following

Definition 2 Let $\mathbf{V}(\omega, \mathbf{x})$ be a vector field in $L^2_{\mathbb{P}}\big(\Omega; L^2(D_{\text{ref}}; \mathbb{R}^d)\big)$. The expansion

$$\mathbf{V}(\omega, \mathbf{x}) = \mathbb{E}[\mathbf{V}](\mathbf{x}) + \sum_{i \in I} \sigma_i X_i(\omega)\boldsymbol{\varphi}_i(\mathbf{x})$$

with $(X_i, X_j)_{L^2_{\mathbb{P}}(\Omega)} = \delta_{ij}$ and $\mathbb{E}[X_i] = 0$ is called *Karhunen-Loève expansion* of $\mathbf{V}(\omega, \mathbf{x})$.

Remark 1 The knowledge of the random vector field $\mathbf{V}(\omega, \mathbf{x})$ is sufficient to compute the related Karhunen-Loève expansion. In practice, however, the random field is often only given in terms of its (empirical) mean $\mathbb{E}[\mathbf{V}]$ and its (empirical) covariance function $\text{Cov}[\mathbf{V}]$. In this case, the orthogonal basis in $L^2_{\mathbb{P}}(\Omega)$ is only determined up to isometry. Therefore, for the use of e.g. a (quasi-) Monte Carlo method, the law of the random variables $\{X_i\}_{i \in I}$ has to be approximated appropriately, for example by a *maximum likelihood estimate*, cf. [35]. This is in contrast to the discretization in the perturbation method, where we do not need to know the random variables' distribution at all.

Without loss of generality, we may assume that $\mathbb{E}[\mathbf{V}](\mathbf{x}) = \mathbf{x}$ is the identity mapping. Otherwise, we replace D_{ref} and $\boldsymbol{\varphi}_k$ by

$$\tilde{D}_{\text{ref}} := \mathbb{E}[\mathbf{V}](D_{\text{ref}}) \quad \text{and} \quad \tilde{\boldsymbol{\varphi}}_k := \sqrt{\det(\mathbb{E}[\mathbf{V}]^{-1})'}\boldsymbol{\varphi}_k \circ \mathbb{E}[\mathbf{V}]^{-1}.$$

3 Shape Sensitivity Analysis

In this section, we summarize results on shape sensitivity analysis for the Poisson equation

$$- \Delta u = f \text{ in } D_{\text{ref}}, \quad u = 0 \text{ on } \Gamma_{\text{ref}} := \partial D_{\text{ref}}. \tag{4}$$

For a more general framework and the details on this topic, we refer the reader to [8, 12, 37] and the references therein.

Assume that D_{ref} is of class C^2. This smoothness assumption guarantees the H^2-regularity of problems (4) and (5), cf. [37, Proposition 2.83]. Moreover, let $\mathbf{V} \in C^2(\mathbb{R}^d; \mathbb{R}^d)$ be a vector field. We may define the family of transformations $\{T_\varepsilon\}_{\varepsilon > 0}$ by the perturbations of identity

$$T_\varepsilon(\mathbf{x}) = \text{Id}(\mathbf{x}) + \varepsilon \mathbf{V}(\mathbf{x}).$$

Then, there exists an $\varepsilon_0 > 0$ such that the transformations T_ε are C^2-diffeomorphisms for all $\varepsilon \in [0, \varepsilon_0]$, cf. [36, Section 1.1]. The related family of domains will be denoted by $D_\varepsilon := T_\varepsilon(D_{\text{ref}})$. We shall consider the Poisson equation on these domains, i.e.

$$- \Delta u_\varepsilon = f \text{ in } D_\varepsilon, \quad u_\varepsilon = 0 \text{ on } \Gamma_\varepsilon := \partial D_\varepsilon. \tag{5}$$

Here, in order to guarantee the well-posedness of the equation, we assume that the right hand side is defined on the hold-all

$$\mathcal{D} = \bigcup_{0 \le \varepsilon \le \varepsilon_0} D_\varepsilon.$$

Now, we have for the weak solution $u_\varepsilon \in H^s(D_\varepsilon)$ with $s \in [0, 2]$ that

$$u^\varepsilon = u_\varepsilon \circ T_\varepsilon \in H^s(D_{\text{ref}})$$

for all $\varepsilon \in [0, \varepsilon_0]$, see e.g. [37]. Especially, we set $\overline{u} := u \in H^s(D_{\text{ref}})$. Then, we may define the *material derivative* of u as in [37, Definition 2.71].

Definition 3 The function $\dot{u}[\mathbf{V}] \in H^s(D_{\text{ref}})$ is called the strong (weak) *material derivative* of $u \in H^s(D_{\text{ref}})$ in the direction $\mathbf{V}$ if the strong (weak) limit

$$\dot{u}[\mathbf{V}] = \lim_{\varepsilon \to 0} \frac{1}{\varepsilon}(u^\varepsilon - \overline{u})$$

exists.

The shape sensitivity analysis considered in this section is based on the notion of the *local shape derivative*. To this end, we consider for $\overline{u} \in H^s(D_{\text{ref}})$ and $u_\varepsilon \in$

$H^s(D_\varepsilon)$ the expression

$$\frac{1}{\varepsilon}\big(u_\varepsilon(\mathbf{x}) - \bar{u}(\mathbf{x})\big).$$

Obviously, this expression is only meaningful if $\mathbf{x} \in D_\varepsilon \cap D_{\text{ref}}$. Nevertheless, according to [12, Remark 2.2.12], there exists an $\varepsilon(\mathbf{x}, \mathbf{V})$ due to the regularity of T_ε such that $\mathbf{x} \in D_\varepsilon \cap D_{\text{ref}}$ for all $0 \leq \varepsilon \leq \varepsilon(\mathbf{x}, \mathbf{V})$. Moreover, in order to define a meaningful functional analytic framework for the limit $\varepsilon \to 0$, one has to consider compact subsets $K \Subset D_{\text{ref}}$, cf. [36]. Hence, we have from [12, Definition 2.2.13] the following

Definition 4 For $K \Subset D_{\text{ref}}$, the function $\delta u[\mathbf{V}] \in H^s(K)$ is called the strong (weak) *local $H^s(K)$ shape derivative* of u in direction $\mathbf{V}$, if the strong (weak) $H^s(K)$ limit

$$\delta u[\mathbf{V}] = \lim_{\varepsilon \to 0} \frac{1}{\varepsilon}(u_\varepsilon - \bar{u})$$

exists. It holds $\delta u \in H^s_{\text{loc}}(D_{\text{ref}})$ strongly (weakly) if the limit exists for arbitrary $K \Subset D_{\text{ref}}$.

Notice that the definition of $\delta u[\mathbf{V}]$ has no meaning on Γ_{ref} in general, cf. [12, Remark 2.2.14]. Nevertheless, since boundary values for $\dot{u}[\mathbf{V}]$ are obtained via the trace operator, cf. [37, Proposition 2.75], we may define the boundary values for $\delta u[\mathbf{V}]$ by employing the relation

$$\dot{u}[\mathbf{V}] = \delta u[\mathbf{V}] + \langle \nabla u, \mathbf{V} \rangle,$$

cf. [37, Definition 2.85]. Therefore, if $f \in H^1(\mathcal{D})$, the local shape derivative for the Poisson equation (5) satisfies the boundary value problem

$$\Delta \delta u = 0 \text{ in } D_{\text{ref}}, \quad \delta u = -\langle \mathbf{V}, \mathbf{n} \rangle \frac{\partial \bar{u}}{\partial \mathbf{n}} \text{ on } \Gamma_{\text{ref}}, \tag{6}$$

cf. [37, Proposition 3.1]. Here, $\mathbf{n}(\mathbf{x})$ denotes the outward normal at the boundary Γ_{ref}.

The representation (6) of $\delta u[\mathbf{V}]$ indicates that it is already sufficient to consider vector fields $\mathbf{V}$ which are compactly supported in a neighbourhood of Γ_{ref}, i.e. $\mathbf{V}|_K \equiv 0$ for all $K \Subset D_{\text{ref}}$, cf. [12, Remark 2.1.6]. More precisely, it holds for two perturbation fields $\mathbf{V}$ and $\tilde{\mathbf{V}}$ that

$$\delta u[\mathbf{V}] = \delta u[\tilde{\mathbf{V}}] \quad \text{if } \mathbf{V}|_{\Gamma_{\text{ref}}} = \tilde{\mathbf{V}}|_{\Gamma_{\text{ref}}},$$

cf. [37, Proposition 2.90]. For example, it is quite common to consider (normal) perturbations of the boundary, see e.g. [12, 25, 29, 30].

Having the local shape derivative of the solution u to (4) at hand, we can linearize the perturbed solution u_ε to (5) in a neighbourhood of D_{ref} in terms of a *shape Taylor expansion*, cf. [10, 11, 21, 31], according to

$$u_\varepsilon(\mathbf{x}) = \overline{u}(\mathbf{x}) + \varepsilon \delta u(\mathbf{x}) + \varepsilon^2 C(\mathbf{x}) \quad \text{for } \mathbf{x} \in K \Subset (D_{\text{ref}} \cap D_\varepsilon), \tag{7}$$

where the function $|C(\mathbf{x})| < \infty$ depends on the distance $\text{dist}(K, \Gamma_{\text{ref}})$ and the load f.

4 Approximation of Mean and Covariance

We shall return now to our model problem, the Poisson equation on a random domain:

$$-\Delta u(\omega, \mathbf{x}) = f(\mathbf{x}) \text{ in } D(\omega), \quad u(\omega, \mathbf{x}) = 0 \text{ on } \Gamma(\omega). \tag{8}$$

We assume that the random domain is described by a random vector field. This means, we have

$$D(\omega) := \mathbf{V}(\omega, D_{\text{ref}}).$$

With respect to the discussion in the end of Sect. 2, it is reasonable to assume that $\mathbf{V}$ is a perturbation of identity. More precisely, we assume that it holds

$$\mathbf{V}(\omega, \mathbf{x}) = \text{Id}(\mathbf{x}) + \mathbf{V}_0(\omega, \mathbf{x})$$

for a vector field $\mathbf{V}_0(\omega) \in C^2(D_{\text{ref}}; \mathbb{R}^d)$ for almost every $\omega \in \Omega$ with $\mathbb{E}[\mathbf{V}_0] = \mathbf{0}$. We shall further assume the uniformity condition $\|\mathbf{V}_0(\omega)\|_{C^2(D_{\text{ref}};\mathbb{R}^d)} \leq c$ for some $c < \infty$ and for almost every $\omega \in \Omega$. Then, in view of (7), the first-order shape Taylor expansion for the solution $u(\omega)$ to (8) with respect to the transformation

$$T_\varepsilon(\omega, \mathbf{x}) = \text{Id}(\mathbf{x}) + \varepsilon \mathbf{V}_0(\omega, \mathbf{x}), \tag{9}$$

is given by

$$u(\omega, \mathbf{x}) = \overline{u}(\mathbf{x}) + \varepsilon \delta u(\mathbf{x})[\mathbf{V}_0(\omega)] + O(\varepsilon^2).$$

In this expansion, $\overline{u}$ is the solution to

$$-\Delta \overline{u} = f \text{ in } D_{\text{ref}}, \quad \overline{u} = 0 \text{ on } \Gamma_{\text{ref}}, \tag{10}$$

while $\delta u[\mathbf{V}_0(\omega)]$ is the solution to

$$\Delta \delta u[\mathbf{V}_0(\omega)] = 0 \text{ in } D_{\text{ref}}, \quad \delta u[\mathbf{V}_0(\omega)] = -\langle \mathbf{V}_0(\omega), \mathbf{n}\rangle \frac{\partial \overline{u}}{\partial \mathbf{n}} \text{ on } \Gamma_{\text{ref}}. \tag{11}$$

As already pointed out in the end of the preceding section, it is sufficient to know $V_0(\omega, \mathbf{x})$ only in a neighbourhood of the boundary Γ_{ref} of D_{ref}. This is in contrast to the domain mapping method where one always has to know the perturbation field for the whole domain D_{ref}.

In order to simplify the notation, we will write $\delta u(\omega)$ instead of $\delta u[V_0(\omega)]$ in the sequel. Having the first-order shape Taylor expansion (9) of $u(\omega)$ at hand, we can approximate the related moments from it.

Theorem 2 *For $\varepsilon > 0$ sufficiently small, it holds for $K \Subset (D_{\text{ref}} \cap D_\varepsilon)$ that*

$$\mathbb{E}[u] = \bar{u} + O(\varepsilon^2) \quad \text{in } K \tag{12}$$

with $\bar{u} \in H_0^1(D_{\text{ref}})$. The covariance of u satisfies

$$\text{Cov}[u] = \varepsilon^2 \, \text{Cov}[\delta u] + O(\varepsilon^3) \quad \text{in } K \times K \tag{13}$$

with the covariance $\text{Cov}[\delta u] \in H^1(D_{\text{ref}}) \otimes H^1(D_{\text{ref}})$. The covariance is given as the solution to the following boundary value problem

$$
\begin{aligned}
(\Delta \otimes \Delta) \, \text{Cov}[\delta u] &= 0 \quad \text{in } D_{\text{ref}} \times D_{\text{ref}}, \\
(\Delta \otimes \gamma_0^{\text{int}}) \, \text{Cov}[\delta u] &= 0 \quad \text{in } D_{\text{ref}} \times \Gamma_{\text{ref}}, \\
(\gamma_0^{\text{int}} \otimes \Delta) \, \text{Cov}[\delta u] &= 0 \quad \text{in } \Gamma_{\text{ref}} \times D_{\text{ref}}, \\
(\gamma_0^{\text{int}} \otimes \gamma_0^{\text{int}}) \, \text{Cov}[\delta u] &= \langle \mathbf{n}(\mathbf{x}), \text{Cov}[\mathbf{V}]\mathbf{n}(\mathbf{y}) \rangle \left(\frac{\partial \bar{u}}{\partial \mathbf{n}} \otimes \frac{\partial \bar{u}}{\partial \mathbf{n}} \right) \quad \text{on } \Gamma_{\text{ref}} \times \Gamma_{\text{ref}}.
\end{aligned}
\tag{14}
$$

Here, $\gamma_0^{\text{int}} \colon H^1(D_{\text{ref}}) \to H^{1/2}(\Gamma_{\text{ref}})$ denotes the interior trace operator.

Proof The equation for the mean is easily obtained by exploiting the linearity of the mean. It remains to show that

$$\mathbb{E}[\delta u] = 0.$$

By Theorem 1, we know that we may interchange the Bochner integral with the Laplace operator. Thus, from (11), we obtain the following boundary value problem for $\mathbb{E}[\delta u]$:

$$\Delta \mathbb{E}[\delta u] = 0 \text{ in } D_{\text{ref}}, \quad \mathbb{E}[\delta u] = -\mathbb{E}\left[\langle \mathbf{V}_0, \mathbf{n} \rangle \frac{\partial \bar{u}}{\partial \mathbf{n}} \right] \text{ on } \Gamma_{\text{ref}}.$$

By the linearity of the Bochner integral, the boundary condition can be written as

$$-\mathbb{E}\left[\langle \mathbf{V}_0, \mathbf{n} \rangle \frac{\partial \bar{u}}{\partial \mathbf{n}} \right] = -\langle \mathbb{E}[\mathbf{V}_0], \mathbf{n} \rangle \frac{\partial \bar{u}}{\partial \mathbf{n}} = 0$$

since $\mathbb{E}[\mathbf{V}_0] = \mathbf{0}$. Thus, $\mathbb{E}[\delta u]$ solves the Laplace equation with homogeneous boundary condition. From this, we infer $\mathbb{E}[\delta u] = 0$.

For the covariance $\mathrm{Cov}[u]$, we obtain

$$\mathrm{Cov}[u] = \mathbb{E}\big[(u - \mathbb{E}[u]) \otimes (u - \mathbb{E}[u])\big]$$

$$= \mathbb{E}\Big[\big(\bar{u} + \varepsilon\delta u(\omega) + O(\varepsilon^2) - \mathbb{E}[u]\big) \otimes \big(\bar{u} + \varepsilon\delta u(\omega) + O(\varepsilon^2) - \mathbb{E}[u]\big)\Big].$$

Since we can estimate $\mathbb{E}[u] - \bar{u} = O(\varepsilon^2)$ in K due to (12), we arrive at

$$\mathrm{Cov}[u] = \mathbb{E}\Big[\big(\varepsilon\delta u(\omega) + O(\varepsilon^2)\big) \otimes \big(\varepsilon\delta u(\omega) + O(\varepsilon^2)\big)\Big]$$

$$= \varepsilon^2 \mathbb{E}[\delta u(\omega) \otimes \delta u(\omega)] + O(\varepsilon^3).$$

In view of $\mathrm{Cov}[\delta u] = \mathbb{E}[\delta u(\omega) \otimes \delta u(\omega)]$, we conclude (13). Finally, by tensorization of (11) and application of the mean together with Theorem 1, one infers that $\mathrm{Cov}[\delta u] \in H^1(D_{\mathrm{ref}}) \otimes H^1(D_{\mathrm{ref}})$ is given by (14). $\qquad\square$

Remark 2 The technique which we used to derive the approximation error for the covariance of u can straightforwardly be applied to obtain a similar result for the k-th moment, i.e.

$$\mathbb{E}\big[\underbrace{(u - \mathbb{E}[u]) \otimes \ldots \otimes (u - \mathbb{E}[u])}_{k\text{-times}} \big].$$

In this case, we end up with the expression

$$\mathbb{E}\big[\big(\varepsilon\delta u + O(\varepsilon^2)\big) \otimes \ldots \otimes \big(\varepsilon\delta u + O(\varepsilon^2)\big)\big] = \varepsilon^k \mathbb{E}[\delta u \otimes \ldots \otimes \delta u] + O(\varepsilon^{k+1}),$$

where the constant depends exponentially on k, see also [7].

5 Discretization of Tensor Product Dirichlet Problems

In the previous section, we have seen that we end up solving the tensor product Dirichlet problem (14) in order to approximate the covariance of the model problem's solution. The treatment of the non-homogenous tensor product Dirichlet boundary condition is non-trivial. Therefore, we shall consider here the discretization by finite elements in detail.

We start with the discretization of univariate Dirichlet problems and then generalize the approach towards the tensor product case. We thus aim at solving the Dirichlet boundary value problem

$$\Delta u = 0 \text{ in } D_{\mathrm{ref}}, \quad u = g \text{ on } \Gamma_{\mathrm{ref}}. \tag{15}$$

By the inverse trace theorem, see e.g. [40], there exists an extension of $u_g \in H^1(D_{\text{ref}})$ with $\gamma_0^{\text{int}} u_g = g$ provided that $g \in H^{1/2}(\Gamma_{\text{ref}})$. Therefore, it remains to determine the function $u_0 = u - u_g \in H_0^1(D_{\text{ref}})$ such that there holds

$$a^D(u_0, v) = -a^D(u_g, v) \quad \text{for all } v \in H_0^1(D_{\text{ref}}). \tag{16}$$

Here and in the sequel, the $H_0^1(D_{\text{ref}})$-elliptic bilinear form related to the Laplace operator is given by

$$a^D(u, v) := (\nabla u, \nabla v)_{L^2(D_{\text{ref}})} \quad \text{for } u, v \in H^1(D_{\text{ref}}).$$

The question arises how to numerically determine a suitable extension u_g of the Dirichlet data. We follow here the approach from [2], see also [13], where the extension is generated by means of an L^2-projection of the given boundary data. To that end, we introduce the nested sequence of finite element spaces

$$V_0 \subset V_1 \subset \cdots \subset V_J \subset H^1(D_{\text{ref}}).$$

Herein, given a uniform triangulation for D_{ref}, the space V_j corresponds to the space of continuous piecewise linear functions with the basis $\{\varphi_{j,k} \in V_j : k \in I_j\}$. Of course, by performing obvious modifications, one can employ the presented framework also for higher order ansatz functions. Notice that we have $\dim V_j \approx 2^{dj}$. In the following, we distinguish between basis functions $\{\varphi_{j,k} \in V_j : k \in I_j^D\}$ which are supported in the interior of the reference domain, i.e. $\varphi_{j,k}|_{\Gamma_{\text{ref}}} \equiv 0$, and boundary functions $\{\varphi_{j,k} \in V_j : k \in I_j^\Gamma\}$ with $\varphi_{j,k}|_{\Gamma_{\text{ref}}} \not\equiv 0$. Notice that $I_j = I_j^D \cup I_j^\Gamma$ and $I_j^D \cap I_j^\Gamma = \emptyset$. The related finite element spaces are then given by

$$V_j^D := \text{span}\{\varphi_{j,k} \in V_j : k \in I_j^D\} \quad \text{and} \quad V_j^\Gamma := \text{span}\{\varphi_{j,k}|_{\Gamma_{\text{ref}}} : \varphi_{j,k} \in V_j, k \in I_j^\Gamma\}.$$

Moreover, we denote the L^2-inner product on Γ_{ref} by

$$a^\Gamma(u, v) := (u, v)_{L^2(\Gamma_{\text{ref}})} \quad \text{for } u, v \in L^2(\Gamma_{\text{ref}}).$$

Then, the L^2-orthogonal projection of the Dirichlet data is given by the solution to the following variational formulation:

$$\text{Find } g_j \in V_j^\Gamma \text{ such that}$$
$$a^\Gamma(g_j, v) = a^\Gamma(g, v) \quad \text{for all } v \in V_j^\Gamma. \tag{17}$$

We are now prepared to formulate the Galerkin discretization of (16). To that end, we introduce the stiffness matrices

$$\mathbf{S}_j^\Lambda := \left[a^D(\varphi_{j,\ell}, \varphi_{j,k})\right]_{k \in I_j^D, \ell \in I_j^\Lambda}, \quad \Lambda \in \{D, \Gamma\}, \tag{18}$$

and the mass matrices with respect to the boundary

$$\mathbf{G}_j := \left[a^\Gamma(\varphi_{j,\ell}, \varphi_{j,k})\right]_{k \in I_j^\Gamma, \ell \in I_{j'}^\Gamma}. \tag{19}$$

The related data vector reads

$$\mathbf{g}_j = \left[a^\Gamma(g, \varphi_{j,k})\right]_{k \in I_j^\Gamma}.$$

In order to compute an approximate solution to this boundary value problem in the finite element space $V_J \subset H^1(D_{\mathrm{ref}})$ for $J \in \mathbb{N}$, we make the ansatz

$$u_J = \sum_{k \in I_J} u_{J,k} \varphi_{J,k} = \sum_{k \in I_J^D} u_{J,k} \varphi_{J,k} + \sum_{k \in I_J^\Gamma} u_{J,k} \varphi_{J,k} = u_J^D + u_J^\Gamma.$$

At first, we determine the boundary part $u_J^\Gamma \in H^1(D)$ such that

$$\mathbf{G}_J u_J^\Gamma = \mathbf{g}_J. \tag{20}$$

Therefore, $u_J^\Gamma|_{\Gamma_{\mathrm{ref}}}$ is the L^2-orthogonal projection of the Dirichlet data g onto the discrete trace space V_J^Γ. Having u_J^Γ at hand, we can compute the domain part $u_J^D \in H_0^1(D)$ from

$$\mathbf{S}_J^D u_J^D = -\mathbf{S}_J^\Gamma u_J^\Gamma. \tag{21}$$

We use the conjugate gradient method to iteratively solve the systems of linear equations (20) and (21). Using a nested iteration, combined with the BPX-preconditioner, cf. [3], in case of (21), results in a linear over-all complexity, see [15]. Moreover, from [2, Theorem 1], we obtain the following convergence result.

Theorem 3 *Let $g \in H^t(\Gamma_{\mathrm{ref}})$ for $0 \le t \le 3/2$. Then, if $g_J \in V_J^\Gamma$ is given by (17), the Galerkin solution u_J to (15) satisfies*

$$\|u - u_J\|_{L^2(D_{\mathrm{ref}})} \lesssim 2^{-J(t+1/2)} \|g\|_{H^t(\Gamma_{\mathrm{ref}})}.$$

Next, we deal with the tensor product boundary value problem (14) and discretize it in $V_J \otimes V_J$. We make the ansatz

$$\text{Cov}[\delta u]_J = \sum_{k \in I_J} \sum_{k' \in I_J} u_{J,k,k'} (\varphi_{J,k} \otimes \varphi_{J,k'})$$

$$= \text{Cov}[\delta u]_J^{D,D} + \text{Cov}[\delta u]_J^{D,\Gamma} + \text{Cov}[\delta u]_J^{\Gamma,D} + \text{Cov}[\delta u]_J^{\Gamma,\Gamma} \qquad (22)$$

with

$$\text{Cov}[\delta u]_J^{\Lambda,\Lambda'} = \sum_{k \in I_J^\Lambda} \sum_{k' \in I_J^{\Lambda'}} u_{J,k,k'} (\varphi_{J,k} \otimes \varphi_{J,k'}) \quad \text{for } \Lambda, \Lambda' \in \{D, \Gamma\}.$$

In complete analogy to the non-tensor product case, we obtain the solution scheme:

1. Solve $(\mathbf{G}_J \otimes \mathbf{G}_J) \mathbf{u}_J^{\Gamma,\Gamma} = \mathbf{g}_J$.
2. Solve $(\mathbf{G}_J \otimes \mathbf{S}_J^D) \mathbf{u}_J^{\Gamma,D} = -(\mathbf{G}_J \otimes \mathbf{S}_J^\Gamma) \mathbf{u}_J^{\Gamma,\Gamma}$ and $(\mathbf{S}_J^D \otimes \mathbf{G}_J) \mathbf{u}_J^{D,\Gamma} = -(\mathbf{S}_J^\Gamma \otimes \mathbf{G}_J) \mathbf{u}_J^{\Gamma,\Gamma}$.
3. Solve $(\mathbf{S}_J^D \otimes \mathbf{S}_J^D) \mathbf{u}_J^{D,D} = -(\mathbf{S}_J^\Gamma \otimes \mathbf{S}_J^\Gamma) \mathbf{u}_J^{\Gamma,\Gamma} - (\mathbf{S}_J^\Gamma \otimes \mathbf{S}_J^D) \mathbf{u}_J^{\Gamma,D} - (\mathbf{S}_J^D \otimes \mathbf{S}_J^\Gamma) \mathbf{u}_J^{D,\Gamma}$.

Herein, we set $\mathbf{u}_J^{\Lambda,\Lambda'} = [u_{J,k,k'}]_{k \in I_J^\Lambda, k' \in I_J^{\Lambda'}}$ for $\Lambda, \Lambda' \in \{D, \Gamma\}$ and

$$\mathbf{g}_J = \left[\left(\langle \mathbf{n}(\mathbf{x}), \text{Cov}[\mathbf{V}]\mathbf{n}(\mathbf{y}) \rangle \left(\frac{\partial \overline{u}}{\partial \mathbf{n}} \otimes \frac{\partial \overline{u}}{\partial \mathbf{n}} \right), \varphi_{J,k} \otimes \varphi_{J,k'} \right)_{L^2(\Gamma_{\text{ref}} \times \Gamma_{\text{ref}})} \right]_{k,k' \in I_J^\Gamma}.$$

The different tensor products of mass matrices and stiffness matrices in this formulation arise from the related tensor products of the bilinear forms $a^D(\cdot, \cdot)$ and $a^\Gamma(\cdot, \cdot)$. Namely, these are

$$a^{\Gamma,\Gamma}(u,v) := (u,v)_{L^2(\Gamma_{\text{ref}} \times \Gamma_{\text{ref}})} \qquad \text{for } u, v \in L^2(\Gamma_{\text{ref}}) \otimes L^2(\Gamma_{\text{ref}}),$$

$$a^{\Gamma,D}(u,v) := ((\text{Id} \otimes \nabla)u, (\text{Id} \otimes \nabla)v)_{L^2(\Gamma_{\text{ref}} \times D_{\text{ref}})} \quad \text{for } u, v \in L^2(\Gamma_{\text{ref}}) \otimes H^1(D_{\text{ref}}),$$

$$a^{D,\Gamma}(u,v) := ((\nabla \otimes \text{Id})u, (\nabla \otimes \text{Id})v)_{L^2(D_{\text{ref}} \times \Gamma_{\text{ref}})} \quad \text{for } u, v \in H^1(D_{\text{ref}}) \otimes L^2(\Gamma_{\text{ref}}),$$

$$a^{D,D}(u,v) := ((\nabla \otimes \nabla)u, (\nabla \otimes \nabla)v)_{L^2(D_{\text{ref}} \times D_{\text{ref}})} \quad \text{for } u, v \in H^1(D_{\text{ref}}) \otimes H^1(D_{\text{ref}}).$$

For the approximation error of the Galerkin solution in $V_J \otimes V_J$, there holds a result similar to Theorem 3, where we set here and in the sequel

$$H_{\text{mix}}^t(D_{\text{ref}} \times D_{\text{ref}}) := H^t(D_{\text{ref}}) \otimes H^t(D_{\text{ref}}),$$

$$H_{\text{mix}}^t(\Gamma_{\text{ref}} \times \Gamma_{\text{ref}}) := H^t(\Gamma_{\text{ref}}) \otimes H^t(\Gamma_{\text{ref}}).$$

Theorem 4 Let $g \in H_{\text{mix}}^t(\Gamma_{\text{ref}} \times \Gamma_{\text{ref}})$ for $0 \leq t \leq 3/2$. Then, if $g_J \in V_J^\Gamma \otimes V_J^\Gamma$ is the L^2-orthogonal projection of the Dirichlet data, the Galerkin solution u_J to the *tensor product Dirichlet problem satisfies*

$$\|u - u_J\|_{L^2(D_{\text{ref}} \times D_{\text{ref}})} \lesssim 2^{-J(t+1/2)} \|g\|_{H^t_{\text{mix}}(\Gamma_{\text{ref}} \times \Gamma_{\text{ref}})}.$$

Proof By a tensor product argument, the proof of this theorem is obtained by summing up the uni-directional error estimates provided by Theorem 3.   □

Unfortunately, the computational complexity of this approximation is of order $(\dim V_J)^2$, which may become prohibitive for increasing level J. A possibility to overcome this obstruction is given by the discretization in *sparse tensor product spaces*. In the following, we shall focus on this approach.

6   Sparse Second Moment Analysis

According to Sect. 4, to leading order, the mean of the solution of the random boundary value problem (8) satisfies the deterministic equation (10). This equation can be discretized straightforwardly by means of finite elements. The resulting system of linear equations may then be solved in optimal complexity, e.g. by a multigrid solver. The solution of the tensor product structured problem (14) is a little more involved and requires another approach in order to maintain the overall complexity.

Instead of discretizing the tensor product boundary value problem (14) in the space $V_J \otimes V_J$, we consider here the discretization in the *sparse tensor product space*

$$\widehat{V_J \otimes V_J} := \sum_{j+j' \leq J} V_j \otimes V_{j'} = \sum_{j+j'=J} V_j \otimes V_{j'} \subset H^1_{\text{mix}}(D_{\text{ref}} \times D_{\text{ref}}). \qquad (23)$$

For the dimension of the sparse tensor product space, we have

$$\dim \widehat{V_J \otimes V_J} \approx \dim V_J \log(\dim V_J)$$

instead of $(\dim V_J)^2$, which is the dimension of $V_J \otimes V_J$, cf. [4]. Thus, the dimension of the sparse tensor product space is substantially smaller than that of the full tensor product space.

The following lemma, proven in [34, 39], tells us that the approximation power in the sparse tensor product space is nearly as good as in the full tensor product space, provided that the given function has some extra regularity in terms of bounded mixed derivatives.

Lemma 1 *For $0 \le t < 3/2$, $t \le q \le 2$ there holds the error estimate*

$$
\inf_{\hat{v}_J \in \widehat{V_J \otimes V_J}} \|v - \hat{v}_J\|_{H^t_{\mathrm{mix}}(D_{\mathrm{ref}} \times D_{\mathrm{ref}})} \lesssim
\begin{cases}
2^{J(t-q)} \sqrt{J} \|v\|_{H^q_{\mathrm{mix}}(D_{\mathrm{ref}} \times D_{\mathrm{ref}})}, & \text{if } q = 2, \\
2^{J(t-q)} \|v\|_{H^q_{\mathrm{mix}}(D_{\mathrm{ref}} \times D_{\mathrm{ref}})}, & \text{otherwise,}
\end{cases}
$$

provided that $v \in H^q_{\mathrm{mix}}(D_{\mathrm{ref}} \times D_{\mathrm{ref}})$.

This lemma gives rise to an estimate for the Galerkin approximation $\widehat{\mathrm{Cov}[\delta u]}_J$ of (14) in the sparse tensor product space $\widehat{V_J \otimes V_J}$, see e.g. [18, Proposition 5].

Corollary 1 *The Galerkin approximate $\widehat{\mathrm{Cov}[\delta u]}_J \in \widehat{V_J \otimes V_J}$ to (14) satisfies the error estimate*

$$
\left\| \mathrm{Cov}[\delta u] - \widehat{\mathrm{Cov}[\delta u]}_J \right\|_{L^2(D_{\mathrm{ref}} \times D_{\mathrm{ref}})} \lesssim 2^{-2J} J \| \mathrm{Cov}[\delta u] \|_{H^2_{\mathrm{mix}}(D_{\mathrm{ref}} \times D_{\mathrm{ref}})}
$$

provided that the given data are sufficiently smooth.

The Galerkin discretization of (14) in the sparse tensor product space is now rather similar to the approach in [18], where *sparse multilevel frames*, cf. [20], have been employed for the discretization. We can considerably improve this approach by combining it with the idea from [22]: Instead of dealing with all combinations which occur in the discretization by a sparse frame for each of the four subproblems on $\Gamma_{\mathrm{ref}} \times \Gamma_{\mathrm{ref}}$, on $D_{\mathrm{ref}} \times \Gamma_{\mathrm{ref}}$, on $\Gamma_{\mathrm{ref}} \times D_{\mathrm{ref}}$ and in $D_{\mathrm{ref}} \times D_{\mathrm{ref}}$, we shall employ the *combination technique*, cf. [14]. Then, we have only to compute combinations of the solution on two consecutive levels instead of all combinations.

The analogue to the ansatz (22) for the Galerkin approximation in the sparse tensor product space reads

$$
\widehat{\mathrm{Cov}[\delta u]}_J = \sum_{j+j' \le J} \sum_{k \in I_j} \sum_{k' \in I_{j'}} \hat{u}_{j,j',k,k'} (\varphi_{j,k} \otimes \varphi_{j',k'})
$$

$$
= \widehat{\mathrm{Cov}[\delta u]}_J^{D,D} + \widehat{\mathrm{Cov}[\delta u]}_J^{D,\Gamma} + \widehat{\mathrm{Cov}[\delta u]}_J^{\Gamma,D} + \widehat{\mathrm{Cov}[\delta u]}_J^{\Gamma,\Gamma}
$$

(24)

with

$$
\widehat{\mathrm{Cov}[\delta u]}_J^{\Lambda,\Lambda'} = \sum_{j+j' \le J} \sum_{k \in I_j^{\Lambda}} \sum_{k' \in I_{j'}^{\Lambda'}} \hat{u}_{j,j',k,k'} (\varphi_{j,k} \otimes \varphi_{j',k'}) \in \widehat{V_J^{\Lambda} \otimes V_J^{\Lambda'}}
$$

(25)

for $\Lambda, \Lambda' \in \{D, \Gamma\}$. The basic idea of our approach is to define *detail spaces* with respect to Galerkin projections in order to remove the redundancy in the sparse frame ansatz for the subproblems (25). We need thus the Galerkin projection $P_j : H^1_0(D_{\mathrm{ref}}) \to V_j^D$, $w \mapsto P_j w$ defined via

$$
\left(\nabla(w - P_j w), \nabla v_j \right)_{L^2(D_{\mathrm{ref}})} = 0 \quad \text{for all } v_j \in V_j^D
$$

and the L^2-orthogonal projection $Q_j \colon L^2(\Gamma_{\text{ref}}) \to V_j^\Gamma$, $w \mapsto Q_j w$, defined via

$$\left((w - Q_j w), v_j \right)_{L^2(\Gamma_{\text{ref}})} = 0 \quad \text{for all } v_j \in V_j^\Gamma.$$

Furthermore, we introduce the related *detail projections*

$$\Theta_j^D := P_j - P_{j-1}, \quad \text{where } P_{-1} := 0,$$

and

$$\Theta_j^\Gamma := Q_j - Q_{j-1}, \quad \text{where } Q_{-1} := 0.$$

With the detail projections at hand, we define the *detail spaces*

$$W_j^D := \Theta_j^D H_0^1(D_{\text{ref}}) = (P_j - P_{j-1}) H_0^1(D_{\text{ref}}) \subset V_j^D$$

and

$$W_j^\Gamma := \Theta_j^\Gamma L^2(\Gamma) = (Q_j - Q_{j-1}) L^2(\Gamma) \subset V_j^\Gamma.$$

Obviously, it holds $V_j^\Lambda = V_{j-1}^\Lambda \oplus W_j^\Lambda$ for $\Lambda \in \{D, \Gamma\}$. This gives rise to the decompositions

$$V_J^\Lambda = W_0^\Lambda \oplus W_1^\Lambda \oplus \ldots \oplus W_J^\Lambda \text{ for } \Lambda \in \{D, \Gamma\}.$$

Especially, these decompositions are orthogonal with respect to their defining inner products.

Lemma 2 *It holds*

$$(\nabla w_j, \nabla w_{j'})_{L^2(D_{\text{ref}})} = 0 \quad \text{for } w_j \in W_j^D, \ w_{j'} \in W_{j'}^D \text{ and } j' \neq j$$

as well as

$$(w_j, w_{j'})_{L^2(D_{\text{ref}})} = 0 \quad \text{for } w_j \in W_j^\Gamma, \ w_{j'} \in W_{j'}^\Gamma \text{ and } j \neq j'.$$

Proof We show the assertion for the spaces W_j^D. The proof for the spaces W_j^Γ is analogous. Without loss of generality, let $j > j'$. Otherwise, due to the symmetry of the inner products, we may interchange the roles of j and j'. Let $w_j = \Theta_j v \in W_j^\Gamma$ for some $v \in H_0^1(D_{\text{ref}})$ and $w_{j'} \in W_{j'}^\Gamma \subset V_{j'}^\Gamma$. Then, since $j - 1 \geq j'$, we have that

$$(\nabla P_j v, \nabla w_{j'})_{L^2(D_{\text{ref}})} = (\nabla v, \nabla w_{j'})_{L^2(D_{\text{ref}})}$$

and

$$(\nabla P_{j-1}v, \nabla w_{j'})_{L^2(D_{\mathrm{ref}})} = (\nabla v, \nabla w_{j'})_{L^2(D_{\mathrm{ref}})}.$$

Thus, we obtain

$$(\nabla w_j, \nabla w_{j'})_{L^2(D_{\mathrm{ref}})} = (\nabla P_j v, \nabla w_{j'})_{L^2(D_{\mathrm{ref}})} - (\nabla P_{j-1}v, \nabla w_{j'})_{L^2(D_{\mathrm{ref}})}$$
$$= (\nabla v, \nabla w_{j'})_{L^2(D_{\mathrm{ref}})} - (\nabla v, \nabla w_{j'})_{L^2(D_{\mathrm{ref}})} = 0.$$

$\square$

Now, we shall rewrite the sparse tensor product spaces given by (23) according to

$$\widehat{V_J^\Lambda \otimes V_J^{\Lambda'}} = \sum_{j+j'=J} V_j^\Lambda \otimes V_{j'}^{\Lambda'} = \sum_{j+j'=J} \left(\bigoplus_{i=0}^{j} W_i^\Lambda \right) \otimes V_{j'}^{\Lambda'} = \bigoplus_{j=0}^{J} W_j^\Lambda \otimes V_{J-j}^{\Lambda'}.$$

Thus, fixing a basis $\psi_{j,k} \in W_j^\Lambda$ for $\Lambda \in \{D, \Gamma\}$, we have for the subproblems (25) the formulation

$$\widehat{\mathrm{Cov}[\delta u]}_J^{\Lambda,\Lambda'} = \bigoplus_{j=0}^{J} \sum_{k \in I_j^\Lambda} \sum_{k' \in I_{J-j}^{\Lambda'}} \hat{u}_{j,J-j,k,k'}(\psi_{j,k} \otimes \varphi_{J-j,k'}) \quad \text{for } \Lambda, \Lambda' \in \{D, \Gamma\}.$$

(26)

Taking further into account the orthogonality described by Lemma 2, we can show that the computation of $\widehat{\mathrm{Cov}[\delta u]}_J^{\Lambda,\Lambda'}$ for $\Lambda, \Lambda' \in \{D, \Gamma\}$ decouples into independent subproblems.

Lemma 3 *Let* $\Lambda, \Lambda' \in \{D, \Gamma\}$. *For* $\hat{v}_j \in W_j^\Lambda \otimes V_{J-j}^{\Lambda'}$ *and* $\hat{v}_{j'} \in W_{j'}^\Lambda \otimes V_{J-j'}^{\Lambda'}$, *there holds*

$$a^{\Lambda,\Lambda'}(\hat{v}_j, \hat{v}_{j'}) = 0 \quad \text{if } j \neq j'.$$

Proof We show the proof for the case $\Lambda = \Gamma$ and $\Lambda' = D$. The other cases are analogous, see also [22, Lemma 6]. Assume that

$$\hat{v}_j = \sum_{i \in I} \alpha_i(\psi_{j,i} \otimes \varphi_{J-j,i}) \quad \text{and} \quad \hat{v}_{j'} = \sum_{i \in I'} \beta_i(\psi_{j',i} \otimes \varphi_{J-j',i})$$

is a representation of $\hat{v}_j \in W_j^\Gamma \otimes V_{J-j}^D$ and $\hat{v}_{j'} \in W_{j'}^\Gamma \otimes V_{J-j'}^D$, respectively, for some finite index sets $\mathcal{I}, \mathcal{I}' \subset \mathbb{N}$. Then, we obtain

$$a^{\Gamma,D}(\hat{v}_j, \hat{v}_{j'})$$

$$= \left((\text{Id} \otimes \nabla) \sum_{i \in \mathcal{I}} \alpha_i(\psi_{j,i} \otimes \varphi_{J-j,i}), (\text{Id} \otimes \nabla) \sum_{i' \in \mathcal{I}'} \beta_i(\psi_{j',i'} \otimes \varphi_{J-j',i'}) \right)_{L^2(\Gamma_{\text{ref}} \times D_{\text{ref}})}$$

$$= \sum_{i \in \mathcal{I}} \sum_{i' \in \mathcal{I}'} \alpha_i \beta_{i'} (\psi_{j,i}, \psi_{j',i'})_{L^2(\Gamma_{\text{ref}})} (\nabla \varphi_{J-j,i}, \nabla \varphi_{J-j',i'})_{L^2(D_{\text{ref}})} = 0$$

whenever $j \neq j'$ due to Lemma 2. $\qquad \square$

This lemma tells us that, given $\widehat{\text{Cov}[\delta u]}_J^{\Gamma,\Gamma}$, the computation of $\widehat{\text{Cov}[\delta u]}_J^{\Lambda,\Lambda'}$ for $\Lambda, \Lambda' \in \{D, \Gamma\}$ decouples into $J + 1$ subproblems. It holds

$$\widehat{\text{Cov}[\delta u]}_J^{\Lambda,\Lambda'} = \sum_{j=0}^J \hat{v}_j,$$

where $\hat{v}_j \in W_j^\Lambda \otimes V_{J-j}^{\Lambda'}$ is the solution to the following Galerkin formulation:

Find $\hat{v}_j \in W_j^\Lambda \otimes V_{J-j}^{\Lambda'}$ such that

$$a^{\Lambda,\Lambda'}(\hat{v}_j, \hat{w}) = \text{rhs}^{\Lambda,\Lambda'}(\hat{w}) \quad \text{for all } \hat{w} \in W_j^\Lambda \otimes V_{J-j}^{\Lambda'}.$$

Herein, we set

$$\text{rhs}^{\Lambda,\Lambda'}(\hat{w}) := \begin{cases} -a^{D,\Gamma}\left(\widehat{\text{Cov}[\delta u]}_J^{\Gamma,\Gamma}, \hat{w}\right), & \Lambda = D, \ \Lambda' = \Gamma, \\ -a^{\Gamma,D}\left(\widehat{\text{Cov}[\delta u]}_J^{\Gamma,\Gamma}, \hat{w}\right), & \Lambda = \Gamma, \ \Lambda' = D, \\ -a^{D,D}\left(\widehat{\text{Cov}[\delta u]}_J^{D,\Gamma} + \widehat{\text{Cov}[\delta u]}_J^{\Gamma,D} + \widehat{\text{Cov}[\delta u]}_J^{\Gamma,\Gamma}, \hat{w}\right), & \Lambda = D, \ \Lambda' = D. \end{cases}$$

$$(27)$$

By taking into account the definition of the detail spaces, we end up with the final representation of the solution to (14) in the sparse tensor product space, which is known as the combination technique.

Theorem 5 *Given* $\widehat{\text{Cov}[\delta u]}_J^{\Gamma,\Gamma}$, *the computation of* $\widehat{\text{Cov}[\delta u]}_J^{\Lambda,\Lambda'}$ *for* $\Lambda, \Lambda' \in \{D, \Gamma\}$ *decouples as follows. It holds*

$$\widehat{\text{Cov}[\delta u]}_J^{\Lambda,\Lambda'} = \sum_{j=0}^J p_{j,J-j} - p_{j-1,J-j}, \qquad (28)$$

where $p_{j,J-j} \in V_j^\Lambda \otimes V_{J-j}^{\Lambda'}$ and $p_{j-1,J-j} \in V_{j-1}^\Lambda \otimes V_{J-j}^{\Lambda'}$ satisfy the following subproblems which are defined relative to full tensor product spaces:

Find $p_{j,j'} \in V_j^\Lambda \otimes V_{j'}^{\Lambda'}$ such that

$$a^{\Lambda,\Lambda'}(p_{j,j'}, q_{j,j'}) = \mathrm{rhs}^{\Lambda,\Lambda'}(q_{j,j'}) \quad \text{for all } q_{j,j'} \in V_j^\Lambda \otimes V_{j'}^{\Lambda'}.$$

Here, the right hand side is given according to (27).

Proof The proof of this theorem is a consequence of the previous lemma together with the definition of the detail spaces W_j^Λ for $\Lambda \in \{D, \Gamma\}$. $\qquad \square$

7 Numerical Implementation

Our numerical realization heavily relies on the sparse frame discretization of the model problem presented in [18]. Nevertheless, in contrast to this work, we make here use of the fact that we already obtain a sparse tensor product representation of the solution if we have the representations in the spaces $V_j \otimes V_{J-j}$ and $V_{j-1} \otimes V_{J-j}$. This means that it is sufficient to compute the diagonal $(j, J-j)$ for $j = 0, \ldots, J$ and the subdiagonal $(j, J-j-1)$ for $j = 0, \ldots, J-1$ of a sparse frame representation. Moreover, each block in this representation corresponds to the solution of a tensor product subproblem as stated in Theorem 5. The corresponding right hand sides are obtained by means of the matrix-vector product in the frame representation. Therefore, in this context, the combination technique can be considered as an improved solver for the approach presented in [18], which results in a remarkable speed-up. In the sequel, we describe this approach in detail.

We start by discretizing the Dirichlet data. The proceeding is as considered in [17]. Setting $\mathcal{J}_0 := I_0^\Gamma$ and $\mathcal{J}_j := I_j^\Gamma \setminus I_{j-1}^\Gamma$ for $j > 0$, the hierarchical basis in $\mathrm{span}\{\varphi_{j,k} \in V_j : k \in I_j^\Gamma\}$ is given by $\bigcup_{j=0}^J \{\varphi_{j,k}\}_{k \in \mathcal{J}_j}$. We replace the normal part of the covariance by its piecewise linear sparse grid interpolant, cf. [4],

$$\langle \mathbf{n}(\mathbf{x}), \mathrm{Cov}[\mathbf{V}]\mathbf{n}(\mathbf{y}) \rangle \approx \left(\sum_{j+j' \le J} \sum_{k \in \mathcal{J}_j} \sum_{k' \in \mathcal{J}_{j'}} \gamma_{(j,k),(j',k')}(\varphi_{j,k} \otimes \varphi_{j',k'}) \right)\bigg|_{\Gamma_{\mathrm{ref}} \times \Gamma_{\mathrm{ref}}}.$$

Thus, the coefficient vector $\mathbf{g}_{j,j'}$ of the Dirichlet data becomes

$$\mathbf{g}_{j,j'} = \sum_{\ell+\ell' \le J} (\mathbf{B}_{j,\ell} \otimes \mathbf{B}_{j',\ell'})[\gamma_{(\ell,k),(\ell',k')}]_{k \in \mathcal{J}_j, k' \in \mathcal{J}_{j'}}, \tag{29}$$

where the matrices $\mathbf{B}_{j,j'}$ are given by

$$\mathbf{B}_{j,j'} = \left[a^\Gamma \left(\frac{\partial \overline{u}}{\partial \mathbf{n}} \varphi_{j,k}, \varphi_{j',k'} \right) \right]_{k \in I_j, k' \in \mathcal{J}_{j'}}, \quad 0 \le j, j' \le J.$$

The expression (29) can be evaluated in optimal complexity by applying the matrix-vector multiplication from [43], i.e. the UNIDIRML algorithm. Nevertheless, for the sake of an easier implementation, we employ here the matrix-vector multiplication from [20], which is optimal up to logarithmic factors. In particular, by using prolongations and restrictions, the matrices $\mathbf{B}_{j,j'}$ have to be provided only for the case $j = j'$. Thus, having all right hand sides at hand, we can solve next

$$(\mathbf{G}_j \otimes \mathbf{G}_{j'})\mathbf{p}_{j,j'}^{\Gamma,\Gamma} = \mathbf{g}_{j,j'}$$

for all indices satisfying $j' = J - j$ or $j' = J - j - 1$. With these coefficients, we determine the right hand sides for the problems on $D_{\mathrm{ref}} \times \Gamma_{\mathrm{ref}}$ and $\Gamma_{\mathrm{ref}} \times D_{\mathrm{ref}}$ according to

$$\mathbf{f}_{j,j'}^{D,\Gamma} = -\sum_{\ell+\ell'\leq J} (\mathbf{S}_{j,\ell}^{\Gamma} \otimes \mathbf{G}_{j',\ell'})\mathbf{p}_{\ell,\ell'}^{\Gamma,\Gamma} \quad \text{and} \quad \mathbf{f}_{j,j'}^{\Gamma,D} = -\sum_{\ell+\ell'\leq J} (\mathbf{G}_{j,\ell} \otimes \mathbf{S}_{j',\ell'}^{\Gamma})\mathbf{p}_{\ell,\ell'}^{\Gamma,\Gamma},$$

where the matrices $\mathbf{S}_{j,j'}^{\Gamma}$ and $\mathbf{G}_{j,j'}$ are given by

$$\left.\begin{array}{l}
\mathbf{S}_{j,j'}^{\Gamma} = \left[a^D(\varphi_{j',\ell}, \varphi_{j,k})\right]_{k\in I_j^D, \ell\in I_{j'}^{\Gamma}} \\
\mathbf{G}_{j,j'} = \left[a^{\Gamma}(\varphi_{j',\ell}, \varphi_{j,k})\right]_{k\in I_j^{\Gamma}, \ell\in I_{j'}^{\Gamma}}
\end{array}\right\} \quad 0 \leq j, j' \leq J.$$

Notice that we have $\mathbf{S}_{j,j}^{\Gamma} = \mathbf{S}_j^{\Gamma}$ and $\mathbf{G}_{j,j} = \mathbf{G}_j$, cf. (18) and (19). Now, we can solve

$$(\mathbf{S}_j^D \otimes \mathbf{G}_{j'})\mathbf{p}_{j,j'}^{D,\Gamma} = \mathbf{f}_{j,j'}^{D,\Gamma} \quad \text{and} \quad (\mathbf{G}_j \otimes \mathbf{S}_{j'}^D)\mathbf{p}_{j,j'}^{\Gamma,D} = \mathbf{f}_{j,j'}^{\Gamma,D}$$

for all indices satisfying $j' = J - j$ or $j' = J - j - 1$.

From the solutions $\mathbf{p}_{j,j'}^{D,\Gamma}$ and $\mathbf{p}_{j,j'}^{\Gamma,D}$, we can finally determine the right hand sides

$$\mathbf{f}_{j,j'}^{D,D} = -\sum_{\ell+\ell'\leq J} (\mathbf{S}_{j,\ell}^{\Gamma} \otimes \mathbf{S}_{j',\ell'}^{\Gamma})\mathbf{p}_{\ell,\ell'}^{\Gamma,\Gamma} + (\mathbf{S}_{j,\ell}^D \otimes \mathbf{S}_{j',\ell'}^{\Gamma})\mathbf{p}_{\ell,\ell'}^{D,\Gamma} + (\mathbf{S}_{j,\ell}^{\Gamma} \otimes \mathbf{S}_{j',\ell'}^D)\mathbf{p}_{\ell,\ell'}^{\Gamma,D},$$

where the matrices $\mathbf{S}_{j,j'}^D$ are given by

$$\mathbf{S}_{j,j'}^D = \left[a^D(\varphi_{j',\ell}, \varphi_{j,k})\right]_{k\in I_j^D, \ell\in I_{j'}^D}, \quad 0 \leq j, j' \leq J.$$

It remains to compute the solutions to

$$(\mathbf{S}_j^D \otimes \mathbf{S}_{j'}^D)\mathbf{p}_{j,j'}^{D,D} = \mathbf{f}_{j,j'}^{D,D}$$

for all indices satisfying $j' = J - j$ or $j' = J - j - 1$.

Appropriate tensorization of the BPX-preconditioner, cf. [3], yields an asymptotically optimal preconditioning for each of the preceding linear systems, cf. [20,

Theorem 7]. Consequently, the computational complexity for their solution is linear, which means it is of the order $O(2^{(j+j')d})$. Moreover, the right hand sides $\mathbf{f}_{j,j'}^{\Lambda,\Lambda'}$ for $\Lambda, \Lambda' \in \{D, \Gamma\}$ can be computed by the UNIDIRML algorithm proposed in [43] with an effort of $O(J2^{dJ})$. We thus obtain the following result:

Theorem 6 *The cost of computing the Galerkin solution* $\widehat{\mathrm{Cov}[\delta u]}_J$ *via the expansion* (28) *is of optimal order* $O(J2^{dJ})$.

Proof For each $0 \le j \le J$ and $\Lambda, \Lambda' \in \{D, \Gamma\}$, the cost to determine $p_{j,J-j}^{\Lambda,\Lambda'}$ and $p_{j-1,J-j}^{\Lambda,\Lambda'}$ is of order $O(2^{dJ})$. Summing over j yields immediately the assertion.   □

8 Numerical Results

To demonstrate the described method, we consider an analytical example on the one hand and a stochastic example on the other hand. In the latter, for a given random domain perturbation described by the random vector field $\mathbf{V}$, we compute the approximate mean $\bar{u}$ in accordance with (10) and the approximate covariance $\mathrm{Cov}[\delta u]$ in accordance with (14). All computations are carried out on a computing server with two Intel(R) Xeon(R) X5550 CPUs with a clock rate of 2.67 GHz and 48 GB of main memory. The computations have been performed single-threaded, i.e. on a single core.

8.1 An Analytical Example

In this analytical example, we want to validate the convergence rates of the combination technique for the sparse tensor product solution of tensor product Dirichlet problems. To that end, consider the tensor product boundary value problem

$$
\begin{aligned}
(\Delta \otimes \Delta)u &= 0 & & \text{in } D_{\mathrm{ref}} \times D_{\mathrm{ref}}, \\
(\Delta \otimes \gamma_0^{\mathrm{int}})u &= 0 & & \text{in } D_{\mathrm{ref}} \times \Gamma_{\mathrm{ref}}, \\
(\gamma_0^{\mathrm{int}} \otimes \Delta)u &= 0 & & \text{in } \Gamma_{\mathrm{ref}} \times D_{\mathrm{ref}}, \\
(\gamma_0^{\mathrm{int}} \otimes \gamma_0^{\mathrm{int}})u &= g_1 \otimes g_2 & & \text{on } \Gamma_{\mathrm{ref}} \times \Gamma_{\mathrm{ref}},
\end{aligned}
\tag{30}
$$

where $D_{\mathrm{ref}} = \{\mathbf{x} \in \mathbb{R}^2 : \|\mathbf{x}\|_2 < 1\}$ is the two-dimensional unit disk. We choose g_1 and g_2 to be the traces of harmonic functions. More precisely, we set

$$
g_1(\mathbf{x}) = x_1^2 - x_2^2 \quad \text{and} \quad g_2(\mathbf{x}) = -\frac{1}{2\pi} \log\left(\sqrt{(x_1 - 2)^2 + (x_2 - 2)^2}\right) \quad \text{for } \mathbf{x} \in \Gamma_{\mathrm{ref}}.
$$

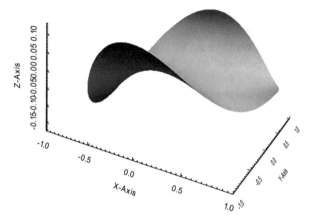

Fig. 1 Trace $u|_{\mathbf{x}=\mathbf{y}}$ of the solution u to (30)

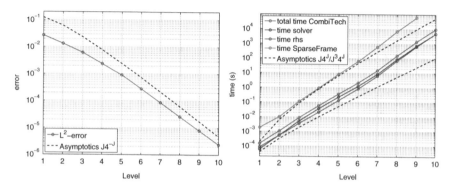

Fig. 2 Relative L^2-error (*left*) and computation times (*right*) of the combination technique in case of the analytic example

Then, the solution u is simply given by the product

$$u(\mathbf{x}, \mathbf{y}) = -\frac{1}{2\pi}(x_1^2 - x_2^2) \log\left(\sqrt{(y_1 - 2)^2 + (y_2 - 2)^2}\right).$$

A visualization of the trace $u|_{\mathbf{x}=\mathbf{y}}$ of this function is found in Fig. 1.

The convergence plot on the left of Fig. 2 shows that the relative L^2-error, indicated by the blue line, exhibits almost the convergence rate predicted in Corollary 1, indicated by the black dashed line. On level 10, there are about 2.1 million degrees of freedom in each spatial variable, which is, up to a logarithmic factor, the number of degrees of freedom appearing in the discretization by the combination technique. Vice versa, a full tensor product discretization on this level would result in about $4.4 \cdot 10^{12}$ degrees of freedom, which is no more feasible.

The plot on the right hand side of Fig. 2 depicts the related computation times. For comparison, we have added here the computation times for the sparse tensor product frame discretization from [18]. The related curve is indicated in green. The computation time consumed by the combination technique is represented by the red curve. Notice that we have set up both methods such that they provide similar accuracies for the approximation of the solution. From level 3–9, the combination technique is in average a factor 30 faster than the frame discretization, where the speed-up is growing when the level increases. Nevertheless, it seems that, from level 7 on, both methods do not achieve the theoretical rate of $J^3 4^J$ anymore.

We present in the plot on the right hand side of Fig. 2 also the time consumed for exclusively computing the respective right hand sides for the combination technique, indicated by the blue line. As can be seen, on the higher levels, this computation takes nearly half of the total computational time. A potential improvement could thus be made by using the matrix-vector product from [43]. Finally, we have plotted the time which is needed for exclusively solving the linear systems by the tensor product solver. Here, it seems that we have the optimal behavior of order $J 4^J$ up to level 7. Then, also this rate deteriorates.

8.2   The Poisson Equation on the Random Unit Disc

For this example, we consider also $D_{\text{ref}} = \{\mathbf{x} \in \mathbb{R}^2 : \|\mathbf{x}\|_2 < 1\}$ as reference domain and the load is set to $f(\mathbf{x}) \equiv 1$. The random vector field $\mathbf{V}$ is provided by its mean $\mathbb{E}[\mathbf{V}](\mathbf{x}) = \mathbf{x}$ and its covariance function

$$\text{Cov}[\mathbf{V}](\mathbf{x}, \mathbf{y}) = \frac{\varepsilon^2}{125} \begin{bmatrix} 5\exp(-4\|\mathbf{x} - \mathbf{y}\|_2^2) & \exp(-0.1\|2\mathbf{x} - \mathbf{y}\|_2^2) \\ \exp(-0.1\|\mathbf{x} - 2\mathbf{y}\|_2^2) & 5\exp(-\|\mathbf{x} - \mathbf{y}\|_2^2) \end{bmatrix}.$$

In Fig. 3, a visualization of the solution $\bar{u}$ to (10) (left) and the variance $\mathbb{V}[\delta u]$ of the solution to (14) (right) is depicted. In order to validate the computational method, we consider a reference solution computed with a quasi-Monte Carlo method based on Halton points. To that end, we have solved the Poisson equation on 10^4 realizations of the random parameter on level $J = 7$ (this corresponds to

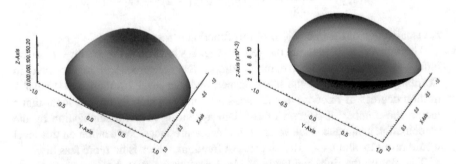

Fig. 3 Solution $\bar{u}$ (*left*) and variance $\mathbb{V}[\delta u]$ (*right*) on the unit disc

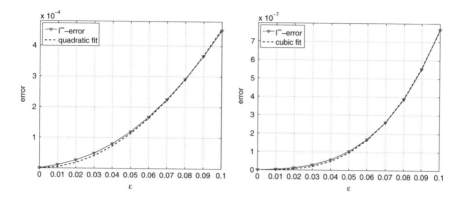

Fig. 4 Error in $\mathbb{E}[u]$ (*left*) and $\mathbb{V}[u]$ (*right*) for increasing values of ε on K

65,536 finite elements). The solutions obtained have then been interpolated on a mesh on level $J = 5$ (this corresponds to 4096 finite elements) for the compactum $K = \{\mathbf{x} \in \mathbb{R}^2 : \|\mathbf{x}\|_2 \leq 0.8\}$. For the combination technique, we set $J = 7$ for the computation of the mean and $J = 9$ for the computation of the variance. The related error plots for the combination technique with respect to different values of ε are shown in Fig. 4, where we used the ℓ^∞-norm to measure the error. As can be seen, the error in the mean, found on the left hand side of the figure, exhibits exactly the expected quadratic behavior, whereas the error in the variance, found on the right hand side of the figure, shows exactly a cubic rate.

References

1. S. Bochner, Integration von Funktionen, deren Werte die Elemente eines Vektorraumes sind. Fundamenta Mathematicae **20**(1), 262–176 (1933)
2. J.H. Bramble, J.T. King, A robust finite element method for nonhomogeneous Dirichlet problems in domains with curved boundaries. Math. Comput. **63**(207), 1–17 (1994)
3. J.H. Bramble, J.E. Pasciak, J. Xu, Parallel multilevel preconditioners. Math. Comput. **55**(191), 1–22 (1990)
4. H.-J. Bungartz, M. Griebel, Sparse grids. Acta Numerica **13**, 147–269 (2004)
5. C. Canuto, T. Kozubek, A fictitious domain approach to the numerical solution of PDEs in stochastic domains. Numerische Mathematik **107**(2), 257–293 (2007)
6. J.E. Castrillon-Candas, F. Nobile, R.F. Tempone, Analytic regularity and collocation approximation for PDEs with random domain deformations (2013). ArXiv e-prints 1312.7845
7. A. Chernov, C. Schwab, First order k-th moment finite element analysis of nonlinear operator equations with stochastic data. Math. Comput. **82**(284), 1859–1888 (2013)
8. C. Delfour, J. Zolésio, *Shapes and Geometries: Metrics, Analysis, Differential Calculus, and Optimization*. Advances in Design and Control, 2nd edn. (Society for Industrial and Applied Mathematics, Philadelphia, 2011)
9. J. Diestel, J.J. Uhl, *Vector Measures*. Mathematical Surveys and Monographs (American Mathematical Society, Providence, 1977)

10. K. Eppler, Boundary integral representations of second derivatives in shape optimization. Discussiones Mathematicae (Differ. Incl. Control Optim.) **20**, 63–78 (2000)
11. K. Eppler, Optimal shape design for elliptic equations via BIE-methods. J. Appl. Math. Comput. Sci. **10**, 487–516 (2000)
12. K. Eppler, Efficient Shape Optimization Algorithms for Elliptic Boundary Value Problems. Habilitation Thesis. Technische Universität Berlin, Germany (2007)
13. G.J. Fix, M.D. Gunzburger, J.S. Peterson, On finite element approximations of problems having inhomogeneous essential boundary conditions. Comput. Math. Appl. **9**(5), 687–700 (1983)
14. M. Griebel, M. Schneider, C. Zenger, A combination technique for the solution of sparse grid problems, in *Iterative Methods in Linear Algebra*, ed. by P. de Groen, R. Beauwens (IMACS, Elsevier, North Holland, 1992), pp. 263–281
15. W. Hackbusch, *Multi-grid Methods and Applications* (Springer, Berlin, 1985)
16. P.R. Halmos, *Measure Theory*. Graduate Texts in Mathematics (Springer, New York, 1976)
17. H. Harbrecht, Finite element based second moment analysis for elliptic problems in stochastic domains, in *Numerical Mathematics and Advanced Applications 2009*, ed. by G. Kreiss, P. Lötstedt, A. Målqvist, M. Neytcheva (Springer, Berlin/Heidelberg, 2010), pp. 433–441
18. H. Harbrecht, A finite element method for elliptic problems with stochastic input data. Appl. Numer. Math. **60**(3), 227–244 (2010)
19. H. Harbrecht, J. Li, First order second moment analysis for stochastic interface problems based on low-rank approximation. ESAIM: Math. Model. Numer. Anal. **47**, 1533–1552 (2013)
20. H. Harbrecht, R. Schneider, C. Schwab, Multilevel frames for sparse tensor product spaces. Numerische Mathematik **110**(2), 199–220 (2008)
21. H. Harbrecht, R. Schneider, C. Schwab, Sparse second moment analysis for elliptic problems in stochastic domains. Numerische Mathematik **109**(3), 385–414 (2008)
22. H. Harbrecht, M. Peters, M. Siebenmorgen, Combination technique based k-th moment analysis of elliptic problems with random diffusion. J. Comput. Phys. **252**, 128–141 (2013)
23. H. Harbrecht, M. Peters, M. Siebenmorgen, Numerical solution of elliptic diffusion problems on random domains. Preprint 2014–08, Mathematisches Institut, Universität Basel (2014)
24. E. Hille, R.S. Phillips, *Functional Analysis and Semi-groups*, vol. 31 (American Mathematical Society, Providence, 1957)
25. A. Kirsch, The domain derivative and two applications in inverse scattering theory. Inverse Probl. **9**(1), 81–96 (1993)
26. M. Loève, *Probability Theory. I+II*, 4th edn. Number 45 in Graduate Texts in Mathematics (Springer, New York, 1977)
27. P.S. Mohan, P.B. Nair, A.J. Keane, Stochastic projection schemes for deterministic linear elliptic partial differential equations on random domains. Int. J. Numer. Methods Eng. **85**(7), 874–895 (2011)
28. C. Pflaum, A. Zhou, Error analysis of the combination technique. Numerische Mathematik **84**(2), 327–350 (1999)
29. O. Pironneau, Optimal shape design for elliptic systems, in *System Modeling and Optimization*, ed. by R. Drenick, F. Kozin. Volume 38 of Lecture Notes in Control and Information Sciences (Springer, Berlin/Heidelberg, 1982), pp. 42–66
30. R. Potthast, Frechét differentiability of boundary integral operators in inverse acoustic scattering. Inverse Probl. **10**(2), 431–447 (1994)
31. R. Potthast, Frechét-Differenzierbarkeit von Randintegraloperatoren und Randwertproblemen zur Helmholtzgleichung und zu den zeitharmonischen Maxwellgleichungen. PHD Thesis. Georg-August-Universität Göttingen, Germany (1994)
32. C. Reisinger, Analysis of linear difference schemes in the sparse grid combination technique. IMA J. Numer. Anal. **33**(2), 544–581 (2013)
33. C. Schwab, R.A. Todor, Sparse finite elements for elliptic problems with stochastic loading. Numerische Mathematik **95**(4), 707–734 (2003)
34. C. Schwab, R.A. Todor, Sparse finite elements for stochastic elliptic problems–higher order moments. Computing **71**(1), 43–63 (2003)

35. C. Schwab, R. Todor, Karhunen-Loève approximation of random fields by generalized fast multipole methods. J. Comput. Phys. **217**, 100–122 (2006)
36. J. Simon, Differentiation with respect to the domain in boundary value problems. Numer. Funct. Anal. Optim. **2**(7–8), 649–687 (1980)
37. J. Sokolowski, J.P. Zolésio, *Introduction to Shape Optimization. Shape Sensitivity Analysis.* Springer Series in Computational Mathematics (Springer, Berlin/Heidelberg, 1992)
38. D.M. Tartakovsky, D. Xiu, Stochastic analysis of transport in tubes with rough walls. J. Comput. Phys. **217**(1), 248–259 (2006)
39. T. von Petersdorff, C. Schwab, Sparse finite element methods for operator equations with stochastic data. Appl. Math. **51**(2), 145–180 (2006)
40. J. Wloka, *Partielle Differentialgleichungen: Sobolevräume und Randwertaufgaben* (Teubner, Stuttgart, 1982)
41. D. Xiu, D.M. Tartakovsky, Numerical methods for differential equations in random domains. SIAM J. Sci. Comput. **28**(3), 1167–1185 (2006)
42. Y. Xu, A. Zhou, Fast Boolean approximation methods for solving integral equations in high dimensions. J. Integral Equ. Appl. **16**(1), 83–110 (2004)
43. A. Zeiser, Fast matrix-vector multiplication in the sparse-grid Galerkin method. J. Sci. Comput. **47**(3), 328–346 (2011)

Adaptive Sparse Grids and Extrapolation Techniques

Brendan Harding

Abstract In this paper we extend the study of (dimension) adaptive sparse grids by building a lattice framework around projections onto hierarchical surpluses. Using this we derive formulas for the explicit calculation of combination coefficients, in particular providing a simple formula for the coefficient update used in the adaptive sparse grids algorithm. Further, we are able to extend error estimates for classical sparse grids to adaptive sparse grids. Multi-variate extrapolation has been well studied in the context of sparse grids. This too can be studied within the adaptive sparse grids framework and doing so leads to an adaptive extrapolation algorithm.

1 Introduction

In [8] sparse grids were studied via projections onto function space lattices. This was a powerful tool which led to the adaptive sparse grid algorithm, in which approximations to sparse grid solutions by the summation of many coarse approximations is iteratively improved by adding additional coarse approximations to the solution space. By viewing the coarse approximations as projections of the true solution onto the coarse solution space, an underlying lattice structure was exploited to provide an update formula for the combination coefficients. This work will be reviewed in Sect. 2.

The hierarchical surplus is an important notion in the study of sparse grids, see for example [1, 3, 5]. The main contribution of this paper is the study of projections onto these hierarchical surpluses within the adaptive sparse grids framework. In Sect. 3.1 we will introduce these projections in relation to the projections onto coarse approximation spaces and derive several well-known properties. These results are then applied in Sect. 3.2 in the study of combination coefficients. Finally, in Sect. 3.3 we show how error analysis of the classical combination technique can be extended to adaptive sparse grids. We derive formulas which may be used to compute a priori error estimates within an adaptive sparse grids algorithm.

B. Harding (✉)
Mathematical Sciences Institute, Australian National University, Canberra, ACT, Australia
e-mail: brendan.harding@anu.edu.au

© Springer International Publishing Switzerland 2016
J. Garcke, D. Pflüger (eds.), *Sparse Grids and Applications – Stuttgart 2014*,
Lecture Notes in Computational Science and Engineering 109,
DOI 10.1007/978-3-319-28262-6_4

79

Extrapolation techniques involve the computation of a more accurate solution from several less accurate ones. Given the many coarse approximations typically used in the computation of adaptive sparse grids, it is natural to see if extrapolation techniques can be used to obtain more accurate approximations. Multi-variate extrapolation has been extensively studied in the context of sparse grids [2, 9–11]. In Sect. 4 we study adaptive sparse grids applied to extrapolations onto coarse approximation spaces using the framework which has been developed. In particular we study the combination coefficients which provide extrapolations. This leads to results which may be applied to an adaptive extrapolation algorithm.

2 Background

Consider function spaces which are tensor products $V = V^1 \otimes \cdots \otimes V^d$ where the components V^k ($k \in \{1, \ldots, d\}$) are nested hierarchical spaces, that is,

$$V_0^k \subset V_1^k \subset \cdots \subset V_{m_k}^k \subseteq V^k,$$

where m_k is possibly ∞. We form coarse approximation spaces of V via the tensor product $V_{\underline{i}} = V_{i_1}^1 \otimes \cdots \otimes V_{i_d}^d$ where $\underline{i}$ is a multi-index $(i_1, \ldots, i_d) \in \mathbb{N}^d$ with $i_k \leq m_k$ for $k = 1 \ldots, d$. We claim that the collection of these $V_{\underline{i}}$ form a function space lattice, that is a partially ordered set $(\mathcal{L}, \leq)$ in which any two members have a unique least upper bound ($\vee$) and unique greatest lower bound ($\wedge$). An example is the spaces of piecewise multi-linear functions $V_{\underline{i}}$ which interpolate function values at points on uniform grids which are successively refined by a factor of 2 in each dimension starting from some initial V_0.

We define an ordering on these spaces given by $V_i^k \leq V_j^k$ iff $V_i^k \subseteq V_j^k$ for $i, j \in \mathbb{N}$. This leads to a natural definition of the greatest lower bound and least upper bound, namely $V_i^k \wedge V_j^k := V_i^k \cap V_j^k$ and $V_i^k \vee V_j^k := V_i^k \cup V_j^k$ respectively, which defines a lattice on $\{V_0^k, V_1^k, \ldots, V_{m_k}^k\}$. Given the nested nature of the V_i^k we observe that $V_i^k \cap V_j^k = V_{\min\{i,j\}}^k$ and $V_i^k \cup V_j^k = V_{\max\{i,j\}}^k$. Notice that $V_i^k \subseteq V_j^k \Leftrightarrow i \leq j$, that is the ordering of the V_i^k is consistent with the natural ordering of $\{0, 1, \ldots, m_k\} \subset \mathbb{N}$. This ordering of $\mathbb{N}$ leads to a lattice on $\{0, 1, \ldots, m_k\}$ via the greatest lower bound and least upper bound defined by $i \wedge j := \min\{i, j\}$ and $i \vee j := \max\{i, j\}$ respectively. Therefore we observe that $V_i^k \wedge V_j^k = V_{i \wedge j}^k$ and $V_i^k \vee V_j^k = V_{i \vee j}^k$ which effectively 'lifts' the lattice structure of $\{0, 1, \ldots, m_k\}$ to $\{V_0^k, V_1^k, \ldots, V_{m_k}^k\}$.

Now consider the partial ordering induced on the tensor product space V. Given $V_{\underline{i}}$ and $V_{\underline{j}}$ we define the partial ordering $V_{\underline{i}} \leq V_{\underline{j}}$ iff $V_{\underline{i}} \subseteq V_{\underline{j}}$, or equivalently $V_{i_k} \leq V_{j_k}$ for $k = 1, \ldots, d$. Additionally, we define $V_{\underline{i}} \wedge V_{\underline{j}} := (V_{i_1}^1 \wedge V_{j_1}^1) \otimes \cdots \otimes (V_{i_d}^d \wedge V_{j_d}^d)$ and similarly $V_{\underline{i}} \vee V_{\underline{j}} := (V_{i_1}^1 \vee V_{j_1}^1) \otimes \cdots \otimes (V_{i_d}^d \vee V_{j_d}^d)$. Equivalently we can write $V_{\underline{i}} \wedge V_{\underline{j}} = V_{\min\{\underline{i},\underline{j}\}}$ and $V_{\underline{i}} \vee V_{\underline{j}} = V_{\max\{\underline{i},\underline{j}\}}$ where min and max are taken component wise over the multi-indices (e.g. $\min\{\underline{i}, \underline{j}\} = (\min\{i_1, j_1\}, \ldots, \min\{i_d, j_d\})$). Observe that whilst $V_{\underline{i}} \wedge V_{\underline{j}} = V_{\underline{i}} \cap V_{\underline{j}}$ it is generally not true that $V_{\underline{i}} \vee V_{\underline{j}}$ is equal to $V_{\underline{i}} \cup V_{\underline{j}}$. As with the V_i^k, this lattice on the $V_{\underline{i}}$ can be viewed as a 'lifting' of the natural lattice structure on $\mathbb{N}^d$. In particular, given the partial ordering of $\underline{i}, \underline{j} \in \mathbb{N}^d$ defined by $\underline{i} \le \underline{j}$ iff $i_k \le j_k$ for $k = 1, \ldots, d$, then the addition of the binary relations $\underline{i} \wedge \underline{j} := \min\{\underline{i}, \underline{j}\}$ and $\underline{i} \vee \underline{j} := \max\{\underline{i}, \underline{j}\}$ describes a lattice. Thus we observe a 'lifting' of the lattice structure via $V_{\underline{i}} \le V_{\underline{j}} \Leftrightarrow \underline{i} \le \underline{j}$, $V_{\underline{i}} \wedge V_{\underline{j}} = V_{\underline{i} \wedge \underline{j}}$ and $V_{\underline{i}} \vee V_{\underline{j}} = V_{\underline{i} \vee \underline{j}}$.

The notion of a covering element is also useful. Given a partial ordered set $(\mathcal{L}, \le)$ then $b \in \mathcal{L}$ is said to cover $a \in \mathcal{L}$ if $a < b$ (where $a < b$ if $a \le b$ and $a \ne b$) and there is no $c \in \mathcal{L}$ such that $a < c < b$. We use the notation $a \prec b$ to denote that b covers a. Thus, given the partial ordering on $\mathbb{N}^d$ described above, $\underline{i} \prec \underline{j}$ if $j_k = i_k + 1$ for exactly one $k \in \{1, \ldots, d\}$ with the remaining components being equal. Similarly, as the partial ordering on $\mathbb{N}^d$ lifts to the $V_{\underline{i}}$, it is immediate that $V_{\underline{i}} \prec V_{\underline{j}} \Leftrightarrow \underline{i} \prec \underline{j}$.

Consider a family of projections $P_{i_k}^k : V^k \to V_{i_k}^k$ from which the tensor product provides the projections $P_{\underline{i}} = \bigotimes_k P_{i_k}^k : V \to V_{\underline{i}}$. The existence of such projections is given by the following proposition [8].

Proposition 1 *For every lattice space generated from a tensor product of nested hierarchical spaces, we have:*

- *there are linear operators $P_{\underline{i}}$ on V with range $R(P_{\underline{i}}) = V_{\underline{i}}$ and $P_{\underline{i}} P_{\underline{j}} = P_{\underline{i} \wedge \underline{j}}$.*
- *Consequently $P_{\underline{i}} P_{\underline{i}} = P_{\underline{i}}$ and $P_{\underline{i}} P_{\underline{j}} = P_{\underline{j}} P_{\underline{i}}$.*

Let I be a subset of the lattice of multi-indices on $\mathbb{N}^d$. We say I is a downset if

$$\underline{i} \in I \,\&\, \underline{j} \le \underline{i} \implies \underline{j} \in I.$$

Given $J \subset \mathbb{N}^d$ we use the notation $J \downarrow$ to denote the smallest downset that contains J. Consider $\mathcal{P}(\mathbb{N}^d)$, i.e. the power set of the set of all multi-indices, and let $\mathcal{D}(\mathbb{N}^d)$ be the subset of the power set containing all of the downsets.

Definition 1 Given $I, J \in \mathcal{D}(\mathbb{N}^d)$ then we define the partial ordering

$$I \le J \Leftrightarrow I \subseteq J$$

Additionally we define the binary relations

$$I \wedge J := I \cap J \quad \text{and} \quad I \vee J := I \cup J.$$

This leads us to the following lemma.

Lemma 1 $\mathcal{D}(\mathbb{N}^d)$ *with the partial ordering and binary operations defined in Definition 1 is a lattice.*

Proof We need only show that given any $I, J \in \mathcal{D}(\mathbb{N}^d)$ then $I \wedge J \in \mathcal{D}(\mathbb{N}^d)$ and $I \vee J \in \mathcal{D}(\mathbb{N}^d)$.

Let $\underline{i} \in I \wedge J = I \cap J$, then $\underline{i} \in I$ and $\underline{i} \in J$. It follows that for each $j \in \mathbb{N}^d$ with $j \leq \underline{i}$ one has $j \in I$ and $j \in J$ since I and J are downsets and therefore $j \in I \wedge J$. As a consequence $I \wedge J$ is a downset.

A similar argument shows that $I \vee J$ is also a downset.   □

For $I, J \in \mathcal{D}(\mathbb{N}^d)$ we also have the cover relation $I \prec J$ iff $J = I \cup \{\underline{i}\}$ for some $\underline{i} \notin I$ for which $j \prec \underline{i} \Rightarrow j \in I$ for all $j \in \mathbb{N}^d$ (or equivalently $j < \underline{i} \Rightarrow j \in I$ for all $j \in \mathbb{N}^d$).

Just as the lattice on $\mathbb{N}^d$ can be lifted to a lattice on $\{V_{\underline{i}}\}_{\underline{i} \in \mathbb{N}^d}$ we can lift the lattice on $\mathcal{D}(\mathbb{N}^d)$ to the so called combination space lattice $\{V_I\}_{I \in \mathcal{D}(\mathbb{N}^d)}$ where

$$V_I := \sum_{\underline{i} \in I} V_{\underline{i}} \,.$$

It is straightforward to show that V_I is a downset if I is itself a downset, and furthermore $V_J \downarrow = V_{J\downarrow}$. Furthermore, we have the induced partial ordering $V_I \leq V_J$ iff $I \leq J$ and the binary relations $V_I \wedge V_J = V_{I \wedge J}$ and $V_I \vee V_J = V_{I \vee J}$ which are lifted from the lattice on $\mathcal{D}(\mathbb{N}^d)$ to define a lattice on $\{V_I\}_{I \in \mathcal{D}(\mathbb{N}^d)}$. Lastly, we also have the covering relation $V_I \prec V_J$ iff $I \prec J$. This brings us to a second proposition [8].

Proposition 2 *Let the lattices $V_{\underline{i}}$ have the projections $P_{\underline{i}}$ as in Proposition 1. Then for $I, J \in \mathcal{D}(\mathbb{N}^d)$ there are linear operators P_I on V with range $R(P_I) = V_I$ such that $P_I P_J = P_{I \cap J}$. Conversely, if P_I is a family of projections with these properties, then $P_{\underline{i}} := P_{\{\underline{i}\}\downarrow}$ defines a family of projections as in Proposition 1.*

The proof given in [8] shows that the linear operators

$$P_I = 1 - \prod_{\underline{i} \in I}(1 - P_{\underline{i}}) \tag{1}$$

are the projections described by Proposition 2. Further, the product may be expanded to the sum

$$P_I = \sum_{\underline{i} \in I} c_{\underline{i}} P_{\underline{i}} \tag{2}$$

where the $c_{\underline{i}} \in \mathbb{Z}$ are commonly referred to as combination coefficients. Unfortunately Proposition 2 doesn't tell us much more about the $c_{\underline{i}}$, this leads us to the following corollary [8].

Corollary 1 *Let $J = I \cup \{\underline{j}\}$ be a covering element of I and let P_I be the family of projections as in Proposition 2 and $P_{\underline{i}} = P_{\{\underline{i}\}\downarrow}$. Then one has:*

$$P_J - P_I = \sum_{\underline{i} \in J} d_{\underline{i}} P_{\underline{i}}$$

where $d_j = 1$ and for $\underline{i} \in I$ we have

$$d_{\underline{i}} = -\sum_{\underline{l} \in I_{\underline{i}|j}} c_{\underline{l}}$$

with $I_{\underline{i}|j} := \{\underline{l} \in I : \underline{j} \wedge \underline{l} = \underline{i}\}$.

Proof Notice that

$$P_J - P_I = P_{\underline{j}} \prod_{\underline{i} \in I} (1 - P_{\underline{i}}) = P_{\underline{j}} - P_{\underline{j}} P_I = P_{\underline{j}} - P_{\underline{j}} \sum_{\underline{i} \in I} c_{\underline{i}} P_{\underline{i}}$$

$$= P_{\underline{j}} - \sum_{\underline{i} \in I} c_{\underline{i}} P_{\underline{j} \wedge \underline{i}} = \sum_{\underline{i} \in J} d_{\underline{i}} P_{\underline{i}}. \quad \square$$

As a result, if we have a solution to a problem using the combination P_I and we add a solution from another function space $V_{\underline{j}}$ such that $V_J = V_{I \cup \{\underline{j}\}}$ covers V_I then the new combination coefficients can be obtained from:

$$P_J = (1)P_{\underline{j}} + \sum_{\underline{i} \in I} (c_{\underline{i}} + d_{\underline{i}}) P_{\underline{i}}$$

where the $d_{\underline{i}}$'s are given by Corollary 1. Using this result, one obtains the adaptive sparse grids algorithm, which takes an approximation $u_I = P_I u$ and produces a better approximation $u_J = P_J u$ by finding the covering element $I \prec J$ which minimises $\mathcal{J}(P_J u)$, where $\mathcal{J}$ is some functional that measures the quality of the approximation $P_J u$.

3 Further Analysis of Adaptive Sparse Grids

3.1 Hierarchical Projections

We introduce new projections which will be an important tool in this section.

Definition 2 For $k = 1, \ldots, d$ and $i_k \in \mathbb{N}$ let $Q_{i_k}^k : V^k \to V^k$ be defined as $Q_{i_k}^k := P_{i_k}^k - P_{i_k-1}^k$ (where $P_{i_k-1}^k := 0$ if $i_k - 1 < 0$).

We now give some basic properties of the $Q_{i_k}^k$.

Lemma 2 $Q_{i_k}^k$ *has the following properties:*

1. $Q_{i_k}^k$ has range in $V_{i_k}^k$,
2. $P_{j_k}^k Q_{i_k}^k = Q_{i_k}^k P_{j_k}^k = 0$ for $j_k < i_k$,
3. $P_{j_k}^k Q_{i_k}^k = Q_{i_k}^k P_{j_k}^k = Q_{i_k}^k$ for $j_k \geq i_k$,
4. $Q_{i_k}^k Q_{i_k}^k = Q_{i_k}^k$,
5. $Q_{j_k}^k Q_{i_k}^k = 0$ for $i_k \neq j_k$,
6. $P_{i_k}^k = \sum_{j_k=0}^{i_k} Q_{j_k}^k$.

Proof The first is immediate as $P_{i_k}^k$ has range $V_{i_k}^k$ and $P_{i_k-1}^k$ has range $V_{i_k-1}^k \subset V_{i_k}^k$. For the second and third, one has

$$P_{j_k}^k Q_{i_k}^k = P_{j_k}^k P_{i_k}^k - P_{j_k}^k P_{i_k-1}^k = \begin{cases} P_{j_k}^k - P_{j_k}^k = 0 & \text{for } j_k < i_k, \\ P_{i_k}^k - P_{i_k-1}^k = Q_{i_k}^k & \text{for } j_k \geq i_k. \end{cases}$$

It follows from these two that

$$Q_{j_k}^k Q_{i_k}^k = (P_{j_k}^k - P_{j_k-1}^k)Q_{i_k}^k = \begin{cases} Q_{i_k}^k - 0 = Q_{i_k}^k & \text{for } j_k = i_k, \\ Q_{i_k}^k - Q_{i_k}^k = 0 & \text{for } j_k > i_k, \\ 0 - 0 = 0 & \text{for } j_k < i_k. \end{cases}$$

The last equality is a result of the telescoping sum

$$\sum_{j_k=0}^{i_k} Q_{j_k}^k = \sum_{j_k=0}^{i_k} P_{j_k}^k - P_{j_k-1}^k = P_{i_k}^k - P_{-1}^k = P_{i_k}^k. \qquad \square$$

As with the P_i, we define $Q_{\underline{i}}$ by the tensor product $Q_{\underline{i}} := \bigotimes_k Q_{i_k}^k$. This leads us to the following lemma

Lemma 3 *Let $Q_{i_k}^k := P_{i_k}^k - P_{i_k-1}^k$ and $Q_{\underline{i}} := \bigotimes_k Q_{i_k}^k$. Then*

$$Q_{\underline{i}} = \sum_{(0\leq)\underline{j}\leq 1} (-1)^{|\underline{j}|} P_{\underline{i}-\underline{j}}$$

with $P_{\underline{i}-\underline{j}} = 0$ if $i_k - j_k < 0$ for some $k \in \{1, \ldots, d\}$. Furthermore,

$$P_{\underline{i}} = \sum_{(0\leq)\underline{j}\leq\underline{i}} Q_{\underline{j}}.$$

Proof It follows from the definition that

$$Q_{\underline{i}} = \bigotimes_{k=1}^{d} Q_{i_k}^k = \bigotimes_{k=1}^{d}(P_{i_k}^k - P_{i_k-1}^k) = \sum_{(0\leq)\underline{j}\leq 1} (-1)^{|\underline{j}|} \bigotimes_{k=1}^{d} P_{i_k-j_k}^k = \sum_{(0\leq)\underline{j}\leq 1} (-1)^{|\underline{j}|} P_{\underline{i}-\underline{j}}.$$

Since $P_{i_k-1}^k := 0$ if $i_k - 1 < 0$ one obtains $P_{\underline{i}-\underline{j}} = 0$ if $i_k - j_k < 0$ for some $k \in \{1, \ldots, d\}$. For the latter equality we have

$$\sum_{(0\leq)\underline{j}\leq\underline{i}} Q_{\underline{j}} = \sum_{j_1=0}^{i_1} \cdots \sum_{j_d=0}^{i_d} \bigotimes_{k=1}^{d}(P_{j_k}^k - P_{j_k-1}^k) = \bigotimes_{k=1}^{d}\sum_{j_k=0}^{i_k}(P_{j_k}^k - P_{j_k-1}^k) = \bigotimes_{k=1}^{d} P_{i_k}^k = P_{\underline{i}}$$

as required. $\square$

A few more properties of the $Q_{\underline{i}}$ will also be useful.

Lemma 4 *With $P_{\underline{i}}$ and $Q_{\underline{i}}$ as previously defined, one has*

1. $Q_{\underline{i}}Q_{\underline{i}} = Q_{\underline{i}}$,
2. $Q_{\underline{i}}Q_{\underline{j}} = Q_{\underline{j}}Q_{\underline{i}} = 0$ *for* $\underline{j} \neq \underline{i}$,
3. $P_{\underline{i}}Q_{\underline{j}} = Q_{\underline{j}}P_{\underline{i}} = Q_{\underline{j}}$ *for* $\underline{j} \leq \underline{i}$,
4. $P_{\underline{i}}Q_{\underline{j}} = Q_{\underline{j}}P_{\underline{i}} = 0$ *for* $\underline{j} \not\leq \underline{i}$.

Proof Each is a direct consequence of the analogous results shown for the $Q_{i_k}^k$ and $P_{i_k}^k$ in Lemma 2. $\qquad\square$

Given a downset $I \in \mathcal{D}(\mathbb{N}^d)$ let us define $Q_I := \sum_{\underline{i} \in I} Q_{\underline{i}}$. Given that $P_{\underline{i}} = \sum_{\underline{j} \leq \underline{i}} Q_{\underline{j}}$ it would be reasonable to expect that $Q_I = P_I$ and indeed this is the case.

Proposition 3 *Let $I \in \mathcal{D}(\mathbb{N}^d)$, then $P_I = Q_I$.*

Proof First we define the maximal elements of I as $\max I := \{\underline{i} \in I : \{\underline{j} \in I : \underline{j} \geq \underline{i}\} = \{\underline{i}\}\}$. Now for each $\underline{j} \in I \setminus \max I$ there exists $\underline{l} \in \max I$ such that $\underline{j} < \underline{l}$ and thus

$$1 - \prod_{\underline{i} \in I}(1 - P_{\underline{i}}) = 1 - (1 - P_{\underline{j}}) \prod_{\underline{i} \in I \setminus \{\underline{j}\}}(1 - P_{\underline{i}})$$

$$= 1 - \prod_{\underline{i} \in I \setminus \{\underline{j}\}}(1 - P_{\underline{i}}) + P_{\underline{j}}(1 - P_{\underline{l}}) \prod_{\underline{i} \in I \setminus \{\underline{j},\underline{l}\}}(1 - P_{\underline{i}})$$

$$= 1 - \prod_{\underline{i} \in I \setminus \{\underline{j}\}}(1 - P_{\underline{i}}).$$

Repeating for all $\underline{j} \in I \setminus \max I$ we have that

$$P_I = 1 - \prod_{\underline{i} \in \max I}(1 - P_{\underline{i}}) = 1 - \sum_{J \subset \max I}(-1)^{|J|}\prod_{\underline{j} \in J} P_{\underline{j}} = \sum_{\substack{J \subset \max I \\ J \neq \emptyset}}(-1)^{|J|+1}\sum_{\underline{j} \leq \wedge J} Q_{\underline{j}},$$

where $|J|$ is the number of elements of J and $\wedge J$ is the greatest lower bound over all elements of J (that is given $J = \{\underline{j}_1, \underline{j}_2, \dots, \underline{j}_{|J|}\}$ then $\wedge J = \underline{j}_1 \wedge \underline{j}_2 \wedge \cdots \wedge \underline{j}_{|J|}$). The right hand side can be expanded to a simple sum over the $Q_{\underline{j}}$, that is

$$P_I = \sum_{\underline{j} \in I} q_{\underline{j}} Q_{\underline{j}}$$

for some coefficients $q_{\underline{j}}$. We are required to show that $q_{\underline{j}} = 1$ for all $\underline{j} \in I$. Since

$$Q_{\underline{i}}P_I = \sum_{\underline{j} \in I} q_{\underline{j}} Q_{\underline{i}} Q_{\underline{j}} = q_{\underline{i}} Q_{\underline{i}}.$$

we need only to show that $Q_i P_I = Q_i$. Note that

$$Q_{\underline{i}} \sum_{\underline{j} \le \wedge J} Q_{\underline{j}} = \begin{cases} Q_{\underline{i}} & \text{if } \underline{i} \le \wedge J \\ 0 & \text{otherwise.} \end{cases}$$

Further, $\underline{i} \le \wedge J$ if and only if $\underline{i} \le \underline{j}$ for all $\underline{j} \in J$. Let $J_{\underline{i}} \subset \max I$ be the largest subset of $\max I$ such that $\underline{i} \le \underline{j}$ for all $\underline{j} \in J_{\underline{i}}$. It follows that

$$Q_{\underline{i}} P_I = \sum_{\substack{J \subset \max I \\ J \neq \emptyset}} (-1)^{|J|+1} Q_{\underline{i}} \sum_{\underline{j} \le \wedge J} Q_{\underline{j}} = \sum_{\substack{J \subset J_{\underline{i}} \\ J \neq \emptyset}} (-1)^{|J|+1} Q_{\underline{i}}.$$

Expanding this yields

$$Q_{\underline{i}} P_I = \sum_{m=1}^{|J_{\underline{i}}|} \sum_{\substack{J \subset J_{\underline{i}} \\ |J|=m}} (-1)^{|J|+1} Q_{\underline{i}} = Q_{\underline{i}} \sum_{m=1}^{|J_{\underline{i}}|} \binom{|J_{\underline{i}}|}{m} (-1)^{m+1}$$

$$= Q_{\underline{i}} \left(1 - \sum_{m=0}^{|J_{\underline{i}}|} \binom{|J_{\underline{i}}|}{m} (-1)^m \right)$$

$$= Q_{\underline{i}} \left(1 - (1-1)^{|J_{\underline{i}}|} \right) = Q_{\underline{i}}.$$

Therefore $q_{\underline{i}} = 1$. Since the choice of $\underline{i} \in I$ was arbitrary the proof is complete. □

3.2   Characterisation of Combination Coefficients

The previous results bring us to a corollary about the nature of the combination coefficients.

Corollary 2 *Given $I \in \mathcal{D}(\mathbb{N}^d)$ with corresponding projection $P_I = 1 - \prod_{\underline{j} \in I}(1 - P_{\underline{j}}) = \sum_{\underline{j} \in I} c_{\underline{j}} P_{\underline{j}}$ then for each $\underline{i} \in I$*

$$c_{\underline{i}} = 1 - \sum_{\underline{i} < \underline{j} \in I} c_{\underline{j}},$$

where $\underline{i} < \underline{j} \in I$ means $\{\underline{j} \in \mathbb{N}^d : \underline{j} \in I \,\&\, \underline{j} > \underline{i}\}$.

Proof From Proposition 3 one has $Q_{\underline{i}} P_I = Q_{\underline{i}}$ and hence

$$Q_{\underline{i}} = Q_{\underline{i}} P_I = Q_{\underline{i}} \sum_{\underline{j} \in I} c_{\underline{j}} P_{\underline{j}} = \sum_{\underline{j} \in I} c_{\underline{j}} Q_{\underline{i}} \sum_{\underline{l} \le \underline{j}} Q_{\underline{l}} = \sum_{\underline{i} \le \underline{j} \in I} c_{\underline{j}} Q_{\underline{i}} = Q_{\underline{i}} \sum_{\underline{i} \le \underline{j} \in I} c_{\underline{j}}.$$

It follows that

$$1 = \sum_{\underline{i} \leq \underline{j} \in I} c_{\underline{j}} = c_{\underline{i}} + \sum_{\underline{i} < \underline{j} \in I} c_{\underline{j}}$$

and re-arranging gives the desired result.   □

Let χ_I be the characteristic function on multi-indices, that is

$$\chi_I(\underline{i}) = \begin{cases} 1 & \text{if } \underline{i} \in I \\ 0 & \text{otherwise.} \end{cases}$$

We now give a useful result regarding the calculation of combination coefficients which appears in [4] without proof.

Proposition 4 *Let $I \in \mathcal{D}(\mathbb{N}^d)$ with corresponding projection $P_I = \sum_{\underline{i} \in I} c_{\underline{i}} P_{\underline{i}}$. Then for each $\underline{i} \in I$ one has*

$$c_{\underline{i}} = \sum_{\underline{i} \leq \underline{j} \leq \underline{i} + \underline{1}} (-1)^{|\underline{j} - \underline{i}|} \chi_I(\underline{j}).$$

Proof From Corollary 2 we have that for $\underline{i} \in I$

$$1 = \sum_{\underline{i} \leq \underline{j} \in I} c_{\underline{j}} = \sum_{\underline{i} \leq \underline{j}} c_{\underline{j}} \chi_I(\underline{j}).$$

It follows that for $\underline{i} \in \mathbb{N}^d$

$$\chi_I(\underline{i}) = \sum_{\underline{i} \leq \underline{j}} c_{\underline{j}} \chi_I(\underline{j}).$$

Substituting this into the right hand side of the desired result we obtain

$$\sum_{\underline{i} \leq \underline{j} \leq \underline{i} + \underline{1}} (-1)^{|\underline{j} - \underline{i}|} \chi_I(\underline{j}) = \sum_{\underline{i} \leq \underline{j} \leq \underline{i} + \underline{1}} (-1)^{|\underline{j} - \underline{i}|} \sum_{\underline{j} \leq \underline{l}} c_{\underline{l}} \chi_I(\underline{l})$$

$$= \sum_{\underline{i} \leq \underline{j} \leq \underline{i} + \underline{1}} \left((-1)^{|j_1 - i_1|} \sum_{l_1 = j_1}^{\infty} \cdots (-1)^{|j_d - i_d|} \sum_{l_d = j_d}^{\infty} \right) c_{\underline{l}} \chi_I(\underline{l})$$

$$= \left(\sum_{l_1 = i_1}^{\infty} - \sum_{l_1 = i_1 + 1}^{\infty} \right) \cdots \left(\sum_{l_d = i_d}^{\infty} - \sum_{l_d = i_d + 1}^{\infty} \right) c_{\underline{l}} \chi_I(\underline{l}) = c_{\underline{i}} \chi_I(\underline{i}),$$

as required.   □

This enables fast computation of combination coefficients given an arbitrary downset I. The following corollary shows that many of these coefficients are typically 0.

Corollary 3 *Let* $I \in \mathcal{D}(\mathbb{N}^d)$ *with corresponding projection* $P_I = \sum_{\underline{i} \in I} c_{\underline{i}} P_{\underline{i}}$. *If* $\underline{i} + \underline{1} \in I$, *then* $c_{\underline{i}} = 0$.

Proof $\underline{i} + \underline{1} \in I$ implies that $\underline{j} \in I$ for all $\underline{j} \leq \underline{i} + \underline{1}$. Therefore

$$c_{\underline{i}} = \sum_{\underline{i} \leq \underline{j} \leq \underline{i} + \underline{1}} (-1)^{|\underline{i} - \underline{j}|} = 0 . \quad \square$$

Corollary 1 provided an update formula when a covering element is added to a downset I. It turns out that the update coefficients have a very particular structure as the next lemma will demonstrate.

Lemma 5 *Let* $I, J \in \mathcal{D}(\mathbb{N}^d)$ *such that* $I \prec J$. *Further, let* $\underline{i}$ *be the multi-index such that* $J = I \cup \{\underline{i}\}$. *Then*

$$P_J - P_I = \sum_{\underline{i} - \underline{1} \leq \underline{j} \leq \underline{i}} (-1)^{|\underline{i} - \underline{j}|} P_{\underline{j}}$$

where $P_{\underline{j}} := 0$ *if* $j_k < 0$ *for any* $k \in \{1, \ldots, d\}$.

Proof Clearly we have $P_J - P_I = Q_{\underline{i}}$. Now we simply note that

$$Q_{\underline{i}} = \sum_{\underline{j} \leq \underline{1}} (-1)^{|\underline{j}|} P_{\underline{i} - \underline{j}} = \sum_{\underline{i} - \underline{1} \leq \underline{j} \leq \underline{i}} (-1)^{|\underline{i} - \underline{j}|} P_{\underline{j}} ,$$

as required. $\square$

This is quite a useful result. For example, in 2 dimensions when a covering element is added with $\underline{i} = (i_1, i_2) \geq (1, 1)$ then only 4 coefficients need to be changed, namely $c_{(i_1, i_2)} \mapsto 1$, $c_{(i_1 - 1, i_2)} \mapsto c_{(i_1 - 1, i_2)} - 1$, $c_{(i_1, i_2 - 1)} \mapsto c_{(i_1, i_2 - 1)} - 1$ and $c_{(i_1 - 1, i_2 - 1)} \mapsto c_{(i_1 - 1, i_2 - 1)} + 1$. Similarly in d dimensions one only needs to change 2^d coefficients.

Another interesting observation to be made is that Proposition 4 and Lemma 5 are in some sense mirror images. The former shows how a coefficient $c_{\underline{i}}$ is determined by the existence of positive neighbours in I whilst the latter shows how adding a covering element $\underline{i}$ to I affects the coefficients of the negative neighbours.

The following lemma is essentially a consistency property for combinations arising from adaptive sparse grids.

Lemma 6 *If* $I \in \mathcal{D}(\mathbb{N}^d)$ *is non-empty, then the coefficients* $c_{\underline{i}}$ *corresponding to the projection* $P_I = \sum_{\underline{i} \in I} c_{\underline{i}} P_{\underline{i}}$ *satisfy*

$$1 = \sum_{\underline{i} \in I} c_{\underline{i}} .$$

Proof We note that $I \neq \emptyset$ implies $\underline{0} \in I$ and one has

$$P_{\underline{0}} = P_{\underline{0}} P_I = P_{\underline{0}} \sum_{\underline{i} \in I} c_{\underline{i}} P_{\underline{i}} = P_{\underline{0}} \sum_{\underline{i} \in I} c_{\underline{i}},$$

from which the desired result follows. □

As adaptive sparse grids are meant to generalise the combination technique we show that the classical coefficients [6] come out for the appropriate downset I.

Lemma 7 *Let* $I = \{\underline{i} \in \mathbb{N}^d : |\underline{i}| \leq n\}$. *Then, the* $c_{\underline{i}}$ *corresponding to* P_I *satisfy*

$$c_{\underline{i}} = (-1)^{n-|\underline{i}|} \binom{d-1}{n-|\underline{i}|}.$$

Proof We know generally that

$$c_{\underline{i}} = \sum_{\underline{i} \leq \underline{j} \leq \underline{i}+\underline{1}} (-1)^{|\underline{j}-\underline{i}|} \chi_I(\underline{j}).$$

Therefore, given $\underline{i} \in I$ such that $|\underline{i}| = n - k$ for some $k \in \{0, \dots, d-1\}$, one has

$$c_{\underline{i}} = \sum_{l=0}^{k} (-1)^l \binom{d}{l}.$$

With an induction argument on k using the identity $\binom{d}{k} - \binom{d-1}{k-1} = \binom{d-1}{k}$ it is easily shown that

$$c_{\underline{i}} = (-1)^k \binom{d-1}{k}.$$

Substituting $k = n - |\underline{i}|$ and recognising that $\binom{d-1}{n-|\underline{i}|} := 0$ for $|\underline{i}| \leq n - d$ and $|\underline{i}| > n$ completes the proof. □

3.3 Error Formula for Adaptive Sparse Grids

From here we are going to restrict ourselves to considering nested hierarchical spaces of piecewise multi-linear functions so that we may formulate some general error formulas. Specifically we let the $V_{\underline{i}}$ be the function spaces of piecewise multi-linear functions on $[0,1]^d$ which interpolate between the function values at grid points $x_{\underline{j}} := (j_1 2^{-i_1}, \dots, j_d 2^{-i_d})$ with $\underline{0} \leq \underline{j} \leq 2^{\underline{i}} = (2^{i_1}, \dots, 2^{i_d})$. Let $u \in V = C([0,1]^d)$ and $I \in \mathcal{D}(\mathbb{N}^d)$. We define $u_{\underline{i}} := P_{\underline{i}} u$ and $u_I := P_I u$ from which one obtains a unique combination formula

$$u_I = \sum_{\underline{i} \in I} c_{\underline{i}} u_{\underline{i}} . \tag{3}$$

If the projections $u_{\underline{i}} = P_{\underline{i}} u$ are exactly the piecewise multi-linear interpolants of u, then u_I is the (linear) sparse grid interpolant of u onto V_I. The following proposition considers the interpolation error of functions $u \in H^2_{0,\text{mix}}(\Omega)$, that is $u : \Omega \subset \mathbb{R}^d \to \mathbb{R}$ for which the norm

$$\|u\|^2_{H^2_{0,\text{mix}}} = \sum_{0 \leq \alpha \leq 2} \left\| \frac{\partial^{|\alpha|}}{\partial x^{\alpha}} u \right\|^2_2 = \sum_{0 \leq \alpha \leq 2} \|D^{\alpha} u\|^2_2$$

is finite and $u|_{\partial \Omega} = 0$.

Proposition 5 *Let $I \subset \mathbb{N}^d$ be a downset, $u \in H^2_{0,\text{mix}}(\Omega)$ and $V_{\underline{i}}, P_{\underline{i}}, u_{\underline{i}}$ as above. Let $c_{\underline{i}} \in \mathbb{R}$ be the combination coefficients corresponding to the projection $P_I = \sum_{\underline{i} \in I} c_{\underline{i}} P_{\underline{i}}$ and let $u_I = P_I u$, then*

$$\|u - u_I\|_2 \leq 3^{-d} \|D^2 u\|_2 \left(3^{-d} - \sum_{1 \leq \underline{i} \in I} 2^{-2|\underline{i}|} \right) . \tag{4}$$

Proof We note that as u is zero on the boundary one has

$$u_I = \sum_{1 \leq \underline{i} \in I} w_{\underline{i}} \quad \text{and} \quad u_{\underline{i}} = \sum_{1 \leq \underline{i}} w_{\underline{i}}$$

where $w_{\underline{i}} = Q_{\underline{i}} u$ are the hierarchical surpluses of u. It follows that

$$\|u - u_I\|_2 = \left\| \sum_{1 \leq \underline{i} \notin I} w_{\underline{i}} \right\|_2 \leq \sum_{1 \leq \underline{i} \notin I} \|w_{\underline{i}}\|_2 .$$

Now applying the standard bound $\|w_{\underline{i}}\|_2 \leq 3^{-d} \|D^2 u\|_2 2^{-2|\underline{i}|}$ for $\underline{i} \geq \underline{1}$ (see for example [1, 3]), one has

$$\|u - u_I\|_2 \leq 3^{-d} \|D^2 u\|_2 \sum_{1 \leq \underline{i} \notin I} 2^{-2|\underline{i}|}$$

and using the fact that

$$\sum_{1 \leq \underline{i} \in I} 2^{-2|\underline{i}|} + \sum_{1 \leq \underline{i} \notin I} 2^{-2|\underline{i}|} = \sum_{1 \leq \underline{i}} 2^{-2|\underline{i}|} = \sum_{i_1, \ldots, i_d = 1}^{\infty} 2^{-2i_1} \cdots 2^{-2i_d} = 3^{-d}$$

one obtains the desired result. $\qquad \qquad \square$

More generally, the projections $u_{\underline{i}} = P_{\underline{i}}u$ will be approximations of u in the function space $V_{\underline{i}}$. Classical analysis of the combination technique assumes that the $u_{\underline{i}}$ satisfy an error splitting. We derive a formula for calculating the error of adaptive sparse grid combinations based on such error splittings. First we provide a result which says that projecting a d dimensional combination onto a $k < d$ dimensional function space yields coefficients corresponding to a k dimensional combination. We note that a result similar to Lemma 8 may be found in [12].

Let $I \in \mathcal{D}(\mathbb{N}^d)$ be non-empty, $1 \le k \le d$ and $\{e_1, \ldots, e_k\} \subseteq \{1, \ldots, d\}$ (with $\{e_{k+1}, \ldots, e_d\} = \{1, \ldots, d\} \backslash \{e_1, \ldots, e_k\}$). We define

$$I_{e_1, \ldots, e_k} := \{(i_{e_1}, \ldots, i_{e_k}) : \underline{i} \in I\}.$$

It is immediate that $I_{e_1, \ldots, e_k} \in \mathcal{D}(\mathbb{N}^k)$.

Lemma 8 *Let $I \in \mathcal{D}(\mathbb{N}^d)$ and P_I be the corresponding projection with $P_I = \sum_{\underline{i} \in I} c_{\underline{i}} P_{\underline{i}}$. Further, let*

$$P_{I_{e_1, \ldots, e_k}} : V^{e_1} \otimes \cdots \otimes V^{e_k} \rightarrow V_{I_{e_1, \ldots, e_k}} := \sum_{\underline{i} \in I_{e_1, \ldots, e_k}} V_{i_1}^{e_1} \otimes \cdots \otimes V_{i_k}^{e_k},$$

which may be written as the combination $P_{I_{e_1, \ldots, e_k}} = \sum_{\underline{j} \in I_{e_1, \ldots, e_k}} \bar{c}_{\underline{j}} \bigotimes_{l=1}^{k} P_{j_l}^{e_l}$. Then, for all $\underline{j} \in I_{e_1, \ldots, e_k}$ one has

$$\bar{c}_{\underline{j}} = \sum_{\underline{i} \in I \text{ s.t. } (i_{e_1}, \ldots, i_{e_k}) = \underline{j}} c_{\underline{i}}.$$

Proof Consider a function $u \in V = V^1 \otimes \cdots \otimes V^d$ which only depends on the coordinates $x_{e_1}, \ldots, x_{e_k}$, that is $u(x_1, \ldots, x_d) = v(x_{e_1}, \ldots, x_{e_k})$ for some $v \in V^{e_1} \otimes \cdots \otimes V^{e_k}$. It follows that $P_{\underline{i}}u = \bigotimes_{l=1}^{k} P_{i_{e_l}}^{e_l} v$ and therefore

$$P_I u = \sum_{\underline{i} \in I} c_{\underline{i}} P_{\underline{i}} u = \sum_{\underline{i} \in I} c_{\underline{i}} \bigotimes_{l=1}^{k} P_{i_{e_l}}^{e_l} v$$

$$= \sum_{(i_{e_1}, \ldots, i_{e_k}) \in I_{e_1, \ldots, e_k}} \left(\bigotimes_{l=1}^{k} P_{i_{e_l}}^{e_l} v \right) \sum_{(i_{e_{k+1}}, \ldots, i_{e_d}) \in I_{e_{k+1}, \ldots, e_d}} \chi_I(\underline{i}) c_{\underline{i}}$$

$$= \sum_{\underline{j} \in I_{e_1, \ldots, e_k}} \left(\bigotimes_{l=1}^{k} P_{j_l}^{e_l} v \right) \sum_{\underline{i} \in I \text{ s.t. } (i_{e_1}, \ldots, i_{e_k}) = \underline{j}} c_{\underline{i}}.$$

Finally, since it is clear that $P_I u = P_{I_{e_1, \ldots, e_k}} v$ and u depending on only $x_{e_1}, \ldots, x_{e_k}$ is arbitrary (i.e. v is arbitrary), one has the desired result. □

This result allows us to write down a general formula regarding error estimates of dimension adaptive sparse grids when an error splitting is assumed.

Proposition 6 *Let* $I \in \mathcal{D}(\mathbb{N}^d)$ *be non-empty. For* $1 \le k \le d$, $\{e_1, \ldots, e_k\} \subseteq \{1, \ldots, d\}$ *and* $I_{e_1, \ldots, e_k}$ *the set defined previously, we define* $\bar{c}_{(i_{e_1}, \ldots, i_{e_k})}$ *to be the coefficient corresponding to the* k *dimensional projection* $P_{I_{e_1, \ldots, e_k}}$ *as in Lemma 8. Let* $u_I = \sum_{i \in I} c_{\underline{i}} u_{\underline{i}}$ *where each* $u_{\underline{i}}$ *satisfies the error splitting*

$$u - u_{\underline{i}} = \sum_{k=1}^{d} \sum_{\substack{\{e_1, \ldots, e_k\} \\ \subseteq \{1, \ldots, d\}}} C_{e_1, \ldots, e_k}(h_{i_{e_1}}, \ldots, h_{i_{e_k}}) h_{i_{e_1}}^{p_{e_1}} \cdots h_{i_{e_k}}^{p_{e_k}}, \tag{5}$$

with $p_1, \ldots, p_d > 0$, $h_l := 2^{-l}$ *and for each* $\{e_1, \ldots, e_k\} \subseteq \{1, \ldots, d\}$ *one has* $|C_{e_1, \ldots, e_k}(h_{i_{e_1}}, \ldots, h_{i_{e_k}})| \le \kappa_{e_1, \ldots, e_k} \in \mathbb{R}^+$ *for all* $i_{e_1}, \ldots, i_{e_k}$. *Then*

$$|u - u_I| \le \sum_{k=1}^{d} \sum_{\substack{\{e_1, \ldots, e_k\} \\ \subseteq \{1, \ldots, d\}}} \kappa_{e_1, \ldots, e_k} \sum_{\underline{i} \in I_{e_1, \ldots, e_k}} |\bar{c}_{(i_{e_1}, \ldots, i_{e_k})}| h_{i_{e_1}}^{p_{e_1}} \cdots h_{i_{e_k}}^{p_{e_k}}.$$

Proof Since $\sum_{i \in I} c_{\underline{i}} = 1$ we have $u - u_I = \sum_{i \in I} c_{\underline{i}}(u - u_{\underline{i}})$. From here one substitutes the error splitting formula obtaining

$$|u - u_I| = \left| \sum_{\underline{i} \in I} c_{\underline{i}} \sum_{k=1}^{d} \sum_{\substack{\{e_1, \ldots, e_k\} \\ \subseteq \{1, \ldots, d\}}} C_{e_1, \ldots, e_k}(h_{i_{e_1}}, \ldots, h_{i_{e_k}}) h_{i_{e_1}}^{p_{e_1}} \cdots h_{i_{e_k}}^{p_{e_k}} \right|$$

$$= \left| \sum_{k=1}^{d} \sum_{\substack{\{e_1, \ldots, e_k\} \\ \subseteq \{1, \ldots, d\}}} \sum_{\underline{j} \in I_{e_1, \ldots, e_k}} \sum_{\substack{\underline{i} \in I \text{ s.t.} \\ (i_{e_1}, \ldots, i_{e_k}) = \underline{j}}} c_{\underline{i}} C_{e_1, \ldots, e_k}(h_{i_{e_1}}, \ldots, h_{i_{e_k}}) h_{i_{e_1}}^{p_{e_1}} \cdots h_{i_{e_k}}^{p_{e_k}} \right|$$

$$\le \sum_{k=1}^{d} \sum_{\substack{\{e_1, \ldots, e_k\} \\ \subseteq \{1, \ldots, d\}}} \sum_{\underline{j} \in I_{e_1, \ldots, e_k}} |C_{e_1, \ldots, e_k}(h_{j_{e_1}}, \ldots, h_{j_{e_k}})| h_{j_{e_1}}^{p_{e_1}} \cdots h_{j_{e_k}}^{p_{e_k}} \left| \sum_{\substack{\underline{i} \in I \text{ s.t.} \\ (i_{e_1}, \ldots, i_{e_k}) = \underline{j}}} c_{\underline{i}} \right|$$

$$\le \sum_{k=1}^{d} \sum_{\substack{\{e_1, \ldots, e_k\} \\ \subseteq \{1, \ldots, d\}}} \kappa_{e_1, \ldots, e_k} \sum_{\underline{j} \in I_{e_1, \ldots, e_k}} h_{j_{e_1}}^{p_{e_1}} \cdots h_{j_{e_k}}^{p_{e_k}} |\bar{c}_{(i_{e_1}, \ldots, i_{e_k})}|. \quad \Box$$

Whilst this result is not simple enough that one could easily write (by hand) the resulting bound for a given I, it does give one a way to quickly compute a bound.

4 Extrapolation Based Adaptive Sparse Grids

4.1 Multivariate Extrapolation

Extrapolation techniques are used to obtain higher rates of convergence of a given method for which the rate of convergence is known. For example, in one spatial dimension, suppose an approximation $u_i \in V_i$ of $u \in V \subseteq C([0, 1])$ is known to satisfy pointwise

$$u - u_i = D \times 2^{-pi} + O(2^{-qi})$$

for some $D \in \mathbb{R}$ (which may depend on x) and $p, q \in \mathbb{N}$ with $1 \leq p < q$, then via classical Richardson extrapolation one obtains

$$u - \left(\frac{2^p}{2^p - 1} u_{i+1} - \frac{1}{2^p - 1} u_i \right) = \frac{2^p}{2^p - 1} (u - u_{i+1}) - \frac{1}{2^p - 1} (u - u_i) = O(2^{-qi}).$$

This may be extended to higher dimensional approximations where, for each $\underline{i} \in \mathbb{N}^d$, one has approximations $u_{\underline{i}} \in V_{\underline{i}}$ of $u \in V \subseteq C([0, 1]^d)$. One may cancel terms in the error expansion of $u - u_{\underline{i}}$ of the form $C_k 2^{-pi_k}$ (univariate terms) for $k = 1, \ldots, d$ by computing

$$\frac{2^p}{2^p - 1} u_{\underline{i}+\underline{1}} - \frac{1}{2^p - 1} u_{\underline{i}}.$$

There are a couple of issues with this approach. First, $u_{\underline{i}+\underline{1}}$ is approximately 2^d times more expensive to compute than $u_{\underline{i}}$ if complexity is linear with the number of unknowns. This makes $u_{\underline{i}+\underline{1}}$ impractical to compute for large d. Second, error expansions in high dimensions typically have terms proportional to $2^{-p(i_{e_1} + \cdots + i_{e_k})}$ for $k \in \{1, \ldots, d\}$ and $\{e_1, \ldots, e_k\} \subseteq \{1, \ldots, d\}$ (multivariate terms). If all terms of $\underline{i}$ are not sufficiently large then some of these terms may be significant such that the cancellation of univariate terms provides little improvement.

Multivariate extrapolation techniques typically attempt to deal with at least one of these issues. Here we use the notation $\underline{i} + 1_k$ for the multi-index $(i_1, \ldots, i_{k-1}, i_k + 1, i_{k+1}, \ldots, i_d)$. One may extrapolate the univariate terms by computing

$$\frac{2^p}{2^p - 1} \left(\sum_{k=1}^{d} u_{\underline{i}+1_k} \right) - \frac{(d-1)2^p + 1}{2^p - 1} u_{\underline{i}}.$$

Each $u_{\underline{i}+1_k}$ costs approximately twice as much as $u_{\underline{i}}$ thus the additional cost is approximately $2d$ compared to 2^d. This type of extrapolation has been extensively studied in the literature [2, 7, 9, 11]. We focus on a different type of extrapolation which focuses on the second problem of classical extrapolation in higher dimensions, that is the cancellation of multivariate terms in the error expansion.

For the remainder of this paper we assume that the $u_{\underline{i}}$ satisfy the generalised pointwise error splitting

$$u - u_{\underline{i}} = \left(\sum_{k=1}^{d} \sum_{\substack{\{e_1,\dots,e_k\} \\ \subseteq\{1,\dots,d\}}} D_{e_1,\dots,e_k} h_{i_{e_1}}^{p_{e_1}} \cdots h_{i_{e_k}}^{p_{e_k}} \right) + R(h_{i_1},\dots,h_{i_d}), \qquad (6)$$

where $p_1,\dots,p_d \in \mathbb{N}_+$, $h_j := 2^{-j}$ for $j \in \mathbb{N}$, and

$$R(h_{i_1},\dots,h_{i_d}) = \sum_{k=1}^{d} \sum_{\substack{\{e_1,\dots,e_k\} \\ \subseteq\{1,\dots,d\}}} E_{e_1,\dots,e_k}(h_{i_{e_1}},\dots,h_{i_{e_k}}) h_{i_{e_1}}^{q_{e_1}} \cdots h_{i_{e_k}}^{q_{e_k}}$$

for some $q_1 > p_1, \dots, q_d > p_d$ and with each $|E_{e_1,\dots,e_k}(h_{i_{e_1}},\dots,h_{i_{e_k}})|$ bounded for all $\underline{i} \in \mathbb{N}^d$. This extends the error splitting considered in [10] by allowing the rate of convergence to differ for each dimension. For $p_1 = \cdots = p_d = 2$ one has the extrapolation formula [10]

$$\tilde{u}_{\underline{i}} = \sum_{\underline{i} \leq \underline{j} \leq \underline{i}+\underline{1}} \frac{(-4)^{|\underline{j}-\underline{i}|}}{(-3)^d} u_{\underline{j}}. \qquad (7)$$

For $p_1 = \cdots = p_d = p$ one may take

$$\tilde{u}_{\underline{i}} = \sum_{\underline{i} \leq \underline{j} \leq \underline{i}+\underline{1}} \frac{(-2^p)^{|\underline{j}-\underline{i}|}}{(1-2^p)^d} u_{\underline{j}}.$$

In the most general case one has the extrapolation formula

$$\tilde{u}_{\underline{i}} = \sum_{\underline{j} \leq \underline{1}} \left(\prod_{m=1}^{d} \frac{(-2^{p_m})^{j_m}}{1-2^{p_m}} \right) u_{\underline{i}+\underline{j}}. \qquad (8)$$

To show these do indeed result in an extrapolation we first need two lemmas.

Lemma 9 *One has*

$$\sum_{\underline{j} \leq \underline{1}} \prod_{m=1}^{d} \frac{(-2^{p_m})^{j_m}}{1-2^{p_m}} = 1.$$

Proof We notice that

$$
\sum_{j \leq 1} \prod_{m=1}^{d} \frac{(-2^{p_m})^{j_m}}{1 - 2^{p_m}} = \sum_{j_1=0}^{1} \cdots \sum_{j_d=0}^{1} \frac{(-2^{p_1})^{j_1}}{1 - 2^{p_1}} \cdots \frac{(-2^{p_d})^{j_d}}{1 - 2^{p_d}}
$$

$$
= \left(\sum_{j_1=0}^{1} \frac{(-2^{p_1})^{j_1}}{1 - 2^{p_1}} \right) \cdots \left(\sum_{j_d=0}^{1} \frac{(-2^{p_d})^{j_d}}{1 - 2^{p_d}} \right)
$$

$$
= \prod_{m=1}^{d} \sum_{j_m=0}^{1} \frac{(-2^{p_m})^{j_m}}{1 - 2^{p_m}} = \prod_{m=1}^{d} \frac{1 - 2^{p_m}}{1 - 2^{p_m}} = 1 . \quad \square
$$

Lemma 10 *Fix $k \in \{1, \ldots, d\}$ and $\{e_1, \ldots, e_k\} \subseteq \{1, \ldots, d\}$, then*

$$
\sum_{j \leq 1} \prod_{m=1}^{d} \frac{(-2^{p_m})^{j_m}}{1 - 2^{p_m}} D_{e_1, \ldots, e_k} h_{j_{e_1}}^{p_{e_1}} \cdots h_{j_{e_k}}^{p_{e_k}} = 0 .
$$

Proof Similar to the previous lemma note we can swap the sum and product

$$
\sum_{j \leq 1} \prod_{m=1}^{d} \frac{(-2^{p_m})^{j_m}}{1 - 2^{p_m}} D_{e_1, \ldots, e_k} h_{j_{e_1}}^{p_{e_1}} \cdots h_{j_{e_k}}^{p_{e_k}}
$$

$$
= D_{e_1, \ldots, e_k} \prod_{m=1}^{d} \sum_{j_m=0}^{1} \frac{(-2^{p_m})^{j_m}}{1 - 2^{p_m}} h_{j_{e_1}}^{p_{e_1}} \cdots h_{j_{e_k}}^{p_{e_k}} .
$$

Now consider $m \in \{e_1, \ldots, e_k\}$, without loss of generality let $m = e_1$, one has

$$
\sum_{j_m=0}^{1} \frac{(-2^{p_m})^{j_m}}{1 - 2^{p_m}} h_{j_{e_1}}^{p_{e_1}} \cdots h_{j_{e_k}}^{p_{e_k}} = h_{j_{e_2}}^{p_{e_2}} \cdots h_{j_{e_k}}^{p_{e_k}} \sum_{j_m=0}^{1} \frac{(-2^{p_m})^{j_m} h_{j_m}^{p_m}}{1 - 2^{p_m}}
$$

$$
= h_{j_{e_2}}^{p_{e_2}} \cdots h_{j_{e_k}}^{p_{e_k}} \sum_{j_m=0}^{1} \frac{(-1)^{j_m}}{1 - 2^{p_m}} = 0 ,
$$

from which the result follows. $\quad \square$

Thus we have the following proposition.

Proposition 7 *If $u_{\underline{i}}$ satisfies the pointwise error splitting (6) then*

$$u - \tilde{u}_{\underline{i}} = \sum_{\underline{j} \le \underline{1}} \left(\prod_{m=1}^{d} \frac{(-2^{p_m})^{j_m}}{1 - 2^{p_m}} \right) \sum_{k=1}^{d} \sum_{\substack{\{l_1,\dots,l_k\} \\ \subset \{1,\dots,d\}}} R(h_{i_1+j_1}, \dots, h_{i_d+j_d}) \,.$$

Proof We need only to show that the $D_{e_1,\dots,e_k}$ terms vanish. Lemma 9 tells us that

$$u - \tilde{u}_{\underline{i}} = u - \sum_{\underline{j} \le \underline{1}} \left(\prod_{m=1}^{d} \frac{(-2^{p_m})^{j_m}}{1 - 2^{p_m}} \right) u_{\underline{i}+\underline{j}} = \sum_{\underline{j} \le \underline{1}} \left(\prod_{m=1}^{d} \frac{(-2^{p_m})^{j_m}}{1 - 2^{p_m}} \right) (u - u_{\underline{j}+\underline{i}}) \,.$$

From here we can substitute in the error splitting (6). We look at a single term, hence fix $k \in \{1, \dots, d\}$ and $\{e_1, \dots, e_k\} \subseteq \{1, \dots, d\}$, then we have

$$\sum_{\underline{j} \le \underline{1}} \left(\prod_{m=1}^{d} \frac{(-2^{p_m})^{j_m}}{1 - 2^{p_m}} \right) D_{e_1,\dots,e_k} h_{j_{e_1}+i_{l_1}}^{p_{e_1}} \cdots h_{j_{e_k}+i_{e_k}}^{p_{e_k}}$$

$$= h_{i_{e_1}}^{p_{e_1}} \cdots h_{i_{e_k}}^{p_{e_k}} \left(\prod_{m=1}^{d} \frac{(-2^{p_m})^{j_m}}{1 - 2^{p_m}} \right) D_{e_1,\dots,e_k} h_{j_{e_1}}^{p_{e_1}} \cdots h_{j_{e_k}}^{p_{e_k}} = 0 \,,$$

from Lemma 10. Thus all terms except the $R(h_{i_1+j_1}, \dots, h_{i_d+j_d})$ terms sum to 0 which yields the result. □

One could substitute $R(h_{i_1+j_1}, \dots, h_{i_d+j_d})$ and obtain a more precise expression for the remainder (as has been done for the case $p_1 = \cdots = p_d = 2$ and $q_1 = \cdots = q_d = 4$, see [10]) but we shall not do so here.

4.2 Extrapolation Within Adaptive Sparse Grids

In [10] the classical combination formula was studied if the approximations $u_{\underline{i}}$ are replaced with the extrapolations $\tilde{u}_{\underline{i}}$ of (7), that is the combination

$$\tilde{u}_n^c = \sum_{k=0}^{d-1} (-1)^k \binom{d-1}{k} \sum_{|\underline{i}|=n-k} \tilde{u}_{\underline{i}} \,.$$

The main result was a careful calculation of an error bound for this particular case with $p_1 = \cdots = p_d = 2$ and $q_1 = \cdots = q_d = 4$. It was also shown that the combination of extrapolations may be expressed as

$$\tilde{u}_n^c = \sum_{k=-d+1}^{d} \sum_{|\underline{i}|=n+k} \tilde{c}_{\underline{i}} u_{\underline{i}},$$

where, if $|\underline{i}|_0$ is the number of non-zero entries in $\underline{i}$, then

$$\tilde{c}_{\underline{i}} = \sum_{k=\max\{0,|\underline{i}|-n-1\}}^{\min\{|\underline{i}|-n+d-1,d-1\}} \frac{(-4)^{|\underline{i}|-n+d-1-k}}{(-3)^d}(-1)^{d-1-k}\binom{d-1}{k}\binom{|\underline{i}|_0}{|\underline{i}|-n+d-1-k}.$$

These coefficients will be the focus of this section and, using the framework of adaptive sparse grids, we develop an adaptive way of combining extrapolations. Whilst we focus on the case with $p_1 = \cdots = p_d = 2$ and $q_1 = \cdots = q_d = 4$, the framework is easily applied to the general case.

We first observe that, if we were to define $u_{\underline{i}} := P_{\underline{i}} u$, then one has

$$\tilde{u}_{\underline{i}} = \left(\sum_{\underline{j} \leq \underline{1}} \frac{(-4)^{|\underline{j}|}}{(-3)^d} P_{\underline{i}+\underline{j}} \right) u.$$

Thus we will define

$$\tilde{P}_{\underline{i}} := \sum_{\underline{j} \leq \underline{1}} \frac{(-4)^{|\underline{j}|}}{(-3)^d} P_{\underline{i}+\underline{j}}$$

such that $\tilde{u}_{\underline{i}} = \tilde{P}_{\underline{i}} u$. The idea now is that rather than substituting this into a classical combination formula we may substitute this into any combination formula obtained via the adaptive sparse grids formulation. That is given a downset $I \in \mathcal{D}(\mathbb{N}^d)$ and the corresponding projection $P_I = \sum_{\underline{i} \in I} c_{\underline{i}} P_{\underline{i}}$ then we instead use the extrapolation projections to obtain

$$\tilde{P}_I := \sum_{\underline{i} \in I} c_{\underline{i}} \tilde{P}_{\underline{i}}.$$

Since the two approaches give reasonable approximations individually, it follows that the above projection would give a reasonable approximation. If the $u_{\underline{i}}$ satisfy the error splitting (6) then for each I one may carry out an analysis similar to that in [10] for the classical case.

There are two questions that immediately come to mind regarding the choice of I and the resulting coefficients. Consider the set $I_{+\underline{1}} := \{\underline{i} : \underline{j} \in I \text{ for some } \underline{i} - \underline{1} \leq \underline{j} \leq \underline{i}\}$, then one can write

$$\tilde{P}_I = \sum_{\underline{i} \in I_{+\underline{1}}} \tilde{c}_{\underline{i}} P_{\underline{i}}.$$

Note that $\tilde{P}_I$ is actually a projection onto $V_{I_{+\underline{1}}}$ rather than V_I. The first question is whether there is a simple way to determine the $\tilde{c}_{\underline{i}}$. The answer is yes and this will be addressed in Proposition 8. The second question is more subtle and relates to whether the $\tilde{P}_I$ can be extended to projections onto V_J for any given downset J (as opposed to only those J for which there exists a downset I such that $J = I_{+\underline{1}}$). This will be addressed later as a consequence of a simple expression for the $\tilde{c}_{\underline{i}}$.

Proposition 8 *Let* $I \in \mathcal{D}(\mathbb{N}^d)$. *Given the combination of extrapolations* $\tilde{P}_I = \sum_{\underline{i} \in I_{+\underline{1}}} \tilde{c}_{\underline{i}} P_{\underline{i}}$, *then the* $\tilde{c}_{\underline{i}}$ *are given by*

$$\tilde{c}_{\underline{i}} = \sum_{-\underline{1} \leq \underline{l} \leq \underline{1}} \frac{(-1)^{|\underline{l}|}}{(-3)^d} 5^{d-|\underline{l}|_0} 4^{d-|\underline{l}+\underline{1}|_0} \chi_I(\underline{i} + \underline{l})$$

where $|\underline{l}|_0$ *is defined to be the number of non-zero elements of* $\underline{l}$.

Note here we allow elements of $\underline{j}$ to be negative and that $\chi_I(\underline{i} + \underline{j}) := 0$ if $i_k + j_k < 0$ for any $k \in \{1, \ldots, d\}$.

Proof We first write $\tilde{P}_I = \sum_{\underline{i} \in I} c_{\underline{i}} \tilde{P}_{\underline{i}}$ for which we know the $c_{\underline{i}}$ are given by

$$c_{\underline{i}} = \sum_{\underline{j} \leq \underline{1}} (-1)^{|\underline{j}|} \chi_I(\underline{i} + \underline{j}). \tag{9}$$

Now we note that a $\tilde{P}_{\underline{j}}$ contains a $P_{\underline{i}}$ term in its sum if $\underline{j} \leq \underline{i} \leq \underline{j} + \underline{1}$. Conversely, for a given $P_{\underline{i}}$, those $\tilde{P}_{\underline{j}}$ which have a $P_{\underline{i}}$ term in its sum are those with $\underline{i} - \underline{1} \leq \underline{j} \leq \underline{i}$ (and $\underline{j} \geq \underline{0}$). Further, the term in

$$\tilde{P}_{\underline{j}} = \sum_{\underline{k} \leq \underline{1}} \frac{(-4)^{|\underline{k}|}}{(-3)^d} P_{\underline{j}+\underline{k}}$$

for which $P_{\underline{j}+\underline{k}} = P_{\underline{i}}$ clearly satisfies $\underline{k} = \underline{i} - \underline{j}$. It follows that

$$\tilde{c}_{\underline{i}} P_{\underline{i}} = \sum_{\substack{\underline{i}-\underline{1} \leq \underline{j} \leq \underline{i} \\ \underline{j} \geq \underline{0}}} c_{\underline{j}} \frac{(-4)^{|\underline{i}-\underline{j}|}}{(-3)^d} P_{\underline{i}}$$

and therefore

$$\tilde{c}_{\underline{i}} = (-3)^{-d} \sum_{\substack{\underline{i}-\underline{1} \leq \underline{j} \leq \underline{i} \\ \underline{j} \geq \underline{0}}} c_{\underline{j}} (-4)^{|\underline{i}-\underline{j}|} .$$

Now substituting (9) for $c_{\underline{j}}$ we obtain

$$\tilde{c}_{\underline{i}} = (-3)^{-d} \sum_{\substack{\underline{i}-\underline{1} \leq \underline{j} \leq \underline{i} \\ \underline{j} \geq \underline{0}}} (-4)^{|\underline{i}-\underline{j}|} \sum_{(\underline{0} \leq) \underline{k} \leq \underline{1}} (-1)^{|\underline{k}|} \chi_I (\underline{j} + \underline{k})$$

$$= (-3)^{-d} \sum_{\substack{(\underline{0} \leq) \underline{j} \leq \underline{1} \\ \underline{j} \leq \underline{i}}} (-4)^{|\underline{j}|} \sum_{(\underline{0} \leq) \underline{k} \leq \underline{1}} (-1)^{|\underline{k}|} \chi_I (\underline{i} - \underline{j} + \underline{k}) .$$

We will now make a substitution $\underline{l} = \underline{k} - \underline{j}$, (allowing $\underline{l}$ to have negative values),

$$\tilde{c}_{\underline{i}} = (-3)^{-d} \sum_{\substack{(\underline{0} \leq) \underline{j} \leq \underline{1} \\ \underline{j} \leq \underline{i}}} (-4)^{|\underline{j}|} \sum_{-\underline{j} \leq \underline{l} \leq \underline{1}-\underline{j}} (-1)^{|\underline{l}+\underline{j}|} \chi_I (\underline{i} + \underline{l})$$

$$= (-3)^{-d} \sum_{\substack{-\underline{1} \leq \underline{l} \leq \underline{1} \\ \underline{l}+\underline{i} \geq \underline{0}}} \sum_{\max\{\underline{0},-\underline{l}\} \leq \underline{j} \leq \min\{\underline{1},\underline{1}-\underline{l}\}} (-4)^{|\underline{j}|} (-1)^{|\underline{l}+\underline{j}|} \chi_I (\underline{i} + \underline{l})$$

$$= (-3)^{-d} \sum_{-\underline{1} \leq \underline{l} \leq \underline{1}} \chi_I (\underline{i} + \underline{l}) \sum_{\max\{\underline{0},-\underline{l}\} \leq \underline{j} \leq \min\{\underline{1},\underline{1}-\underline{l}\}} (-4)^{|\underline{j}|} (-1)^{|\underline{l}+\underline{j}|} .$$

The second last equality here is a change of order of summation, and the min and max over the multi-indices are component wise. We now consider the inner summation. Noting that $\underline{j} \geq \underline{0}$ and $\underline{l} + \underline{j} \geq \underline{0}$, we can write this as

$$\sum_{\max\{\underline{0},-\underline{l}\} \leq \underline{j} \leq \min\{\underline{1},\underline{1}-\underline{l}\}} (-4)^{j_1+\cdots+j_d} (-1)^{l_1+j_1+\cdots+l_d+j_d}$$

$$= (-1)^{l_1+\cdots+l_d} \sum_{\max\{\underline{0},-\underline{l}\} \leq \underline{j} \leq \min\{\underline{1},\underline{1}-\underline{l}\}} 4^{j_1} \dots 4^{j_d}$$

$$= (-1)^{l_1+\cdots+l_d} \left(\sum_{j_1=\max\{0,-l_1\}}^{\min\{1,1-l_1\}} 4^{j_1} \right) \cdots \left(\sum_{j_d=\max\{0,-l_d\}}^{\min\{1,1-l_d\}} 4^{j_d} \right)$$

$$= (-1)^{l_1+\cdots+l_d} 5^{d-|\underline{l}|_0} 4^{d-|\underline{l}+\underline{1}|_0} ,$$

where $d - |\underline{l}|_0$ and $d - |\underline{l} + \underline{1}|_0$ essentially count the number of elements of $\underline{l}$ which are 0 and -1 respectively. Finally we note that $(-1)^{l_1+\cdots+l_d} = (-1)^{|\underline{l}|}$ since for any

$l_k = -1$ we have $(-1)^{l_k} = (-1)^{l_k+2} = (-1)^{-l_k}$. Substituting these back yields the desired result.   □

As a result of this proposition, we can compute coefficients for an adaptive extrapolation approach very quickly.

Lemma 11 *Let $I \in \mathcal{D}(\mathbb{N}^d)$ (non-empty) and $\tilde{P}_I$ be the combination of extrapolations $\tilde{P}_I = \sum_{\underline{i} \in I_{+1}} \tilde{c}_{\underline{i}} P_{\underline{i}}$, then for $\underline{i} \in I$ one has $\sum_{\underline{j} \geq \underline{i}} \tilde{c}_{\underline{j}} = 1$.*

Proof We note that as $\tilde{P}_I = \sum_{\underline{j} \in I} c_{\underline{j}} \tilde{P}_{\underline{j}}$ and the $\tilde{P}_{\underline{j}}$ sums over $P_{\underline{k}}$ with $\underline{j} \leq \underline{k} \leq \underline{j}+1$ then

$$\sum_{\underline{j} \geq \underline{i}} \tilde{c}_{\underline{j}} = \sum_{\underline{j} \geq \underline{i}} c_{\underline{j}} \sum_{\underline{k} \leq \underline{1}} \frac{(-4)^{|\underline{k}|}}{(-3)^d}.$$

Further, the sum of coefficients in each $\tilde{P}_{\underline{i}}$ sum to 1 (Lemma 9), that is

$$\sum_{\underline{j} \leq \underline{1}} \frac{(-4)^{|\underline{j}|}}{(-3)^d} = 1.$$

Thus $\sum_{\underline{j} \geq \underline{i}} \tilde{c}_{\underline{j}} = \sum_{\underline{j} \geq \underline{i}} c_{\underline{j}} = 1$ as a consequence of Lemma 2.   □

Corollary 4 *If $I \in \mathcal{D}(\mathbb{N}^d)$ is non-empty then $\sum_{\underline{i} \in I_{+1}} \tilde{c}_{\underline{i}} = 1$.*

Proof Simply apply the previous lemma to $\underline{i} = \underline{0}$.   □

Note that when a covering element $\underline{i}$ is added to a downset I for which $\underline{i}$ has one or more 0 coefficients then we actually need to compute a few additional grids in order to compute $\tilde{P}_{I \cup \{\underline{i}\}}$. In fact this is connected to our second question regarding the downsets I_{+1} for which our extrapolations operate on. Let us rewrite the $\tilde{c}_{\underline{i}}$ of $\tilde{P}_I$ by shifting $\underline{l}$ by $\underline{1}$ as follows

$$\tilde{c}_{\underline{i}} = \sum_{-\underline{1} \leq \underline{l} - \underline{1} \leq \underline{1}} \frac{(-1)^{|\underline{l}-\underline{1}|}}{(-3)^d} 5^{d-|\underline{l}-\underline{1}|_0} 4^{d-|\underline{l}|_0} \chi_I(\underline{i} + \underline{l} - \underline{1})$$

$$= \sum_{\underline{0} \leq \underline{l} \leq \underline{2}} \frac{(-1)^{|\underline{l}|}}{3^d} 5^{d-|\underline{l}-\underline{1}|_0} 4^{d-|\underline{l}|_0} \chi_{I_{+1}}(\underline{i} + \underline{l}).$$

Now given a non-empty downset J we consider the combination

$$\tilde{\tilde{P}}_J = \sum_{\underline{i} \in J} \tilde{\tilde{c}}_{\underline{i}} P_{\underline{i}} u$$

where the coefficients are given by

$$\tilde{\tilde{c}}_{\underline{i}} = \sum_{\underline{0} \le \underline{l} \le \underline{2}} \frac{(-1)^{|\underline{l}|}}{3^d} 5^{d-|\underline{l}-\underline{1}|_0} 4^{d-|\underline{l}|_0} \chi_J(\underline{i} + \underline{l}). \tag{10}$$

Our second question can now be rephrased as determining for which J is the above approximation reasonable. By reasonable we specifically mean two things. First, the coefficients should sum to 1 to provide consistency. Second, the 2nd order error terms should sum to zero when the $P_{\underline{i}} u$ satisfy the error splitting. Observations and experiments seem to indicate that the resulting combination is reasonable if $\underline{1} \in J$ (and thus $\{\underline{1}\} \downarrow\subset J$). This is not too surprising as the multivariate extrapolation applied to a single $u_{\underline{i}}$ requires those $u_{\underline{j}}$ with $\underline{i} \le \underline{j} \le \underline{i} + \underline{1}$. We suspect this is a both a sufficient and necessary condition. This is relatively straightforward to show in 2 dimensions by breaking the problem up into the few cases that can occur. Whilst a case by case argument could also be used for higher dimensions the number of cases grows quickly.

The formula (10) lends itself to a more general adaptive scheme for extrapolations, and analogous to the regular adaptive sparse grids we could look at $\tilde{\tilde{P}}_J - \tilde{\tilde{P}}_I$ where J covers I.

Lemma 12 *Let* $I \in \mathcal{D}(\mathbb{N}^d)$ *with* $\underline{1} \in I$ *and let* $\underline{i}$ *be a covering element of* I. *Let* $J = I \cup \{\underline{i}\}$, *then*

$$\tilde{\tilde{P}}_J - \tilde{\tilde{P}}_I = \sum_{\substack{\underline{i}-\underline{2} \le \underline{j} \le \underline{i} \\ \underline{j} \ge \underline{0}}} \frac{(-1)^{|\underline{i}-\underline{j}|}}{3^d} 5^{d-|\underline{i}-\underline{j}-\underline{1}|_0} 4^{d-|\underline{i}-\underline{j}|_0} P_{\underline{j}}.$$

Proof Simply note that the $\tilde{\tilde{c}}_{\underline{i}}$ that are affected are those with $\underline{i} + \underline{l} = \underline{j}$ for some $\underline{0} \le \underline{l} \le \underline{2}$ and that (10) provides the update for each such $\underline{i}$. □

For the general extrapolation formula (8) one may follow the same procedure to obtain the formulas

$$\tilde{\tilde{c}}_{\underline{i}} = \sum_{-\underline{1} \le \underline{l} \le \underline{1}} (-1)^{|\underline{l}|} \chi_I(\underline{i} + \underline{l}) \left(\prod_{m=1}^{d} \frac{\delta_{-1,l_m} 2^{p_m} + \delta_{0,l_m}(1 + 2^{p_m}) + \delta_{1,l_m}}{1 - 2^{p_m}} \right),$$

(with $\delta_{a,b} = 1$ if $a = b$ and $\delta_{a,b} = 0$ otherwise), and

$$\tilde{\tilde{P}}_J - \tilde{\tilde{P}}_I = \sum_{\substack{\underline{i}-\underline{2} \le \underline{j} \le \underline{i} \\ \underline{j} \ge \underline{0}}} (-1)^{|\underline{i}-\underline{j}|} \left(\prod_{m=1}^{d} \frac{\delta_{i_m-2,j_m} + \delta_{i_m-1,j_m}(1 + 2^{p_m}) + \delta_{i_m,j_m} 2^{p_m}}{2^{p_m} - 1} \right) P_{\underline{j}}$$

for Proposition 8 and Lemma 12 respectively.

5 Conclusions

We have extended the early work on adaptive sparse grids by studying the projections onto the hierarchical surpluses. Specifically we have derived exact expressions for the coefficient updates and provided error bounds. Further, we have applied this framework to the study of multivariate extrapolations. By studying the coefficients of extrapolation formulas within the adaptive sparse grids framework we derive several results which could be applied in an adaptive extrapolation algorithm. The effectiveness of this approach for numerical applications is something to be tested in the future.

References

1. H.J. Bungartz, M. Griebel, Sparse grids. Acta Numerica **13**, 147–269 (2004)
2. H.J. Bungartz, M. Griebel, U. Rüde, Extrapolation, combination, and sparse grid techniques for elliptic boundary value problems. Comput. Methods Appl. Mech. Eng. **116**, 243–252 (1994)
3. J. Garcke, Sparse grids in a nutshell, in *Sparse Grids and Applications*, ed. by J. Garcke, M. Griebel. Lecture Notes in Computational Science and Engineering, vol. 88 (Springer, Berlin/New York, 2013), pp. 57–80
4. T. Gerstner, M. Griebel, Numerical integration using sparse grids. Numer. Algorithms **18**(3–4), 209–232 (1998). Kluwer Academic
5. M. Griebel, H. Harbrecht, On the convergence of the combination technique, in *Sparse Grids and Applications*, ed. by J. Garcke, D. Pflüger. Lecture Notes in Computational Science and Engineering, vol. 97 (Springer, Cham/New York, 2014), pp. 55–74
6. M. Griebel, M. Schneider, C. Zenger, A combination technique for the solution of sparse grid problems, in *Iterative Methods in Linear Algebra*, ed. by P. de Groen, R. Beauwens (IMACS, Elsevier, North Holland, 1992), pp. 263–281
7. B. Harding, M. Hegland, Robust solutions to PDEs with multiple grids, in *Sparse Grids and Applications*, ed. by J. Garcke, D. Pflüger. Lecture Notes in Computational Science and Engineering, vol. 97 (Springer, Cham/New York, 2014), pp. 171–193
8. M. Hegland, Adaptive sparse grids. Anziam J. **44**(April), 335–353 (2003)
9. C.B. Liem, T. Lü, T.M. Shih, *The Splitting Extrapolation Method: A New Technique in Numerical Solution of Multidimensional Problems*. Series on Applied Mathematics, vol. 7 (World Scientific, Singapore/River Edge, 1995)
10. C. Reisinger, Numerische Methoden für hochdimensionale parabolische Gleichungen am Beispiel von Optionspreisaufgaben. Universität Heidelberg (2004)
11. U. Rüde, Extrapolation and related techniques for solving elliptic equations. Bericht I-9135, Institut für Informatik, TU München, Sept 1991
12. M. Wong, Theory of the sparse grid combination technique. Australian National University (2015, submitted)

A Cache-Optimal Alternative to the Unidirectional Hierarchization Algorithm

Philipp Hupp and Riko Jacob

Abstract The sparse grid combination technique provides a framework to solve high-dimensional numerical problems with standard solvers by assembling a sparse grid from many coarse and anisotropic full grids called component grids. *Hierarchization* is one of the most fundamental tasks for sparse grids. It describes the transformation from the nodal basis to the hierarchical basis. In settings where the component grids have to be frequently combined and distributed in a massively parallel compute environment, hierarchization on component grids is relevant to minimize communication overhead.

We present a cache-oblivious hierarchization algorithm for component grids of the combination technique. It causes $|\mathbf{G}_\ell| \cdot \left(\frac{1}{B} + O\left(\frac{1}{\sqrt[d]{M}} \right) \right)$ cache misses under the tall cache assumption $M = \omega\left(B^d\right)$.[1] Here, $\mathbf{G}_\ell$ denotes the component grid, d the dimension, M the size of the cache and B the cache line size. This algorithm decreases the leading term of the cache misses by a factor of d compared to the unidirectional algorithm which is the common standard up to now. The new algorithm is also optimal in the sense that the leading term of the cache misses is reduced to scanning complexity, i.e., every degree of freedom has to be touched once. We also present a variant of the algorithm that causes $|\mathbf{G}_\ell| \cdot \left(\frac{2}{B} + O\left(\frac{1}{\sqrt[d-1]{M \cdot B^{d-2}}} \right) \right)$ cache misses under the assumption $M = \omega(B)$. The new algorithms have been implemented and outperform previously existing software. In several cases the measured performance is close to the best possible.

---

[1] The dimension d is assumed to be constant in the O-notation.

P. Hupp (✉)
ETH Zürich, Zürich, Switzerland
e-mail: philipp.hupp@inf.ethz.ch; philipp@huppweb.de

R. Jacob
IT University of Copenhagen, Rued Langgaards Vej 7, DK-2300 København S, Denmark
e-mail: rikj@itu.dk

© Springer International Publishing Switzerland 2016
J. Garcke, D. Pflüger (eds.), *Sparse Grids and Applications – Stuttgart 2014*,
Lecture Notes in Computational Science and Engineering 109,
DOI 10.1007/978-3-319-28262-6_5

1 Introduction

The gap between peak performance and memory bandwidth on modern processors is already large and still increasing. In many situations, this phenomenon can be counteracted by using caches, i.e., a small but fast additional memory close to the processor. Now, the expensive communication is between the memory and the cache, and this kind of communication efficiency is crucial for high performance code. All areas of computer science acknowledge this phenomenon, but call it and the methods to design such algorithms slightly differently. The algorithms that reduce memory traffic are called, e.g., I/O efficient algorithms [1, 22, 31], communication avoiding algorithms [2, 13, 27], and blocked algorithms [30]. Still, all these efforts aim to increase temporal locality (reuse over time) and spatial locality (use of several items of a cache line) to reduce the amount of data that is transferred between the different levels of the memory hierarchy.

Sparse grids [3, 40, 41] are a numerical discretization scheme that allows to solve high-dimensional numerical problems by lessening the curse of dimensionality from $O\left(h_n^{-d}\right)$ to $O\left(h_n^{-1} \cdot |\log_2 h_n|^{d-1}\right)$ for dimension d and minimum mesh size $h_n = 2^{-n}$. Crucial for the reduction in the degrees of freedom is a change of basis from the nodal basis to the hierarchical basis and the selection of the most important basis functions of the hierarchical basis. This change of basis is called *hierarchization* and is one of the most fundamental algorithms for sparse grids. The reduction in the degrees of freedom for sparse grids comes at the cost of a less regular structure and more complicated data access patterns for sparse grid algorithms. In consequence, communication efficient algorithms are in particular important and less obvious for sparse grids. Because hierarchization is among the most simple algorithmic tasks that are based on the hierarchical structure of the sparse grids, we consider it prototypical in the sense that algorithmic ideas that work for it are also applicable to more complicated tasks.

The sparse grid combination technique [18] assembles the sparse grid from a linear combination of many coarse and anisotropic, i.e. refined differently in different dimensions, full grids called component grids. This allows to solve the numerical problem on the full component grids with standard solvers while taking advantage of the reduced number of degrees of freedom of the sparse grid. For time dependent problems, the combination technique can be applied as depicted in Fig. 1: a standard solver is employed to each of the (regular) component grids. Then, a reduce step assembles the sparse grid solution as a linear combination of the component grid solutions. This is followed by a broadcast step that distributes the joint solution back to the component grids. The change of basis from the regular grid basis to the hierarchical basis can facilitate the reduce and the broadcast step. In this situation, hierarchization is on the performance critical path of the solver. Current approaches to master large simulations of hot fusion plasmas are a prominent example [37].

The task of hierarchization we consider here has as input an array of values representing the sampled function in the nodal basis, i.e., as function values

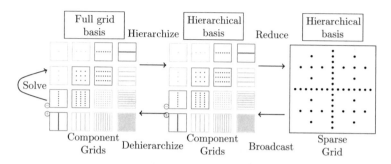

Fig. 1 (De-)hierarchization as pre- and postprocessing steps for the reduce/broadcast step of the combination technique. The combination technique computes a solution in the sparse grid space by a suitable linear combination (*blue*: $+1$; *red*: -1) of the component grid solutions

sampled at the grid point, and as output the same function represented in the hierarchical basis as explained later. The grid points form a regular anisotropic (not all dimensions are refined equally) grid, and the coefficient values are laid out in lexicographic order, i.e., in a generalized row major layout. The algorithm is formulated and analyzed in the cache oblivious model [10]. In this model the algorithm is formulated to work on a random access memory where every memory position holds an element, i.e., an input value, an intermediate value or an output value. It is analyzed in the I/O model with a fast memory (cache) that holds M elements, the transfer to the slower memory is done in blocks or cache lines with B elements, and the cache replacement strategy is optimal. This makes sure that elements that participate in algebraic operations reside in cache. The performance is measured as the number of cache misses (also called I/Os) the algorithm incurs. Note that in this model the CPU operations are not counted as if the CPU was infinitely fast. Hence, this model focuses on one level of the memory hierarchy, usually the biggest relevant one. As our experiments demonstrate, these assumptions capture the most important aspects of our test machine. More precisely, it turns out that the shared level 3 cache of the CPU is usually the bottleneck, because the four cores of the CPU together are fast enough to keep the memory connection busy all the time. Hence it is reasonable as a theoretical model to regard the CPU(s) as infinitely fast.

The unidirectional principle is the dominating design pattern for sparse grid algorithms. The unidirectional principle exploits the tensor product structure of the underlying basis and decomposes the global operator into d sweeps over the grid. In each sweep it works locally on all one-dimensional subproblems, called poles, of the current work direction [3]. By working in d sweeps the unidirectional hierarchization algorithm (Algorithm 1) needs only $3d$ arithmetic operations to hierarchize piecewise d-linear basis functions. In contrast, a direct hierarchization algorithm, i.e., calculating the hierarchical surplus (the coefficient in the hierarchical basis) of each grid point in one go, like formulating the task as a multiplication with a sparse matrix, would require $c \cdot 3^d$ arithmetic operations for $1 \leq c \leq 2$. Therefore, the unidirectional algorithm is a good choice with respect to the number of arithmetic operations. As such, the unidirectional algorithm is, however, inherently cache inefficient in the sense that it performs d sweeps over the data and therefore causes at least $d \cdot \frac{|\mathbf{G}_\ell|}{B} - (d-1) \cdot \frac{M}{B} = d \cdot \frac{1}{B} \cdot \left(|\mathbf{G}_\ell| - \frac{d-1}{d} \cdot M\right)$ cache misses. Here, $\mathbf{G}_\ell$ denotes the input grid, M the size of the internal memory or cache and B the block or cache line size. Furthermore, the unidirectional hierarchization algorithm has been implemented for component grids such that it is within a factor of 1.5 of this unidirectional memory bound [24]. In consequence, any significant further improvements have to avoid the unidirectional principle on a global scale.

This paper presents a cache-oblivious [10] hierarchization algorithm (Algorithm 2) that avoids the unidirectional principle on a global scale but applies it (recursively) to smaller subproblems that fit into cache. It actually computes precisely the same intermediate values at the same memory locations as the unidirectional algorithm, but it computes them in a different order. By doing so, the algorithm avoids the d global passes of the unidirectional algorithm. For component grids and the piecewise-linear basis this algorithm causes $|\mathbf{G}_\ell| \cdot \left(\frac{1}{B} + O\left(\frac{1}{\sqrt[d]{M}}\right)\right)$ cache misses, i.e., it works with scanning complexity (touching every grid point once) plus a lower order term. For the second term to be of lower order a strong tall cache assumption of $M = \omega\left(B^d\right)$ is needed. It reflects that we, as is usual (e.g. [10, 38]), consider the asymptotics of increasing M, and here in particular demand that M grows faster than B^d. With this strong tall cache assumption, the leading term of this complexity result is optimal, as every algorithm needs to scan the input. In addition, the presented algorithm reduces the leading term of the cache misses by at least a factor of d compared to any unidirectional algorithm. For the situation that the cache is not that tall but only of size $M = \omega(B)$, we give a variant of the algorithm that causes at most $|\mathbf{G}_\ell| \cdot \left(\frac{2}{B} + O\left(\frac{1}{\sqrt[d-1]{M \cdot B^{d-2}}}\right)\right)$ cache misses. Depending on the size of the cache, the leading term of the cache misses is therefore reduced by a factor of d or $d/2$ compared to the unidirectional algorithm. The presented algorithm is cache-oblivious, works on a standard row major layout, relies on a least recently used (LRU) cache replacement strategy, is in-place, performs the same arithmetic operations as the unidirectional algorithm and works for anisotropic component grids.

To ease readability and in agreement with common usage in the sparse grid literature, this paper generally assumes for the O-notation that the dimension d is constant: In numerics, the dimension d is a parameter inherent to the problem under consideration. If a more accurate solution is required, the refinement level of the discretization is increased while the dimension of the problem stays constant. For completeness, we state the complexity of the divide and conquer hierarchization algorithm for component grids also including the constant d at the end of the relevant section.

The rest of the paper is organized as follows: Sect. 2 considers related work, Sect. 3 explains the relevant concepts of sparse grids and how they are presented in this paper, Sect. 4 formulates the algorithm and analyses it, Sect. 5 reports on run time experiments of an implementation on current hardware, and Sect. 6 discusses conclusions and directions for future work.

2   Related Work

Hong and Kung started the analysis of the I/O-complexity of algorithms with their red-blue pebble game [22] assuming an internal memory of size M and basic blocks of size $B = 1$.[2] To use data for computations, it has to reside in internal memory. Aggarwal and Vitter extended this basic model to the cache-aware external memory model [1] with arbitrary block or cache line size B to account for spatial locality. Frigo et al. generalized this to the cache-oblivious model [10] in which the parameters M and B are not known to the algorithm and the cache replacement strategy is assumed to be the best possible. As the parameters M and B are not known when a cache-oblivious algorithm is designed, a cache-oblivious algorithm is automatically efficient for several layers of the memory hierarchy simultaneously. All mentioned theoretical models assume a fully associative cache and so does the analysis presented in this paper.

Sparse grids [3, 40, 41] and the sparse grid combination technique [18] have been developed to solve high-dimensional numerical problems. They have been applied to a variety of high-dimensional numerical problems, including partial differential equations (PDEs) from fluid mechanics [17], financial mathematics [4, 21] and physics [29], real-time visualization applications [6, 7], machine learning problems [11, 12, 36], data mining problems [5, 12] and so forth. In a current project [37], the combination technique is used as depicted in Fig. 1 to simulate hot fusion plasmas as they occur in plasma fusion reactors like the international flagship project ITER. In this project the fusion plasmas are modeled using the gyrokinetic approach which results in a high-dimensional PDE, i.e., five space and velocity dimensions plus time. Furthermore, the convergence of the combination technique has been studied for several special cases [34, 35, 39] as well as general operator equations [15].

Due to the coarse grain parallelism of the component grids the combination technique is ideal for high performance computing [14]. This coarse grain parallelism also allows to incorporate algorithm based fault tolerance into the combination technique [20]. It was discovered early that, for time dependent PDEs, the component grid solutions need to be synchronized after few time steps [16] and that the communication needed for this synchronization can be reduced if the component grid solutions are represented in the hierarchical basis [19]. Recently, communication schemes that use the hierarchical representation of the component grid solutions to minimize communication in this synchronization step were derived, implemented and tested for the setting of the gyrokinetic approach [23, 25, 26].

The problem considered in this work, namely finding efficient hierarchization and dehierarchization algorithms for sparse grids, has been investigated in many occasions [6, 8, 9, 24, 28, 32, 33, 36]. All these algorithms implement the unidirectional algorithm and hence sweep d times over the whole data set. The unidirectional

---

[2]We use the terms internal memory and cache as well as cache line size and block size synonymously.

hierarchization algorithm for component grids has been implemented such that it is within a factor of 1.5 of the unidirectional memory bound [24]. Therefore, any significant further improvements have to avoid the global unidirectional principle. This paper extends the first algorithm that avoids the unidirectional principle on a global scale [23]. In contrast to this initial version of the algorithm, the algorithm presented in this paper works in-place and performs the same arithmetic operations as the unidirectional algorithm. Also, the first lower bound for the hierarchization task was proven in [23].

3   Sparse Grid Definitions and the Unidirectional Hierarchization Algorithm

This section describes the necessary notation and background to discuss the sparse grid hierarchization algorithm. For a thorough description of sparse grids we refer to the survey by Bungartz and Griebel [3].

Let us begin with a conventional level ℓ discretization of the 1-dimensional space $\Omega := [0, 1]$. The grid points x of $\mathbf{G}_\ell$ are

$$\mathbf{G}_\ell = \left\{ x = \frac{i}{2^\ell} \in \Omega : i \in \{0, 1, \ldots, 2^\ell\} \right\} .$$

The corresponding nodal basis functions are the piecewise linear hat functions with peak at the grid point and support of the form $]\frac{i-1}{2^\ell}, \frac{i+1}{2^\ell}[$. The 1-dimensional sparse grid of level $n = \ell$ has the same grid points. We use the notation $x_{k,i} = \frac{i}{2^k}$ for $0 \le k \le \ell$ and $i \in \{0, \ldots, 2^k\}$. Two distinct pairs (k, i) and (k', i') describe the same grid point if the coordinates of the grid points are identical, i.e., if $x_{k,i} = x_{k',i'}$. For a level-index pair (k, i) the reduced pair (k', i') is defined to have the smallest i', i.e., i' is odd, with $x_{k',i'} = x_{k,i}$. For odd i and $1 \le k$, the interval $I_{k,i} =]x_{k,i} - 2^{-k}, x_{k,i} + 2^{-k}[$, is the support of the corresponding hierarchical basis function $\phi_{k,i}(x)$ with $x_{k,i}$ as its midpoint, i.e. $\phi_{k,i}(x) = \max(0, 1 - |x - x_{k,i}| \cdot 2^k)$ as is also depicted in Fig. 2. The reduced level index pairs of the two endpoints (allowing $(0, 0)$ as a special case) define the functions $\mathcal{L}$ and $\mathcal{R}$ by $I_{k,i} =]x_{\mathcal{L}(k,i)}, x_{\mathcal{R}(k,i)}[$. The two grid points $x_{\mathcal{L}(k,i)}$ and $x_{\mathcal{R}(k,i)}$ are called the left and respectively right hierarchical predecessor. We additionally define the interval $I_{0,0} = [0, 1]$. $I_{0,0}$ differs from $I_{1,1} =]0, 1[$ only by its two endpoints which are called global boundary points. These two global boundary points are the only grid points that have no hierarchical predecessors. The closure $\overline{I_{k,i}}$ of an interval $I_{k,i} =]x_{\mathcal{L}(k,i)}, x_{\mathcal{R}(k,i)}[$ is defined in the usual way as $\overline{I_{k,i}} := [x_{\mathcal{L}(k,i)}, x_{\mathcal{R}(k,i)}]$.

We say that $I_{k,i}$ has the two children intervals $I_{k+1,2i-1} =]x_{\mathcal{L}(k,i)}, x_{k,i}[$ and $I_{k+1,2i+1} =]x_{k,i}, x_{\mathcal{R}(k,i)}[$. We extend this notion (by transitive closure) to descendants,

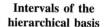

**Intervals of the
hierarchical basis**

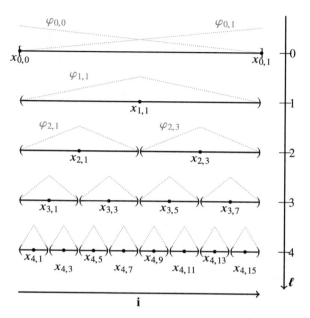

Fig. 2 The intervals $I_{k,i}$ and the corresponding grid points $\mathbf{x}_{k,i}$ as well as the piecewise linear basis functions $\varphi_{k,i}$ of the hierarchical basis

and observe that an interval $I_{k',i'}$ is a descendant of $I_{k,i}$ if and only if $I_{k',i'} \subset I_{k,i}$. This immediately leads to the following statement:

Lemma 1 *If a grid point $x_{k',i'}$ is element of the interval $I_{k,i}$ (with odd i and i') then $k' \geq k$ and the hierarchical predecessors of $x_{k',i'}$ are in the closure of $I_{k,i}$, i.e.,*
$$(x_{k',i'} \in I_{k,i}) \Rightarrow \left(\{x_{\mathcal{L}(k',i')}, x_{\mathcal{R}(k',i')}\} \subset \overline{I_{k,i}}\right).$$

Hierarchization is a change of basis of a piecewise linear function from the nodal basis of level ℓ, given as the function values y_i at position $i2^{-\ell} \in \{0, 2^{-\ell}, \ldots, 1\}$, into the hierarchical basis of the sparse grid. For odd i, $k \leq \ell$ and $I_{k,i} =]x_{\mathcal{L}(k,i)}, x_{\mathcal{R}(k,i)}[$, we get the hierarchical surplus as $\alpha_{k,i} = y_{k,i} - \frac{1}{2}(y_{\mathcal{L}(k,i)} + y_{\mathcal{R}(k,i)})$ for $k \geq 1$, and for the global boundary points we have $\alpha_{0,0} = y_{0,0}$ and $\alpha_{0,1} = y_{0,1}$. The grid points $\{x_{k,i}, x_{\mathcal{L}(k,i)}, x_{\mathcal{R}(k,i)}\}$ form the 3-point stencil of $x_{k,i}$.

Let us now address the d-dimensional case and the discretization of the space $\Omega := [0, 1]^d$. In general, vectors are written in bold face and operations on them are meant component wise. The conventional anisotropic grid $\mathbf{G}_\ell$ with mesh-width $h_{\ell_r} := 2^{-\ell_r}$ and discretization level ℓ_r in dimension $r \in \{1, \ldots, d\}$ has the grid points

$$\mathbf{G}_\ell = \left\{\mathbf{x} = \frac{\mathbf{i}}{2^\ell} \in \Omega : i_r \in \{0, 1, \ldots, 2^{\ell_r}\} \,\forall r \in \{1, \ldots, d\}\right\} .$$

The corresponding basis functions are the tensor products of the one-dimensional basis functions. Hence, the grid $\mathbf{G}_\ell \subset [0,1]^d$ is completely defined by its level vector $\ell \in \mathbb{N}_0^d$ describing how often dimension $r \in \{1,\ldots,d\}$ has been refined. A grid of refinement level ℓ_r consists of $2^{\ell_r} + 1$ grid points in dimension r, the outermost two of which are called global boundary points, i.e., the points with $i_r \in \{0, 2^{\ell_r}\}$.

The d-dimensional sparse grid results from a tensor product approach. To express this, a level and index is replaced by a d-fold level- and index-vector in the above definition of the sparse grid. The grid points have the form $(x_{k_1,i_1}, \ldots, x_{k_d,i_d})$ corresponding to the basis function with support $I_{k_1,i_1} \times \cdots \times I_{k_d,i_d}$. The regular sparse grid of level n consists of the grid points with $|\mathbf{k}|_1 \le n + d - 1$, i.e., $k_1 + \cdots + k_d \le n + d - 1$. The anisotropic component grid with level vector ℓ consists of the grid points with $k_r \le \ell_r$.

In this case hierarchization can be performed using the unidirectional principle using $d - 1$ intermediate results at every grid point. More precisely, we define $d + 1$ variables $\alpha_{\ell,\mathbf{i}}^{(j)}$. For $j = 0$, the variable $\alpha_{\ell,\mathbf{i}}^{(0)}$ is the function value at position described by $(\mathbf{i}, \ell)$. The final value $\alpha_{\ell,\mathbf{i}}^{(d)}$ is the hierarchical surpluses, i.e., the coefficient of the hierarchical basis functions that represent the function with the prescribed values at the grid points. For $j > 0$, the variable $\alpha_{\ell,\mathbf{i}}^{(j)}$ is what we call "hierarchized up to dimension j", also referred to as the coefficient at position $(\ell, \mathbf{i})$ being in state j, and it is computed from the variables $\alpha_*^{(d-1)}$ by applying the 3-point stencil in direction j. More precisely, for a level index vector $(\ell, \mathbf{i})$ define the left hierarchical predecessor in direction r as $\mathcal{L}_r(\ell, \mathbf{i}) := (\ell', \mathbf{i}')$, with $(\ell_r', i_r') := \mathcal{L}(\ell_r, i_r)$ and $(\ell_s', i_s') = (\ell_s, i_s)$ for $s \ne r$. $\mathcal{R}_r$ is defined analogously for the right hierarchical predecessor in direction r. With this we define $\alpha_{\ell,\mathbf{i}}^{(j)} = \alpha_{\ell,\mathbf{i}}^{(j-1)} - \frac{1}{2}\left(\alpha_{\mathcal{L}_j(\ell,\mathbf{i})}^{(j-1)} + \alpha_{\mathcal{R}_j(\ell,\mathbf{i})}^{(j-1)} \right)$, and for the boundary points in direction r with $(k_r, i_r) = (0,0)$ or $(k_r, i_r) = (0,1)$ we have $\alpha_{\ell,\mathbf{i}}^{(j)} = \alpha_{\ell,\mathbf{i}}^{(j-1)}$. If boundary points are not part of the task, for the sake of uniformity, we consider the modification of the boundary points as applying a 3-point stencil, too, only that the non-existent hierarchical predecessor variables are considered being 0. For a set of grid points U and a direction r let $\mathcal{H}_r(U)$ denote the set of hierarchical predecessors in direction r.

It is well known that one-dimensional hierarchization can be performed in-place by performing the hierarchization from high level to low level. It follows immediately that also high-dimensional hierarchization can be performed in-place by using the unidirectional principle. This is expressed in Algorithm 1, the classical unidirectional hierarchization algorithm. This formulation of the algorithm uses the notion of a pole, i.e., the grid points that are an axis-aligned one-dimensional grid in dimension k. In our notation, a pole in direction r is expressed as $\mathbf{G}_{\ell,I}$ where the interval I is such that $I_r = [0,1]$ for the direction r and all other components of I are single (grid) coordinates. We also use $\pi_r(\mathbf{G}_\ell)$ for the projection of $\mathbf{G}_\ell$ along dimension r, i.e., replacing the r-th coordinate by 0. Therefore, $\pi_r(\mathbf{G}_\ell)$ contains exactly one grid point of each pole in direction r and can be used to loop over all poles in this direction.

Algorithm 1: The unidirectional hierarchization algorithm

---

1 **Function** unidirHierarchize $(\mathbf{G}_\ell)$
2   **for** $r \leftarrow 1$ **to** d **do** // *unidirectional loop over dimensions*
3   **forall** $\mathbf{x}_{k,i} \in \pi_r(\mathbf{G}_\ell)$ **do** // *loop over all poles in dimension* r
4   **for** *level* $\leftarrow \ell_r$ down to 1 **do** // *update pole bottom up*
5             $\mathbf{k}_r = level$
6   **forall** *index* $\in \{1, \ldots, 2^{k_r} - 1\}$ *and index odd* **do**
7                $\mathbf{i}_r = index$
8                $\mathbf{x}_{k,i} = \mathbf{x}_{k,i} - 0.5 * \left(\mathbf{x}_{\mathcal{L}_r(k,i)} + \mathbf{x}_{\mathcal{R}_r(k,i)}\right)$

---

4 Divide and Conquer Hierarchization

This section first derives the basic version of the divide and conquer hierarchization algorithm (Algorithm 2), proves its correctness and then analyzes its complexity. Subsequently, this algorithm is used as basic building block to derive hybrid algorithms which trade a weaker tall cache assumption for an increase in cache misses. The section ends with a sketch of parallelization possibilities for the derived algorithms.

In the new algorithm, Algorithm 2, we perform hierarchization in-place as in the unidirectional algorithm, but we do not follow the unidirectional principle globally. Like in the unidirectional algorithm We still have one intermediate result per grid point at any time, but the dimension up to which a grid point is hierarchized depends on its location.

The new algorithm divides the grid spatially and works on d-dimensional intervals (generalized axis parallel rectangle). The constituting one-dimensional interval may consist of a single grid point or be the support of a basis function corresponding to a grid point. More precisely, we call the d-dimensional interval $I = I_1 \times \cdots \times I_d$ a valid grid interval for the grid $\mathbf{G}_\ell$ with respect to level vector $\boldsymbol{\ell}$ if all intervals are of the form $I_r = I_{k_r,i_r}$ or a single grid point $I_r = \{x_{k_r,i_r}\} = [x_{k_r,i_r}, x_{k_r,i_r}]$, with $k_r \leq \ell_r$. Now the subgrid of $\mathbf{G}_\ell$ corresponding to the interval I is defined as

$$\mathbf{G}_{\ell,I} = \mathbf{G}_\ell \cap I$$

Such valid grid intervals have few hierarchical predecessors outside of the interval itself which will ensure that the algorithm is efficient: if I_r is a singleton of the form $I_r = \{x_{k,i}\}$, then $I'_r := \{x_{\mathcal{L}(k,i)}, x_{\mathcal{R}(k,i)}\}$. Otherwise ($I_r$ is an open interval) set $I'_r := \overline{I_r}$. Complete the definition of I' by setting $I'_s = I_s$ for $s \neq r$. With that, the notion of hierarchical predecessors is extended to valid grid intervals by defining $\mathcal{H}_r(\mathbf{G}_{\ell,I}) := I'$. Subgrids $\mathbf{G}_{\ell,I}$ can be defined for arbitrary intervals I and are not restricted to valid grid intervals. Furthermore, we write the shorthand $\mathbf{G}_{\ell,\mathcal{H}_r(I)} := \mathcal{H}_r(\mathbf{G}_{\ell,I})$. To describe the hierarchical predecessors that are part of the 3-point stencil, it is convenient to define the boundary of an interval as $\mathcal{B}_r(I) = \mathcal{H}_r(I) \setminus I$, i.e., the hierarchical predecessors of I which are outside of I. Observe that these boundary points in different directions are disjoint, i.e., $\mathcal{B}_r(I) \cap \mathcal{B}_s(I) = \emptyset$ if $r \neq s$.

The number of grid points $s_r(\mathbf{G}_{\ell,I})$ (size) in dimension r of grid $\mathbf{G}_{\ell,I}$, $s_r(\mathbf{G}_{\ell,I})$ is the number of different coordinates in dimension r that occur for grid points in $\mathbf{G}_{\ell,I}$. For a valid grid interval I of $\mathbf{G}_\ell$ we have

$$
s_r(\mathbf{G}_{\ell,I}) = \begin{cases} 2^{\ell_r - k_r + 1} - 1 & \text{if } I_r = I_{k_r, i_r} \text{ (with } i_r \text{ odd),} \\ 2^{\ell_r} + 1 & \text{if } I_r = [0, 1], \\ 1 & \text{if } I_r \text{ is a single grid point.} \end{cases}
$$

Next, we define the $\texttt{chooseDim}(\mathbf{G}_{\ell,I})$ function for a grid $\mathbf{G}_{\ell,I}$. It returns the dimension for which the grid $\mathbf{G}_{\ell,I}$ has the most grid points (ties can be broken arbitrarily, e.g., choose the smallest dimension).

$$
\texttt{chooseDim}(\mathbf{G}_{\ell,I}) = \arg\max_{1 \leq r \leq d} \{s_r(\mathbf{G}_{\ell,I})\} .
$$

To split the multidimensional interval $I = I_1 \times \cdots \times I_d$ in direction r into three parts we define the following functions. The split functions rely upon $\mathbf{G}_{\ell,I}$ containing more than one grid point in direction r, i.e. $k_r < \ell_r$. For all dimensions $s \neq r$ we set $I_s^* = I_s$ (for $* \in \{0, int, 1, \text{left}, \text{mid}, \text{right}\}$). The case where $I_r = [0, 1]$ is the only situation where an outer boundary should be split off, hence we set $I_r^0 = \{0\}$, $I_r^{int} =\,]0, 1[$ and $I_r^1 = \{1\}$, and write

$$
\texttt{boundarySplit}(r, I) := \left(I_r^0, I_r^{int}, I_r^1\right) .
$$

Otherwise $I_r = I_{k,i}$, and we set $I_r^{\text{left}} = I_{k+1,2i-1} =\,]x_{\mathcal{L}(k,i)}, x_{k,i}[$, $I_r^{\text{mid}} = \{x_{k,i}\}$ and $I_r^{\text{right}} = I_{\ell+1,2i-1} =\,]x_{k,i}, x_{\mathcal{R}(k,i)}[$. We write

$$
\texttt{interiorSplit}(r, I) := \left(I_r^{\text{left}}, I_r^{\text{mid}}, I_r^{\text{right}}\right) .
$$

Both splits are depicted in Fig. 3. Clearly, the three parts are a partitioning of the subgrid, and because only non-trivial dimensions are split, they are all non-empty.

With these definitions, Algorithm 2 is well defined. Its call structure is illustrated in Fig. 4. Next we show that a call $\texttt{hierarchizeRec}(0, d, \mathbf{G}_\ell, [0, 1]^d)$ hierarchizes the grid correctly. Subsequently, the complexity of the algorithm is analyzed.

Fig. 3 *Left:* Applying the $\texttt{boundarySplit}$ to the interval $[0, 1]$. *Right:* Applying the $\texttt{interiorSplit}$ to an interior grid interval $]x_{k,j}, x_{k',j'}[$

Algorithm 2: Divide and conquer hierarchization algorithm

---

1 Function hierarchizeRec $(s, t, \mathbf{G}_\ell, I)$
    // I is a valid grid interval for $\mathbf{G}_\ell$. Initially, all variables
    // of $\mathbf{G}_{\ell,I}$ store the values $\alpha^{(s)}$, in the end they store $\alpha^{(t)}$.
    // This function changes only variables of $\mathbf{G}_{\ell,I}$.
    // It assumes that $\mathbf{G}_{\ell, \mathcal{B}_r(I)}$, i.e., the boundary points of I in
    // direction r, stores the values $\alpha^{(r)}$.

2   **if** $\mathbf{G}_{\ell,I} = \{x_{\mathbf{k},\mathbf{i}}\}$ **then** // grid consists of a single grid point
3   **for** $r \leftarrow (s+1)$ **to** t **do**
4 $x_{\mathbf{k},\mathbf{i}} = x_{\mathbf{k},\mathbf{i}} - 0.5 * \left(x_{\mathcal{L}_r(\mathbf{k},\mathbf{i})} + x_{\mathcal{R}_r(\mathbf{k},\mathbf{i})}\right)$

5   **else** // i.e., $\left|\mathbf{G}_{\ell,I}\right| > 1$
6 $r = \texttt{chooseDim}\left(\mathbf{G}_{\ell,I}\right)$
     // split G into subgrids in dimension r
7   **if** $I_r = [0, 1]$ **then** // case of global boundary
8 $\left(I^0, I^{int}, I^1\right) = \texttt{boundarySplit}\left(r, I\right)$
9   hierarchizeRec$\left(s, (r-1), \mathbf{G}_\ell, I^0\right)$
10   hierarchizeRec$\left(s, (r-1), \mathbf{G}_\ell, I^1\right)$
11   hierarchizeRec$\left(s, t, \mathbf{G}_\ell, I^{int}\right)$
12   hierarchizeRec$\left((r-1), t, \mathbf{G}_\ell, I^0\right)$
13   hierarchizeRec$\left((r-1), t, \mathbf{G}_\ell, I^1\right)$
14   **else** // $I_r \subset (0, 1)$, i.e., no global boundary
15 $\left(I^{left}, I^{mid}, I^{right}\right) = \texttt{interiorSplit}\left(r, I\right)$
16   hierarchizeRec$\left(s, (r-1), \mathbf{G}_\ell, I^{mid}\right)$
17   hierarchizeRec$\left(s, t, \mathbf{G}_\ell, I^{left}\right)$
18   hierarchizeRec$\left(s, t, \mathbf{G}_\ell, I^{right}\right)$
19   hierarchizeRec$\left((r-1), t, \mathbf{G}_\ell, I^{mid}\right)$

4.1 Correctness

To prove the correctness of Algorithm 2 it is sufficient to show that whenever we apply the 3-point stencil in direction r, all three participants store the value $\alpha^{(r-1)}$ and that, in the end, the 3-point stencils in all d dimensions have been applied for all grid points. The presented argument does not rely on the regular structure of the component grids and hence works for regular and adaptively refined sparse grids identically.

When an interval $I_{\mathbf{k},\mathbf{i}}$ is split, the boundary of the resulting subintervals is either part of the boundary of $I_{\mathbf{k},\mathbf{i}}$ or lies exactly in the middle of $I_{\mathbf{k},\mathbf{i}}$:

Lemma 2 *Let $I = I_{\mathbf{k},\mathbf{i}}$ for $\mathbf{k} \geq 1$ be a valid interval for some grid and assume I is split by $\left(I^{left}, I^{mid}, I^{right}\right) := \texttt{interiorSplit}(r, I)$.*

1. In directions different from r, the boundary remains: If $s \neq r$ we have $\mathcal{B}_s(I^{left}) \subset \mathcal{B}_s(I), \mathcal{B}_s(I^{right}) \subset \mathcal{B}_s(I)$, and $\mathcal{B}_s(I^{mid}) \subset \mathcal{B}_s(I)$.

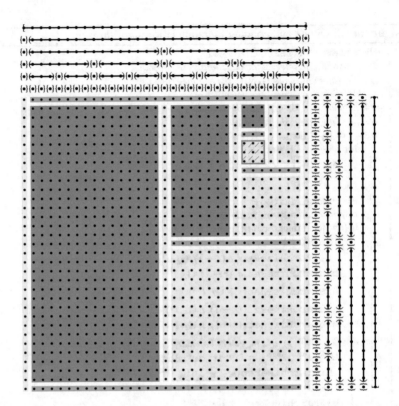

Fig. 4 Hierarchizing a 2-dimensional grid with the divide and conquer hierarchization algorithm (Algorithm 2). The *yellow* (or *blue* or *red*) grid points are hierarchized up to dimension 0 (or 1 or 2), i.e., store the original functional values $\alpha_{k,i}^{(0)}$ (or the values $\alpha_{k,i}^{(1)}$ or the hierarchical surpluses $\alpha_{k,i}^{(2)}$, respectively). The subgrid that is currently updated is hatched (in *red*). The algorithm progresses by hierarchizing complete grid intervals (depicted on the sides)

2. *The boundary of full-dimensional parts is the boundary or the split-plane:* $\mathcal{B}_r(I^{left}) \subset (\mathcal{B}_r(I) \cup I^{mid})$, *and* $\mathcal{B}_r(I^{right}) \subset (\mathcal{B}_r(I) \cup I^{mid})$.
3. *The hierarchical predecessors of the split plane is the old boundary:* $\mathcal{B}_r(I^{mid}) = \mathcal{B}_r(I)$.

Proof Follows from the definitions and Lemma 1   □

For the correctness of Algorithm 2 observe that the purpose of the call is to hierarchize the points inside I in direction $s + 1, \dots, t$. Accordingly, all points inside I are assumed to be already hierarchized in directions up to s, i.e., the variables present $\alpha^{(s)}$. We show that the following invariants hold for the recursion.

For a call of $\texttt{hierarchizeRec}(s, t, \mathbf{G}_\ell, I)$ we formulate the following states of the grid $\mathbf{G}_\ell$.

Definition 1 (Precondition($\mathbf{G}_\ell, I, s$))

1. a variable in $\mathbf{G}_{\ell,I}$ holds the value $\alpha^{(s)}$.
2. a variable in $\mathbf{G}_\ell \cap \mathcal{B}_r(I)$ holds the value $\alpha^{(r)}$.

Definition 2 (Postcondition($\mathbf{G}_\ell, \mathbf{G}'_\ell, I, t$))

1. a variable in $\mathbf{G}'_{\ell,I}$ holds the value $\alpha^{(t)}$.
2. all other variables have the same value in $\mathbf{G}'_\ell$ as in $\mathbf{G}_\ell$.

Lemma 3 *If* hierarchizeRec $(s, t, \mathbf{G}_\ell, I)$ *is called in a situation as described by Definition 1, then the grid $\mathbf{G}'_\ell$ after the call is as described by Definition 2.*

Proof By induction on size of the current subgrid, i.e., the number of grid points in $\mathbf{G}_{\ell,I}$. First, consider the case that the current subgrid $\mathbf{G}_{\ell,I}$ contains a single grid point. Then the for-loop in Line 3 changes the state from the first item of the precondition to the first item of the postcondition. In this, the stencil in Line 4 is correct because of the second item of the precondition.

Second, if $|\mathbf{G}_{\ell,I}| > 1$, then the algorithm performs a split operation. By Lemma 2, Item 1, the invariant on the boundary in directions different from r is transferred from the call to all recursive calls. Because the split partitions $\mathbf{G}_{\ell,I}$ into three subgrids, Item 1 of the precondition is also transferred. The algorithm distinguishes between 2 further cases with $|\mathbf{G}_{\ell,I}| > 1$, namely $I_r = [0, 1]$ or $I_r \subset (0, 1)$.

In the first case we have $I_r = [0, 1]$. In this case the boundary of I^0 and I^1 in direction r is empty, i.e., $\mathcal{B}_r(I^0) = \mathcal{B}_r(I^1) = \emptyset$. Hence, Item 2 of the precondition holds trivially for the calls in Lines 9, 10 and 12, 13. As $\mathcal{B}_r(I^{int}) = (I^0 \cup I^1)$, it follows from the postcondition of the first two calls that Item 2 of the precondition for the call in Line 11 is fulfilled. Item 1 of the precondition for the call in Line 12 follows from Item 1 of the postcondition of the call in Line 9. A similar argument works for Line 13.

Otherwise I_r is an open interval that contains at least 3 grid points of $\mathbf{G}_\ell$. Hence, the split operation partitions the subgrid into three non-empty subgrids. Item 2 of the precondition for the call in Line 16 follows directly from Lemma 2, Item 3. This establishes by Lemma 2, Item 2 the precondition for the calls in Item 17 and Item 18. Item 1 of the precondition for the call in Line 19 follows from Item 1 of the postcondition of the call in Line 16.

The postcondition on the final grid $\mathbf{G}'_\ell$ for the whole call now follows from the postconditions of the recursive calls. □

Therefore, the call hierarchizeRec$(0, d, \mathbf{G}_\ell, [0, 1]^d)$ hierarchizes the whole grid $\mathbf{G}_\ell$ correctly as the precondition holds trivially and the postcondition means that all grid points are correctly hierarchized.

4.2 Complexity Analysis

For each grid point Algorithm 2 performs exactly d updates and for each update precisely 3 grid points are needed. As at most 3 cache lines are needed in cache simultaneously, the hierarchization of any grid point does not cause more than $3d$ cache misses, i.e., 3 cache misses per update, for any reasonable cache replacement strategy. If the subgrid the algorithm works on is sufficiently small, i.e., the subgrid and all hierarchical predecessors of the stencil fit it into memory, the algorithm is much more efficient. In that case the whole subgrid can be hierarchized by loading it and its hierarchical predecessors into memory once and performing the updates in memory. As the subgrid fits into cache and no other cache lines are accessed in between, a LRU strategy ensures that the subgrid stays in cache as long as it is needed. The analysis builds upon these observations.

Note that for all d-dimensional grids, i.e., for all grids that have more than 1 grid point in each dimension, the call $\texttt{hierarchizeRec}(s, t, \mathbf{G}_{\ell, I})$ always happens with $s = 1$ and $t = d$. The parameters s and t are only altered for the grids of the form $\mathbf{G}_{\ell, I_r^0}$, $\mathbf{G}_{\ell, I_r^1}$ and $\mathbf{G}_{\ell, I_r^{mid}}$ which have at least one dimension with a single grid point.

The presented analysis assumes that the grid $\mathbf{G}_\ell$ is significantly larger than the cache. In particular, it is assumed that all directions are refined such that there exist isotropic subgrids, i.e., subgrids with the same number of grid points in each dimension, that do not fit into cache. For component grids, this is the case if $\left(2^{\min_r \ell_r} - 1\right)^d \geq M$, where M is the memory size.

We analyze the performance of the algorithm by focusing on certain calls to $\texttt{hierarchizeRec}(0, d, \mathbf{G}_\ell, I_i)$ on the same level of the recursion, given by the family of intervals $I_i, i \in F$, where $F = \{(\ell, \mathbf{i}) \mid \ell = (m, \ldots, m)\}$. All I_i have the same shape and size. We choose them in a way that the level of the subgrid in each dimension is m. With our particular definition of $\texttt{chooseDim}(\mathbf{G}_{\ell, I})$ and the sufficiently large grids we consider, these calls are actually performed. The intervals I_i ($i \in F$) almost partition the domain, namely they are disjoint and the union of their closures is the complete domain, i.e., $\cup_i \bar{I_i} = [0, 1]^d$. Note that $\mathbf{G}_{\ell, \bar{I_i} \setminus I_i}$ contains the boundary points in all directions and a few more points.

The analysis of such a call is based on the number of grid points of the subgrid in its interior $N(m) = |\mathbf{G}_{\ell, I_i}|$ and on its boundary $Q(m) = |\mathbf{G}_{\ell, \bar{I_i} \setminus I_i}|$. More precisely, we need a good lower bound on $N(m)$ (progress), and good upper bounds on $N(m) + Q(m)$ (base cost and memory requirement) and $Q(m)$ (additional cost). For the base cost and memory requirement we additionally, have to take into consideration the layout of the grid and how it interacts with the blocks of the (external) memory.

Once we identified an m such that the whole grid $\mathbf{G}_{\ell, \bar{I_i}}$ fits into memory, we can estimate the overall number of cache misses in the following way: the call $\texttt{hierarchizeRec}(0, d, \mathbf{G}_\ell, I_i)$ hierarchizes $\mathbf{G}_{\ell, I_i}$ and costs loading the subgrid and its boundary. The number of cache misses is $(N(m) + Q(m))/B$ plus an additional term that is less then $Q(m)$ to account for blocks that are not completely filled (assuming a row major layout). Hierarchizing the boundary $\mathbf{G}_{\ell, \bar{I_i} \setminus I_i}$ incurs at most $3d$ cache misses per boundary point. The number of calls can be estimated by $|F| < |\mathbf{G}_\ell|/N(m)$. Hence, the total number of cache misses is at most

$$\frac{|\mathbf{G}_\ell|}{N(m)}\left(\frac{N(m)+Q(m)}{B}+(3d+1)Q(m)\right) \leq |\mathbf{G}_\ell|\left(\frac{1}{B}+\frac{(3d+2)Q(m)}{N(m)})\right)$$

Hence, in the following we analyze the asymptotic behavior of the additional term $\frac{(3d+2)Q(m)}{N(m)}$, and show that it is $o(1/B)$.

Lemma 4 *Hierarchizing a component grid $\mathbf{G}_\ell$ with $\ell_r \geq \frac{1}{d}\log_2 M$ ($\forall r$) using Algorithm 2 takes*

$$|\mathbf{G}_\ell|\left(\frac{1}{B}+O\left(\frac{1}{\sqrt[d]{M}}\right)\right)$$

cache misses in the cache-oblivious model with the tall cache assumption $B = o(\sqrt[d]{M})$ and an LRU cache replacement strategy.

Proof In the setting of component grids we have $N(m) = (2^m - 1)^d$, and $Q(m) \leq 2d(2^m + 1)^{d-1}$. We choose

$$m = \log_2\left(\sqrt[d]{M/2} - 1\right).$$

leading to

$$\frac{1}{N(m)} = \frac{1}{(2^m-1)^d} = O\left(\frac{1}{M}\right) \text{ and } Q(m) \leq 2d\left(\sqrt[d]{M/2}\right)^{d-1} = O\left(M^{\frac{d-1}{d}}\right).$$

From the tall cache assumption we conclude that for any constant c we have $c \cdot B^d \leq M$ for large enough M and B, which we use as $B \leq \frac{1}{2} \cdot \sqrt[d]{M/2}$. Assuming a row major layout, the number of occupied cache lines is upper bounded by

$$\left(\left\lceil\frac{2^m+1}{B}\right\rceil + 1\right) \cdot (2^m+1)^{d-1} \leq \frac{(2^m+1)^d}{B} + 2 \cdot (2^m+1)^{d-1} =$$

$$= \frac{M}{2 \cdot B} + 2^{1/d} \cdot M^{\frac{d-1}{d}} \leq \frac{M}{2 \cdot B} + 2^{1/d} \cdot M^{\frac{d-1}{d}} \cdot \underbrace{\frac{M^{1/d}}{2^{\frac{d+1}{d}}B}}_{\geq 1} = \frac{M}{B}.$$

Hence, the choice of m is as required in the preceding discussion. In that case the additional term is

$$\frac{(3d+2)Q(m)}{N(m)} = O\left(\frac{M^{\frac{d-1}{d}}}{M}\right) = O\left(\frac{1}{\sqrt[d]{M}}\right) = o\left(\frac{1}{B}\right),$$

where the last equality is the tall cache assumption. □

When the constant d is made explicit in the last equation the lower order term reads $O\left(d^2/\sqrt[d]{M}\right)$.

4.3 Hybrid Algorithms

One aspect of Algorithm 2 and its analysis that might limit its applicability is the fairly strong tall cache assumption $B = o\left(\sqrt[d]{M}\right)$. Algorithm 1, in contrast, can be modified to work with $|\mathbf{G}_\ell|\left(\frac{d}{B} + O\left(\frac{1}{M}\right)\right)$ cache misses if $B = o\left(M\right)$. In fact, these two algorithms mark the corners of a whole spectrum of algorithms that become more cache efficient as the cache gets taller. Instead of working subsequently in all d directions as Algorithm 1, or merging all d phases as Algorithm 2, these hybrid algorithms merge $c \in \mathbb{N}$, $1 \leq c \leq d$ phases of the unidirectional principle. To discuss these hybrid algorithms it is first assumed that the component grid $\mathbf{G}_\ell$ is stored in a block aligned fashion, i.e., every pole in direction 1 is padded with dummy elements such that every pole starts at the beginning of a cache line. As a result, all poles are split into cache lines in the very same way. After discussing this aligned case, the hybrid algorithms are also sketched for the case that the alignment is not possible.

Lemma 5 *For every $c \in \mathbb{N}$, $1 \leq c \leq d$ and c divides d there is an algorithm that, assuming a tall cache with $M = \omega\left(B^c\right)$, performs hierarchization on a block aligned component grid $\mathbf{G}_\ell$ with $|\mathbf{G}_\ell|\left(\frac{d}{c}\cdot\frac{1}{B} + O\left(\frac{1}{\sqrt[c]{M}}\right)\right)$ cache misses.*

Proof Let us first consider the case of $c = 1$, i.e., the mentioned modification of Algorithm 1. To hierarchize a pole, replace Algorithm 4 to Algorithm 8 by the call hierarchizeRec $(r-1, r, \mathbf{G}_\ell, I)$ for the one-dimensional poles e.g. $I = [0, 1] \times \{x_{k_2,i_2}\} \times \cdots \times \{x_{k_d,i_d}\}$.

For $r = 1$ and the considered row major layout, the poles are contiguous in memory. Therefore, each pole can be considered as a 1-dimensional subgrid to which Algorithm 2 is applied. Therefore, Lemma 4 yields that this modification of Algorithm 1 needs $|\mathbf{G}_\ell|\left(\frac{1}{B} + O\left(\frac{1}{M}\right)\right)$ cache misses for the first unidirectional pass, i.e., to hierarchize the first dimension. In that case, the tall cache assumption of Lemma 4 is $B = o\left(M\right)$.

For $r > 1$ the poles worked on are not stored contiguously in memory, and working on a single pole at a time would access a whole cache line to only work with a single element. This can be avoided by the following kind of blocking that works with the poles in direction r that share the same cache line. These poles are by layout neighboring in direction 1. For the sake of the formulation of the algorithm and its complexity analysis, this allows us to consider the cache lines instead of the grid points as the atomic elements, which results in cache line size $B' = 1$, internal memory size $M' = M/B$ and grid size $\mathbf{G}'_\ell = \mathbf{G}_\ell/B$ (all in cache lines). As we now consider $B' = 1$, the memory can be filled with the new (meta-)poles without polluting the internal memory with other grid points.

Therefore the analysis is identical to the case of $r = 1$ which yields that transferring the grid from state $\alpha^{(r-1)}$ to state $\alpha^{(r)}$ needs $|G_\ell| \left(\frac{1}{B} + O\left(\frac{1}{M}\right) \right)$ cache misses and a tall cache assumption of $B = o(M)$.

For $c > 1$ the one-dimensional poles are replaced by c-dimensional planes. The hybrid algorithm works in d/c phases and each phases hierarchizes the grid from state $\alpha^{((p-1)\cdot c)}$ to state $\alpha^{(p\cdot c)}$ for some $p \in \mathbb{N}$. If $c = 2$, the interval is for example $I = [0,1] \times [0,1] \times \{x_{k_3,i_3}\} \times \cdots \times \{x_{k_d,i_d}\}$ and this I can be regarded as a 2-dimensional pole which can be hierarchized in dimension 1 and 2 by the call hierarchizeRec $(0, 2, G_\ell, I)$. The hybrid algorithm performs this call for all such intervals, bringing the complete grid to the state $\alpha^{(c)}$.

For $p = 1$, the intervals are in contiguous memory and the c-dimensional analysis of Lemma 4 applies. This shows that transforming the grid from state $\alpha^{(0)}$ to state $\alpha^{(c)}$ takes $|G_\ell| \left(\frac{1}{B} + O\left(\frac{1}{\sqrt[c]{M}}\right) \right)$ cache misses and requires the tall cache assumption $B = o(\sqrt[c]{M})$.

For $p > 1$, i.e., if the intervals are orthogonal to the direction of the layout, we can again use a version of the algorithm that works on B intervals simultaneously, i.e., regards the cache lines as the atomic elements instead of the grid points (i.e., cache line size $B' = 1$, internal memory size $M' = M/B$ and grid size $G'_\ell = G_\ell/B$ (all in cache lines)). As we consider the case $B' = 1$, the same analysis as in the case of hierarchizing the first c dimensions, i.e., $p = 1$, can be applied. This yields that transforming the grid from state $\alpha^{((p-1)\cdot c)}$ to state $\alpha^{(p\cdot c)}$ for any $p > 1$ causes

$$\frac{|G_\ell|}{B} \left(\frac{1}{1} + O\left(\sqrt[c]{\frac{B}{M}} \right) \right) = |G_\ell| \left(\frac{1}{B} + O\left(\frac{1}{\sqrt[c]{M \cdot B^{c-1}}} \right) \right)$$

cache misses. The tall cache assumption required to use the analysis of the $p = 1$ case is $B = o(M)$. This assumption also guarantees that the second term is of lower order.

As this hybrid algorithm performs d/c sweeps over the complete grid and the lower order term for $p = 1$ dominates that of $p > 1$, i.e., $\frac{1}{\sqrt[c]{M \cdot B^{c-1}}} = O\left(\frac{1}{\sqrt[c]{M}}\right)$ given $B = o(M)$, the statement of the lemma follows.   □

Lemma 6 *For $1 \leq u < d$ and assuming a tall cache of $M = \omega(B^u)$, the number of cache misses to hierarchize a block aligned component grid G_ℓ is*

$$|G_\ell| \left(\frac{2}{B} + O\left(\frac{1}{\sqrt[u]{M}} \right) + O\left(\frac{1}{\sqrt[d-u]{M \cdot B^{d-u-1}}} \right) \right).$$

Proof Consider a hybrid algorithm which works in 2 passes, each pass hierarchizing a different number of dimensions: the first pass hierarchizes the first u dimensions, i.e., $c = u$ and $p = 1$. The second pass hierarchizes the last $(d - u)$ dimensions, i.e., $c = (d - u)$ and $p > 1$. It follows from the proof of Lemma 5 that the first pass requires a tall cache assumption of $B = o\left(\sqrt[u]{M}\right)$ and causes $|G_\ell| \left(\frac{1}{B} + O\left(\frac{1}{\sqrt[u]{M}}\right) \right)$ cache misses. Also by the proof of Lemma 5, the second pass requires a tall cache assumption of $B = o(M)$ and causes $|G_\ell| \left(\frac{1}{B} + O\left(\frac{1}{\sqrt[d-u]{M \cdot B^{d-u-1}}} \right) \right)$ cache misses.   □

For $u \geq d/2$, it holds that $\frac{1}{d - \sqrt[u]{M \cdot B^{d-u-1}}} = O\left(\frac{1}{\sqrt[u]{M}}\right)$ such that the first lower order term in Lemma 6 dominates. As the tall cache assumption $M = \omega(B^u)$ just becomes stronger as u increases, choosing $u > d/2$ is therefore not advantageous. For $u < d/2$, it depends on the actual size of the cache whether the first or the second lower order term dominates. In particular, for $u = 1$, Lemma 6 becomes:

Lemma 7 *Assuming a cache of size* $M = \omega(B)$, *a block aligned component grid* $\mathbf{G}_\ell$ *can be hierarchized with* $|\mathbf{G}_\ell| \left(\frac{2}{B} + O\left(\frac{1}{d - \sqrt[u]{M \cdot B^{d-2}}}\right)\right)$ *cache misses.*

If for some reason a block aligned layout of the component grid is not feasible, the hybrid algorithms can block b poles together. When b is sufficiently larger than B (i.e., $B = o(b)$), then there are at most two cache lines which contain also unused grid points for every $\lfloor b/B \rfloor - 1$ full cache lines. This changes the term $1/B$ to $(1/B + 3/b)$ in the above analysis, adding another lower order term, and the effective size of the cache to $M' = M/b$.

4.4 Parallelization

To achieve high performance on modern machines, it is important that an algorithm can use many parallel processors. In Algorithm 2 this is possible by executing the two recursive calls in Line 17 and Line 18 in parallel. This is still a correct algorithm because I_r^{left} and I_r^{right} are disjoint and the precondition for both calls is already established after Line 16. On the level of grid points, i.e. $B = 1$, the resulting algorithm implements an "exclusive write" police, i.e., two different processors never write to the same memory location simultaneously. Without further synchronization it requires the possibility of "concurrent read" because both parallel calls read the variables in $\mathbf{G}_{\ell, I_r^{mid}}$.

Considering cache-lines, it is possible that two different processors write to the same cache line. To avoid this, the algorithm performs the two calls serially if and only if the split was done in dimension 1, the direction in which a cache line extends. Hierarchization of the boundaries only needs $O(|\mathbf{G}_\ell|/M)$ cache misses, even if it would be performed in serial. On a system with P processors, each having a private cache of size M, the above version of Algorithm 2 achieves that the number of parallel cache misses is $|\mathbf{G}_\ell| \left(\frac{1}{PB} + O\left(\frac{1}{\sqrt[d]{M}}\right)\right)$ as long as $P \leq \prod_{r=2}^{d} 2^{\ell_r}$.

5 Experimental Evaluation

In this section we report on the run times of our implementation of the described recursive algorithm, its variants and alternatives. The experiments confirm that the main bottleneck of the task is the memory access. We conclude this from the observation that the measured running times are generally close to what the I/O model predicts. Hence, further improvement can only be expected when implementing a different I/O algorithm. Still, it should be noted that this is only true once the implementation is sufficiently carefully optimized in other aspects, most notably multicore parallelism and vectorization, but also branch mispredictions, overhead for (recursive) function calls, and the creation of parallel tasks. We also observe that with increasing dimension the gap between the prediction and the measurements increases, which we suppose has several reasons: Our analysis is not particularly careful with respect to higher dimensions and constant memory size. The I/O-model ignores additional memory effects like the TLB, i.e., the cache used to perform virtual memory translation.

5.1 Setup and Systems

We implemented the algorithm in C++, using openMP for parallelization and hand coded AVX-vectorization. In the implementation we use the C-style numbering of the dimensions starting with 0, but in the description here we translate this to the usual numbering from 1 to d. The experiments are performed for the case without boundary points, i.e., where all global boundary points are implicitly 0.0. For the recursive algorithms, this is actually more complicated than if all boundaries are present because the recursive calls create some cases with boundaries and hence it is necessary to keep track of the existence of boundaries in the different directions. This allows a direct comparison with [24].

The main focus of the experiments is wall-clock run-time as measured by the chrono timer provided in C++11. This is usually taken relative to the time it takes to touch the grid once, as calculated from the measured performance by the stream benchmark (using as many cores as helpful), multiplied with the size of the grid. The stream benchmark measures the speed at which the CPU can access large amounts of data that is stored contiguously. It turned out to be sensitive to the used compiler, so we always took the highest performance reported.

The experiments do not empty the cache in order to provide cold cache measurements, but by the size of the grid and the structure of the algorithm, the influence of the content of the cache when the measurement starts is small.

The implementation is compiled with gcc in the version available on the architecture (see below), using the flags `-mavx -Wa,-q -Wall -fopenmp -std=c++11 -march=native -O3`. We use openMP to run the code on several cores in parallel, using the static scheduling to distribute the work of for loops, and the concept of tasks for the recursive algorithm. Our code compiles with icc, but a small set of test runs showed that for the bigger grids we are interested in, there were hardly any differences to gcc.

5.1.1 System 1: Rechenteufel

Most of the experiments were performed on this system. It is a standalone workstation (called Rechenteufel) with an IvyBridge Intel(R) Xeon(R) CPU E3-1240 V2 3.40 GHz. with 8 MB shared L3 cache. (From IvyBridge Specs: private L2 Cache of 256 KB, private L1 Cache of 64 KB.) It has 1 CPU with 4 cores (no hyperthreading) and 32 GB DDR3 main memory. The stream benchmark using icc version 13.1.3 (gcc version 4.7.0 compatibility) gives a performance of 21.9 GB/s = $21.9 \cdot 10^9$ byte/s. The used gcc has version 4.8.3. For the reported experiments the maximum grid sizes are roughly 8 GB (a quarter of the main memory).

5.1.2 System 2: Hornet

This system is used for the experiments with GENE. It is one node of a supercomputer called hornet. The CPU is an Intel Haswell E5-2680v3 2,5 GHz with 12 Cores, hyperthreading off, and 30 MB shared L3 cache. (from haswell sepcs: L2 cache: 256 KB per core, L1 cache 64 KB per core), and it has 64 GB DDR4 main memory. One node has two such CPUs, but our experiments only used one of them.

The stream Benchmark with cc (cray compiler) version 8.3.6, using 12 cores, gives $57.286 \cdot 10^9$ Bytes/s.

5.2 Compared Algorithms

Our experimental evaluation considers the task of hierarchization without boundary points.

5.2.1 Unidirectional Algorithm

The basic unidirectional algorithm has been implemented very efficiently as described in [24]. It has a natural lower bound of d times scanning, which is almost achieved in many cases.

5.2.2 Recursive Algorithm

This is an implementation of Algorithm 2, as explained and analyzed in Sect. 4. Hence, for sufficiently big cache (compared to the dimension), this algorithm scans the data set once.

To reduce the overhead of recursive execution, we use as base case regions (sub-)poles in dimension 1 of level `recTile`. In our data layout, such a region is consecutive in memory. Hierarchization in dimension 1 needs two additional values and is done iteratively and without vectorization. In contrast, each hierarchization in a dimension different from 1 needs two other such regions, and the application of the stencil is done vectorized using AVX instructions on 4 doubles.

Multi-core parallelism is implemented as described in Sect. 4 using openMP tasks. To avoid the task creation overhead for very small tasks, we do not create tasks if the level sum of the current recursive call is too small. With a focus on the shared level 3 cache, we also do not parallelize the tasks if the level sum is too big. These limits are called minSpawnLevel and maxSpawnLevel.

Given that this algorithm has the three parameters recTile, minSpawnLevel and maxSpawnLevel, we conducted a parameter study that lead to a reasonable heuristic to choose these parameters. This algorithm turned out to be the fastest for problems with 2, 3 or 4 dimensions.

5.2.3 Hybrid Algorithm: Twice Rec

This is an implementation of the hybrid Algorithm described in Sect. 4.3 that performs two scans over the data set. The first phase considers meta-poles in dimensions 1 to d_{split}, i.e. $1 \leq d_{\text{split}} < d$. These are small complete d_{split} dimensional component grids, presumably small enough to fit into cache. They are hierarchized iteratively using the optimized unidirectional algorithm of [24]. The loop over these subgrids is parallelized.

The second phase is Algorithm 2 operating on vectors that constitute the hierarchized d_{split} dimensional subgrids. Here, the base case is a single vector, and the application of the stencil is vectorized. To increase the number of vectors that can fit into cache, we split the vectors into chunks of BlockSize elements. But, to amortize the overhead of the recursive call structure to sufficiently big base cases, we should not choose BlockSize too small. Hence for the second phase we have the parameters BlockSize, minSpawnLevel and maxSpawnLevel, and for the overall algorithm the additional d_{split}. Again, a parameter study lead to a heuristic to choose good values for these parameters, namely to use BlockSize $= 1024$ and choose d_{split} in a way that the level sum of the subgrids is at least 14.

This algorithm currently gives the best performance for dimensions 5 and 6.

5.3 Parameter Study and Heuristics

The study of the parameters minSpawnLevel, maxSpawnLevel, and recTile for the recursive and BlockSize and d_{split} for the hybrid Algorithm needs to consider a big parameter space. Hence, we did not include further parameters and always used 8 openMP threads and grids of size 8 GB. Further, we did not repeat the individual runs. Accordingly, the heuristic to choose the parameters might not be perfect, but as we will see in the next sections, the performance achieved with this heuristic is usually already pretty good.

5.3.1 Recursive

The parameters of the recursive algorithms are all depending on the level sum of the current rectangle. Hence, they can be at most $\texttt{maxLevel} = \sum_{i=2}^{d} \ell_i$. In the parameter study we vary $\texttt{maxSpawnLevel}$ between 4 and $\texttt{maxLevel}$, and $\texttt{minSpawnLevel}$ between 3 and $\texttt{maxSpawnLevel} - 1$. The parameter $\texttt{recTile}$ varies between 2 and ℓ_1.

The following heuristic yields performance that is close to the best choice of parameters: We chose $\texttt{recTile}$ to be ℓ_1 if this is smaller than 14, else we choose it to be 5. Further we choose $\texttt{maxSpawnLevel} = \texttt{maxLevel} - 2$, $\texttt{minSpawnLevel} = \texttt{maxLevel} - 7$. This choice of parameters is reasonable in the following sense: A real split in the first dimension is expensive as on the boundary we access a whole cache line to use a single grid point. Therefore, if the first dimension is small, it is better to not split it at all. If the first dimension is very large, then the $\texttt{recTile}$ is chosen rather small such that the tiles are more quadratic and the interior to boundary ratio is better than for tiles with a very long first dimension. Only recursive calls between $\texttt{maxSpawnLevel}$ and $\texttt{minSpawnLevel}$ are parallelized. Hence, this difference needs to be at least the binary log of the intended number of parallel tasks, explaining the difference of 5. Keeping $\texttt{maxSpawnLevel}$ slightly away from $\texttt{maxLevel}$ leads to all threads working on some (still big) subgrid, which seems beneficial, perhaps because of caching effects in the virtual address translation (TLB). Generally we observe that the performance was not very sensitive to the choice of the spawn levels.

5.3.2 Twice Recursive

The algorithm is only meaningful if the parameter split dimension is in the interval $1 \leq d_{\text{split}} < d$, and the parameter study explores this whole range. The block size $\texttt{BlockSize}$ is set to all powers of 2 between 4 and $\left(\prod_{i=1}^{d_{\text{split}}} 2^{\ell_i} \right) - 2$.

This lead to the heuristic of choosing d_{split} as the smallest dimension i such that $l = \sum_{j=1}^{i} \ell_j > 13$, and choosing $\texttt{BlockSize} = 1024$. With this heuristic, the subgrids of the first phase are at least $2^{14+3} bytes = 128\,\text{KB}$ big (and not too much bigger). Hence, they can fit into the private L2 cache of 256 KB if $l \leq 15$, and will fit into the 8 MB big L3 cache even if all four cores are active if $l \leq 18$. The choice of $\texttt{BlockSize}$ is a reasonable compromise between keeping the memory requirement small and having enough work to amortize the overhead of the recursion.

5.3.3 Data and Results for Parameter Study

The results of the parameter study for the recursive algorithm are shown in Table 1, that of the twice recursive algorithm in Table 2. Comparing the running times

Table 1 Best runtime of the "Recursive" Algorithm over the searched parameter space and runtime given the parameters chosen by the heuristic. 8 OMP threads on a 4 core CPU (rechenteufel)

$d = 2$	Runtime (best) [s]	Parameters (best)	Runtime (heuristic) [s]	Parameters (heuristic)	Runtime (heuristic) / Runtime (best)
$\ell = (15, 15)$	0.98	SpawnLevel $= (13, 14)$ recTile $= 5$	1.00	SpawnLevel $= (8, 13)$ recTile $= 5$	1.01
$\ell = (20, 10)$	0.99	SpawnLevel $= (7, 8)$ recTile $= 5$	1.00	SpawnLevel $= (3, 8)$ recTile $= 5$	1.01
$\ell = (10, 20)$	0.88	SpawnLevel $= (16, 17)$ recTile $= 7$	0.90	SpawnLevel $= (13, 18)$ recTile $= 10$	1.02
$d = 3$	Runtime (best) [s]	Parameters (best)	Runtime (heuristic) [s]	Parameters (heuristic)	Runtime (heuristic) / Runtime (best)
$\ell = (10, 10, 10)$	1.21	SpawnLevel $= (17, 20)$ recTile $= 7$	1.21	SpawnLevel $= (13, 18)$ recTile $= 10$	1.00
$\ell = (15, 8, 7)$	1.57	SpawnLevel $= (12, 14)$ recTile $= 5$	1.58	SpawnLevel $= (8, 13)$ recTile $= 5$	1.01
$\ell = (8, 7, 15)$	1.12	SpawnLevel $= (19, 20)$ recTile $= 8$	1.12	SpawnLevel $= (15, 20)$ recTile $= 8$	1.00
$d = 4$	Runtime (best) [s]	Parameters (best)	Runtime (heuristic) [s]	Parameters (heuristic)	Runtime (heuristic) / Runtime (best)
$\ell = (8, 8, 7, 7)$	1.77	SpawnLevel $= (18, 21)$ recTile $= 6$	1.80	SpawnLevel $= (15, 20)$ recTile $= 8$	1.02
$\ell = (12, 6, 6, 6)$	2.14	SpawnLevel $= (4, 17)$ recTile $= 7$	2.21	SpawnLevel $= (11, 16)$ recTile $= 12$	1.03
$\ell = (6, 6, 6, 12)$	1.86	SpawnLevel $= (15, 16)$ recTile $= 6$	1.98	SpawnLevel $= (17, 22)$ recTile $= 6$	1.06
d, ℓ	Runtime (best) [s]	Parameters (best)	Runtime (heuristic) [s]	Parameters (heuristic)	Runtime (heuristic) / Runtime (best)
$d = 5$ $\ell = (6, 6, 6, 6, 6)$	2.76	SpawnLevel $= (19, 20)$ recTile $= 5$	3.04	SpawnLevel $= (17, 22)$ recTile $= 6$	1.10
$d = 6$ $\ell = (5, 5, 5, 5, 5, 5)$	3.81	SpawnLevel $= (10, 22)$ recTile $= 3$	4.61	SpawnLevel $= (18, 23)$ recTile $= 5$	1.21

between the tables shows that the recursive algorithm is clearly faster for up to 3 dimensions, for 4 dimensions it is slightly faster, and for 5 and 6 dimensions twice recursive is faster. This is coherent with the theoretical analysis that the interior to boundary ratio and the tall cache requirement become bad for the recursive algorithm, whereas the twice recursive algorithm can in both phases be close to the scanning bound. The anisotropic grids are only reported for the generally faster algorithm.

The heuristic works well, it manages to get within 5 % of the running time with the best parameters for the recursive algorithm, and within 22 % for the twice recursive algorithm. In one case the running time of the heuristic is actually reported to be faster than that of the best parameters, which is an artifact of repeating the run and means that the heuristic is optimal up to measurement accuracy. In the following, we will always use the heuristic to chose the parameters.

Table 2 Best runtime of the "Twice Recursive" Algorithm over the searched parameter space and runtime given the parameters chosen by the heuristic

d, ℓ	Runtime (best) [s]	Parameters (best)	Runtime (heuristic) [s]	Parameters (heuristic)	Runtime (heuristic)/Runtime (best)
$d = 2$ $\ell = (15, 15)$	1.77	$d_{\text{split}} = 2$ BlockSize $= 8192$	1.85	$d_{\text{split}} = 2$ BlockSize $= 1024$	1.05
$d = 3$ $\ell = (10, 10, 10)$	2.12	$d_{\text{split}} = 2$ BlockSize $= 256$	2.90	$d_{\text{split}} = 3$ BlockSize $= 1024$	1.37
$d = 4$ $\ell = (8, 8, 7, 7)$	2.03	$d_{\text{split}} = 3$ BlockSize $= 1024$	1.99	$d_{\text{split}} = 3$ BlockSize $= 1024$	0.98
$d = 5$	Runtime (best) [s]	Parameters (best)	Runtime (heuristic) [s]	Parameters (heuristic)	Runtime (heuristic)/Runtime (best)
$\ell = (6, 6, 6, 6, 6)$	2.42	$d_{\text{split}} = 3$ BlockSize $= 1024$	2.45	$d_{\text{split}} = 4$ BlockSize $= 1024$	1.01
$\ell = (10, 5, 5, 5, 5)$	1.97	$d_{\text{split}} = 3$ BlockSize $= 2048$	2.22	$d_{\text{split}} = 3$ BlockSize $= 1024$	1.13
$\ell = (5, 5, 5, 5, 10)$	1.41	$d_{\text{split}} = 4$ BlockSize $= 1024$	1.50	$d_{\text{split}} = 4$ BlockSize $= 1024$	1.06
$d = 6$	Runtime (best) [s]	Parameters (best)	Runtime (heuristic) [s]	Parameters (heuristic)	Runtime (heuristic)/Runtime (best)
$\ell = (5, 5, 5, 5, 5, 5)$	1.66	$d_{\text{split}} = 4$ BlockSize $= 1024$	1.72	$d_{\text{split}} = 4$ BlockSize $= 1024$	1.04
$\ell = (8, 5, 5, 4, 4, 4)$	2.23	$d_{\text{split}} = 4$ BlockSize $= 8192$	2.30	$d_{\text{split}} = 4$ BlockSize $= 1024$	1.03
$\ell = (5, 5, 4, 4, 4, 8)$	1.55	$d_{\text{split}} = 4$ BlockSize $= 512$	1.89	$d_{\text{split}} = 4$ BlockSize $= 1024$	1.22

5.4 Strong Scaling

A classical experiment is that of strong scaling, i.e., comparing the runtime for the same task with different number of threads, depicted in Fig. 5. In all cases we see perfect scaling between one and two threads, and in most cases a constant performance for four or more threads, which reflects that the machine has four cores. We also see that two threads already achieve more than half of the best performance. For the 6 dimensional case we observe that the second phase of the twice recursive algorithm has a somewhat unstable performance, and that only for 11 and 12 threads it is close to scanning time. For the 5 dimensional case the performance is stable but the first phase takes two times scanning, reflecting that the subgrids are bigger than they should ideally be. All in all, twice recursive does not quite achieve the possible performance of scanning twice, but it still outperforms the unidirectional scanning bound by a factor of two for six dimension, and almost that for five dimensions.

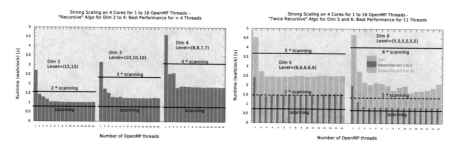

Fig. 5 Scaling of the two algorithms on a four core single CPU machine

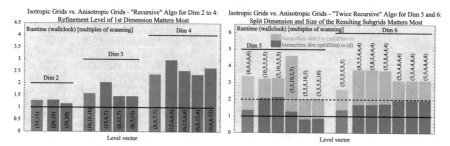

Fig. 6 Influence of anisotropy on the performance

5.5 Anisotropic Grids

In the context of the combination technique, many component grids are anisotropic. We address this by considering grids of different dimensions with level sum 30, i.e., roughly 8 GB of data, as reported Fig. 6. In all tested cases, the unidirectional bound is beaten, in many cases quite clearly. Three cases achieve almost the best possible performance. The recursive algorithm suffers somewhat from the first dimension being refined further. For the twice recursive algorithm, changing the most refined dimension actually changes the split between the two phases, which has a strong influence on performance. Further, the performance is better if the refined dimension is handled in the second phase.

5.6 Speedup over Unidirectional ICCS Code

So far we mainly compared the recursive implementations with the scanning bound, which provides lower bounds, directly for the recursive algorithm, multiplied by 2 for twice recursive and multiplied by d for the unidirectional algorithm. Here we compare this performance with the unidirectional implementation presented in [24]. As we can see in Fig. 7, the new implementation is superior for grids with more than a million points, i.e., roughly 100 MB size.

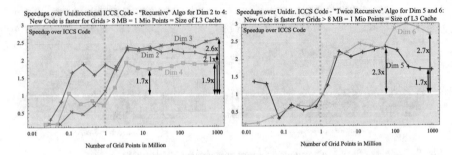

Fig. 7 Comparison of the unidirectional implementation and the recursive ones

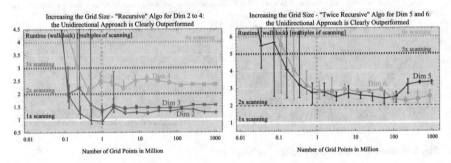

Fig. 8 Performance of the recursive algorithms with respect to grid size. Reports average running time per point (relative to scanning bound) over 10 runs, errorbars show min and max

5.7 Increase Grid Size

Another important aspect of the code is how it scales with the size of the grids as depicted in Fig. 8. We see that for grids with at least a million points (8 MB size), the performance is stable, and it seems to converge to a constant depending on the dimension. This is in line with the analysis in Sect. 4.2. In all cases we see the performance well below the bound of the unidirectional algorithm, and for dimensions 2 and 3 the recursive algorithm is close to scanning once. In dimensions 5 and 6 the performance is reasonable close to scanning twice, as expected for the twice recursive algorithm.

5.8 GENE

One important example for the combination technique, as mentioned in the introduction, is the case of using GENE to simulate a fusion reactor, as reported in Fig. 9. The peculiar situation here is that the first two dimensions are in phase space and should hence not be hierarchized. This can easily be accommodated by using the twice recursive algorithm with $d_{\text{split}} = 2$. The grid sizes stem from a pilot study performed

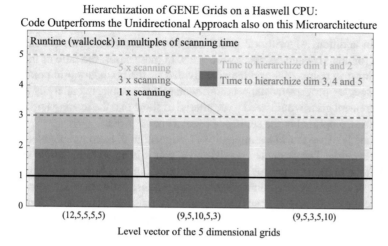

Fig. 9 Grid sizes relevant for the fusion reactor simulation using GENE; haswell system running with 12 threads on 12 cores. The grids have size 32 GB and scanning once is roughly one second

by Mario Heene and Dirk Pflüger in Stuttgart, and we conducted these experiments on the haswell system (as described earlier). We see that the hierarchization of dimensions 3 to 5 takes less time than scanning the grid twice. These measurements show that also on this architecture the implementation performs well and even the heuristic for choosing the parameters (taken from rechenteufel) is not too system specific.

6 Conclusions

This paper has introduced a novel approach to sparse grid algorithms by deriving a cache-oblivious hierarchization algorithm (Algorithm 2) that avoids the d global phases of the unidirectional principle but applies it recursively to smaller subproblems that fit into cache. For the piecewise linear basis and the component grids of the sparse grid combination technique, a discretization scheme to solve high-dimensional numerical problems, the cache complexity of this algorithm is optimal as the leading term of the cache misses is reduced to scanning complexity. For optimality, Algorithm 2 relies on the tall cache assumption $M = \omega\left(B^d\right)$. The general idea of divide and conquer, however, can also be used to derive hybrid algorithms that merge several but not all phases of the unidirectional principle. These hybrid algorithms trade a weaker tall cache assumption off against a slightly increased complexity. One such algorithm only needs a cache of size $M = \omega\left(B\right)$ and, basically, scans the grid twice.

As sparse grids are inherently hierarchical, the divide and conquer approach can also be generalized to other kinds of basis functions and sparse grid tasks such as dehierarchization, i.e., the inverse transformation from the hierarchical basis to the nodal basis, and up-down schemes used to solve PDEs directly in the sparse grid space. In addition, Algorithm 2 is not limited to component grids but can also be applied for adaptive and regular sparse grids. In these cases, the ratio of the number of interior grid points of a grid interval divided by the boundary grid points, which can be seen as progress/costs, becomes worse. As a result, the analysis presented for component grids would need to be altered to show that the leading term of cache misses is also optimal in the setting of regular sparse grids.

The analysis of the I/O complexity of Algorithm 2 is complemented with an implementation. The presented results show, that it is possible to handle additional factors that influence runtime such as vectorization and branch predictions well enough that the memory connection is used almost fully. In particular, the new implementation clearly outperforms previously existing implementations, and in several cases it comes close to optimal performance (as is possible by the memory system).

In the experiments, we see that it is advantageous if the base case is square with respect to cache lines, a case that should be possible to analyze theoretically as well. Then, perhaps, a weaker tall cache assumption might be sufficient.

Acknowledgements We would like to thank Dirk Pflüger and Mario Heene for support and discussions, in particular for enabling the experiments on Hornet. We also thank two anonymous referees for detailed feedback on an earlier draft.

References

1. A. Aggarwal, J.S. Vitter, The input/output complexity of sorting and related problems. Commun. ACM **31**(9), 1116–1127 (1988)
2. G. Ballard, J. Demmel, O. Holtz, O. Schwartz, Minimizing communication in numerical linear algebra. SIAM J. Matrix Anal. Appl. **32**(3), 866–901 (2011)
3. H.-J. Bungartz, M. Griebel, Sparse grids. Acta Numer. **13**, 147–269 (2004)
4. H.-J. Bungartz, A. Heinecke, D. Pflüger, S. Schraufstetter, Option pricing with a direct adaptive sparse grid approach. J. Comput. Appl. Math. **236**(15), 3741–3750 (2011). Online Okt. 2011
5. H.-J. Bungartz, D. Pflüger, S. Zimmer, Adaptive sparse grid techniques for data mining, in *Modelling, Simulation and Optimization of Complex Processes 2006, Proceedings of the International Conference on HPSC*, Hanoi, ed. by H. Bock, E. Kostina, X. Hoang, R. Rannacher (Springer, 2008), pp. 121–130
6. G. Buse, R. Jacob, D. Pflüger, A. Murarasu, A non-static data layout enhancing parallelism and vectorization in sparse grid algorithms, in *Proceedings of the 11th International Symposium on Parallel and Distributed Computing (ISPDC)*, Munich, 25–29 June 2012 (IEEE, 2012), pp. 195–202
7. D. Butnaru, D. Pflüger, H.-J. Bungartz, Towards high-dimensional computational steering of precomputed simulation data using sparse grids, in *Proceedings of the International Conference on Computational Science (ICCS)*, Tsukaba. Volume 4 of Procedia CS (Springer, 2011), pp. 56–65
8. P. Butz, Effiziente verteilte Hierarchisierung und Dehierarchisierung auf vollen Gittern, Bachelor's thesis, University of Stuttgart, 2014, http://d-nb.info/1063333806

9. C. Feuersänger, Sparse grid methods for higher dimensional approximation, PhD thesis, Universität Bonn, 2010
10. M. Frigo, C. E. Leiserson, H. Prokop, S. Ramachandran, Cache-oblivious algorithms, in *Proceedings of the 40th Annual Symposium on Foundations of Computer Science (FOCS'99)*, New York (IEEE Computer Society Press, 1999), pp. 285–297
11. J. Garcke, Maschinelles Lernen durch Funktionsrekonstruktion mit verallgemeinerten dünnen Gittern, PhD thesis, Universität Bonn, 2004
12. J. Garcke, M. Griebel, On the parallelization of the sparse grid approach for data mining, in *Large-Scale Scientific Computing*, ed. by S. Margenov, J. Waśniewski, P. Yalamov. Volume 2179 of Lecture Notes in Computer Science (Springer, Berlin/Heidelberg, 2001), pp. 22–32
13. E. Georganas, J. González-Domínguez, E. Solomonik, Y. Zheng, J. Touriño, K. Yelick, Communication avoiding and overlapping for numerical linear algebra, in *Proceedings of the International Conference on High Performance Computing, Networking, Storage and Analysis (SC'12)*, Salt Lake City (IEEE Computer Society Press, Los Alamitos, 2012), pp. 100:1–100:11
14. M. Griebel, The combination technique for the sparse grid solution of PDE's on multiprocessor machines. Parallel Process. Lett. **2**, 61–70 (1992)
15. M. Griebel, H. Harbrecht, On the convergence of the combination technique, in *Sparse Grids and Applications*. Volume 97 of Lecture Notes in Computational Science and Engineering (Springer, Cham/New York, 2014), pp. 55–74
16. M. Griebel, W. Huber, Turbulence simulation on sparse grids using the combination method, in ed. by N. Satofuka, J. Periaux, A. Ecer, *Proceedings Parallel Computational Fluid Dynamics, New Algorithms and Applications (CFD'94)*, Kyoto, Wiesbaden Braunschweig (Vieweg, 1995), pp. 75–84
17. M. Griebel, W. Huber, C. Zenger, Numerical turbulence simulation on a parallel computer using the combination method, in *Flow Simulation on High Performance Computers II, Notes on Numerical Fluid Mechanics 52*, pp. 34–47 (Vieweg, Wiesbaden 1996) DOI:10.1007/978-3-322-89849-4_4
18. M. Griebel, M. Schneider, C. Zenger, A combination technique for the solution of sparse grid problems, in *Iterative Methods in Linear Algebra* (IMACS/Elsevier, Amsterdam 1992), pp. 263–281
19. M. Griebel, V. Thurner, The efficient solution of fluid dynamics problems by the combination technique. Int. J. Numer. Methods Heat Fluid Flow **5**, 51–69 (1995)
20. B. Harding, M. Hegland, A robust combination technique, in *CTAC-2012*. Volume 54 of ANZIAM Journal, 2013, pp. C394–C411
21. M. Holtz, *Sparse Grid Quadrature in High Dimensions with Applications in Finance and Insurance*. Volume 77 of Lecture Notes in Computational Science and Engineering (Springer, Heidelberg, 2011)
22. J.-W. Hong, H.-T. Kung, I/O complexity: The red-blue pebble game, in *Proceedings of STOC'81*, New York (ACM, 1981), pp. 326–333
23. P. Hupp, Communication efficient algorithms for numerical problems on full and sparse grids, PhD thesis, ETH Zurich, 2014
24. P. Hupp, Performance of unidirectional hierarchization for component grids virtually maximized, in *International Conference on Computational Science*. Volume 29 of Procedia Computer Science (Elsevier, Amsterdam 2014), pp. 2272–2283
25. P. Hupp, M. Heene, R. Jacob, D. Pflüger, Global communication schemes for the numerical solution of high-dimensional PDEs. Parallel Comput. (2016). DOI:10.1016/j.parco.2015.12.006
26. P. Hupp, R. Jacob, M. Heene, D. Pflüger, M. Hegland, Global communication schemes for the sparse grid combination technique. in *Parallel Computing – Accelerating Computational Science and Engineering (CSE)*. Volume 25 of Advances in Parallel Computing (IOS Press, 2014), pp. 564–573
27. D. Irony, S. Toledo, A. Tiskin, Communication lower bounds for distributed-memory matrix multiplication. J. Parallel Distrib. Comput. **64**(9), 1017–1026 (2004)

28. R. Jacob, Efficient regular sparse grid hierarchization by a dynamic memory layout, in *Sparse Grids and Applications 2012*, Munich, ed. by J. Garcke, D. Pflüger. Volume 97 of Lecture Notes in Computational Science and Engineering (Springer, Cham/New York, 2014) pp. 195–219

29. C. Kowitz, M. Hegland, The sparse grid combination technique for computing eigenvalues in linear gyrokinetics. Procedia Comput. Sci. **18**, 449–458 (2013). International Conference on Computational Science.

30. M.D. Lam, E.E. Rothberg, M.E. Wolf, The cache performance and optimizations of blocked algorithms. SIGPLAN Not. **26**(4), 63–74 (1991)

31. A. Maheshwari, N. Zeh, A survey of techniques for designing I/O-efficient algorithms, in *Algorithms for Memory Hierarchies*. ed. by U. Meyer, P. Sanders, J. Sibeyn. Volume 2625 of Lecture Notes in Computer Science, pp. 36–61 (Springer, Berlin/Heidelberg, 2003)

32. A. Murarasu, J. Weidendorfer, G. Buse, D. Butnaru, D. Pflüger, Compact data structure and scalable algorithms for the sparse grid technique, in *Proceedings of the 16th ACM Symposium on Principles and Practice of Parallel Programming (PPoPP)*, San Antonio (ACM, 2011), pp. 25–34

33. A. F. Murarasu, G. Buse, D. Pflüger, J. Weidendorfer, A. Bode, fastsg: A fast routines library for sparse grids. Procedia CS **9**, 354–363 (2012)

34. C. Pflaum, Convergence of the combination technique for second-order elliptic differential equations. SIAM J. Numer. Anal. **34**(6), 2431–2455 (1997)

35. C. Pflaum, A. Zhou, Error analysis of the combination technique. Numer. Math. **84**(2), 327–350 (1999)

36. D. Pflüger, Spatially adaptive sparse grids for high-dimensional problems, PhD thesis, Institut für Informatik, Technische Universität München, 2010

37. D. Pflüger, H.-J. Bungartz, M. Griebel, F. Jenko, T. Dannert, M. Heene, A. Parra Hinojosa, C. Kowitz, and P. Zaspel, Exahd: An exa-scalable two-level sparse grid approach for higher-dimensional problems in plasma physics and beyond, in *Euro-Par 2014: Parallel Processing Workshops*. Volume 8806 of Lecture Notes in Computer Science (Springer, Cham 2014), pp. 565–576

38. H. Prokop, Cache-oblivious algorithms, Master's thesis, Massachusetts Institute of Technology, 1999

39. C. Reisinger, Analysis of linear difference schemes in the sparse grid combination technique. IMA J. Numer. Anal. **33**(2), 544–581 (2013)

40. S. Smolyak, Quadrature and interpolation formulas for tensor products of certain classes of functions. Sov. Math. Dokl. **4**, 240–243 (1963)

41. C. Zenger, Sparse grids, in *Parallel Algorithms for Partial Differential Equations*. Volume 31 of Notes on Numerical Fluid Mechanics (Vieweg, Wiesbaden 1991), pp. 241–251

Spatially-Dimension-Adaptive Sparse Grids for Online Learning

Valeriy Khakhutskyy and Markus Hegland

Abstract This paper takes a new look at regression with adaptive sparse grids. Considering sparse grid refinement as an optimisation problem, we show that it is in fact an instance of submodular optimisation with a cardinality constraint. Hence, we are able to directly apply results obtained in combinatorial optimisation research concerned with submodular optimisation to the grid refinement problem. Based on these results, we derive an efficient refinement indicator that allows the selection of new grid indices with finer granularity than was previously possible. We then implement the resulting new refinement procedure using an averaged stochastic gradient descent method commonly used in online learning methods. As a result we obtain a new method for training adaptive sparse grid models. We show both for synthetic and real-life data that the resulting models exhibit lower complexity and higher predictive power compared to currently used state-of-the-art methods.

1 Introduction

Sparse grids have been successfully used for different data mining and machine learning tasks including regression, classification, clustering, and density estimation [10, 13, 14, 23, 24, 26]. As the demand for mining larger datasets with more and more features increased over the years, so did the number of different sparse

With the support of the Technische Universität München – Institute for Advanced Study, funded by the German Excellence Initiative (and the European Union Seventh Framework Programme under grant agreement nr 291763).

V. Khakhutskyy (✉)
Technische Universität München, München, Germany
e-mail: khakhutv@in.tum.de

M. Hegland
The Australian National University, Canberra ACT, Australia,
e-mail: markus.hegland@anu.edu.au

© Springer International Publishing Switzerland 2016
J. Garcke, D. Pflüger (eds.), *Sparse Grids and Applications – Stuttgart 2014*,
Lecture Notes in Computational Science and Engineering 109,
DOI 10.1007/978-3-319-28262-6_6

grid techniques. These new techniques are based on the application of dimension-adaptive generalised sparse grids [15] and the combination technique [13] to data mining problems, the development of spatially-adaptive sparse grids [27], and the implementation of different methods for parallelised training of sparse grids models, i.e. parallel CG [16] and ADMM [18].

Replacing a full grid by a sparse grid leads to a substantial reduction in time complexity with just slightly lower accuracy, consequently delaying the onset of the curse of dimensionality [9]. As a result, Garcke [13] and Pflüger [26] were able to solve high-dimensional data mining problems including the 18-dimensional Data-Mining-Cup, the 22-dimensional mushroom classification problem, the 64-dimensional optical digits recognition problem, and the 166-dimensional Musk-1 benchmark.

This paper focuses on the procedures for refinement and fitting of sparse grid models. In particular, we suggest to fit models using online gradient descent methods instead of the conjugate gradient method used before for this purpose.

Online learning is an actively developing area of machine learning with many different facets. Moreover, the term "online learning" is used alongside and sometimes interchangeably with the terms "stream learning" and "one-pass learning". In the following, we will use the term "online learning" to imply "training a predictive model incrementally, by updating the model using one data point at a time". Other aspects of online learning with sparse grids have been discussed elsewhere: for example, Peherstorfer et al. presented a classification method based on the Offline/Online splitting [25] and Strätling analysed incremental updates to handle concept drifts [31].

The online gradient descent method – a stochastic approximation of the gradient – enables online learning of predictive models [5] and thus renders regression problems with large numbers of data instances feasible [4]. In this work we focus on the averaged stochastic gradient descent (ASGD) method, which is a stabilised version of the online gradient descent method [28]. While some studies investigate the setting where every training example can be considered only once (stream or one-pass learning), we allow a revision of the samples (multi-pass learning). Although ASGD is not a new method, to the best of our knowledge, its effectiveness in the context of sparse grid regression has not been studied previously.

Furthermore, we show how the grid refinement can be analysed in the framework of submodular optimisation and develop new optimality criteria arising from this analysis. Based on this we are able to present a new procedure for training sparse grids with lower computational costs.

The remainder of this paper is organised as follows: Sect. 2 introduces the basic theory and notation used throughout this paper. The new method is described in Sect. 4. We compare and discuss the performance of the new method in Sect. 14. Finally, Sect. 3 concludes this work with a summary and a discussion of open questions.

2 Theoretical Background

We begin with the description of how regression can be cast into the regularised least squares problem and how adaptive sparse grid models can be used to solve this problem.

A typical regression problem in data mining goes as follows: We are given a dataset, a finite set $S = \{(x_{i1}, \ldots, x_{id}, y_i)\}_{i=1}^N \subset [0, 1]^d \times \mathbb{R}\}$ drawn from a probability distribution S with often unknown probability density function p. We call $\mathbf{x}_i = (x_{i1}, \ldots, x_{id})^T$ the input variables and corresponding y_i the dependent or target variables. In general, $\mathbf{x}_i$ may take any value in $\mathbb{R}^d$. However, for any finite dataset we can normalise the input variables to $[0, 1]^d$.[1]

We further assume the existence of an unknown function that generates the targets from the input variables. Our goal is to find an approximation to this function in a function space V.

In practice we are computing the minimiser of the regularised empirical risk with squared loss function

$$R(f; S) := \frac{1}{N} \sum_{i=1}^N (f(\mathbf{x}_i) - y_i))^2 + \lambda \|f\|_V^2. \tag{1}$$

The regularisation term $\|f\|_V^2$ measures the model complexity and the parameter λ controls the trade-off between data fitting and smoothness.

Since the sample size is limited, the minimisation of (1) can be ill-posed if V is infinite-dimensional. In this work we use a finite-dimensional Hilbert space

$$V_G := \mathrm{span}\{\varphi_g \mid g \in G\},$$

where G is a finite set whose elements index the basis functions.

Let f and u be two functions from V_G with $f(\mathbf{x}) = \sum_{g \in G} w_g \varphi_g(\mathbf{x})$ and $u(\mathbf{x}) = \sum_{g \in G} v_g \varphi_g(\mathbf{x})$. Let $\mathbf{w} = (w_g)_{g \in G}$ and $\mathbf{v} = (v_g)_{g \in G}$ be the vectors indexed by the set G with components w_g and v_g respectively. Since the elements of G can uniquely identify the components of $\mathbf{w}$ and $\mathbf{v}$, we denote the space of all possible values of these vectors by $\mathbb{R}^G$. The inner product of V_G is defined as

$$\langle f, u \rangle_{V_G} = \sum_{g \in G} w_g v_g. \tag{2}$$

---

[1]To normalise a finite dataset we would need to calculate minimal and maximal values of the input in all dimensions. This can be done even by passing through the dataset: first we initialise two variables $\mathbf{x}_{\min}$ and $\mathbf{x}_{\max}$ with the first element and then update the components of these two variables if the new input patterns would have smaller/larger vales than stored.

Every $f \in V_G$ is uniquely identified by its coefficients $\mathbf{w} \in \mathbb{R}^G$. Hence, the regularised expected risk functional (1) for finite-dimensional function spaces can be rewritten in terms of $\mathbf{w}$ and G:

$$J(\mathbf{w}; G) := \sum_{i=1}^{N} \left(\sum_{g \in G} w_g \phi_g(\mathbf{x}) - y_i \right)^2 + \lambda N \sum_{g \in G} w_g^2. \tag{3}$$

Here we have also multiplied the expression by N, which simplifies the notation that follows and does not change the minimiser of J.

Given that the function space V_G is fixed, to solve the regression problem we compute the minimiser

$$\mathbf{w}^\star := \arg \min_{\mathbf{w} \in \mathbb{R}^G} J(\mathbf{w}; G). \tag{4}$$

This cost function can be further split into the sum of functions of individual data points:

$$J(\mathbf{w}; G) = \sum_{i=1}^{N} \left(\sum_{g \in G} w_g \varphi_g(\mathbf{x}_i) - y_i \right)^2 + \lambda N \sum_{g \in G} w_g^2$$

$$= \sum_{i=1}^{N} \left(\left(\sum_{g \in G} w_g \varphi_g(\mathbf{x}_i) - y_i \right)^2 + \lambda \sum_{g \in G} w_g^2 \right).$$

We denote the term corresponding to the data point $(\mathbf{x}, y)$ by

$$J_{\text{point}}(\mathbf{w}, \mathbf{x}, y; G) := \left(\sum_{g \in G} w_g \varphi_g(\mathbf{x}) - y \right)^2 + \lambda \sum_{g \in G} w_g^2, \tag{5}$$

such that

$$J(\mathbf{w}; G) = \sum_{i=1}^{N} J_{\text{point}}(\mathbf{w}, \mathbf{x}_i, y_i; G).$$

Numerically, one may determine the minimiser of $J(\mathbf{w}; G)$ using the gradient descent method which defines a sequence $\mathbf{w}^t$ by

$$\mathbf{w}^{t+1} = \mathbf{w}^t - \gamma^t \nabla_{\mathbf{w}} J(\mathbf{w}^t; G). \tag{6}$$

Depending on the particular choice of γ^t and the adjustment of the update direction with respect to the past updates, the update step (6) leads to a number of different methods, such as the Newton's method, the Broyden–Fletcher–Goldfarb–Shanno algorithm, or the conjugate gradients (CG) method [22]. Borrowing a term from the neural network literature, we call this class of algorithms optimisation in *batch mode*, since the information from the whole "batch" S is taken into account at every update step [4].

The complexity of these descent methods is dominated by function evaluations. And since the intermediate results are rarely stored explicitly, every gradient descent iteration would have a time complexity in $O(N \cdot |G|)$. For k update steps the algorithms also perform k passes through the data points. This yields a time complexity of the whole gradient descent procedure in $O(kN \cdot |G|)$. Therefore, in Sect. 14 we will evaluate the results using the number of data passes as a complexity measure.

For some problems the batch gradient descent method is infeasible, for example, when the dataset is too large to fit into the main memory at once. For some problems the batch gradient descent is impractical, for example, when the training patterns arrive continually in a data stream. To overcome this problem, one can consider a stochastic approximation of the gradient term in (6) by the gradient at one (random) point [5], which yields an update of the form

$$\mathbf{w}^{t+1} = \mathbf{w}^t - \gamma^t \nabla_{\mathbf{w}} J_{\text{point}}(\mathbf{w}^t, \mathbf{x}_t, y_t; G). \tag{7}$$

This class of methods is called online gradient descent or optimisation in *online mode* [4].

In practice, additional measures are taken to obtain robust results and faster convergence [2, 28, 30]. Bottou suggested an averaging scheme for training neural networks [7]. We use his averaging scheme in this work as discussed in Sect. 2.3.

2.1 Classification by Regression

If the target variable y can have only two values, e.g. 0 or 1, we call this problem binary classification. This problem is simpler than general multi-class classification problems. We can solve it by training a regression function and establishing a threshold for decision making:

$$\text{class}(\mathbf{x}) := \begin{cases} 0 & f(\mathbf{x}) < 0.5 \\ 1 & f(\mathbf{x}) \geq 0.5. \end{cases} \tag{8}$$

We will use this property in Sect. 2.5 to evaluate the suggested refinement algorithm.

2.2 Adaptive Sparse Grids for Regression

In the past decade a number of works have focused on solving the problem (4) using sparse grid discretisation techniques [10, 14, 15, 27]. We build upon these works and, in particular, we use piecewise linear basis functions constructed using the following principles:

$$\phi_{\mathbf{l},\mathbf{i}}(\mathbf{x}) := \prod_{t=1}^{d} \phi_{l_t,i_t}(x_t), \text{ with } \mathbf{l}, \mathbf{i} \in \mathbb{N}^d \tag{9}$$

with

$$\phi_{l,i}(x) := \phi(x \cdot 2^l - i) \tag{10}$$

for the linear basis and with

$$\phi_{l,i}(x) := \begin{cases} 1 & \text{if } l = 1 \wedge i = 1, \\ \begin{cases} 2 - 2^l \cdot x & \text{if } x \in [0, 2^{1-l}] \\ 0 & \text{else} \end{cases} & \text{if } l > 1 \wedge i = 1, \\ \begin{cases} 2^l \cdot x + 1 - i & \text{if } x \in [1 - 2^{1-l}, 1] \\ 0 & \text{else} \end{cases} & \text{if } l > 1 \wedge i = 2^l - 1, \\ \phi(x \cdot 2^l - i) & \text{else} \end{cases}$$

for the modified linear basis and

$$\phi(x) := \max\{1 - |x|, 0\}.$$

These functions have proved to offer a good trade-off between approximation accuracy and fast evaluation. The models the with linear basis functions are restricted to the value 0 on the boundaries, while the models with modified linear basis do not have this restriction.

Figure 1 illustrates these functions for a one-dimensional case. An index $(\mathbf{l}, \mathbf{i})$ with $\mathbf{l} := (l_1, \ldots, l_d)^T$ and $\mathbf{i} := (i_1, \ldots, i_d)^T$ uniquely identifies a basis function. Let G be a set of indices $(\mathbf{l}, \mathbf{i})$, called *sparse grid index-set*. Hence, the set of all sparse grid basis functions spans the function space V_G defined through the one-dimensional basis function construction rule (9) and the inner product (2).

An important property of a sparse grid model is its ability to adapt the grid structure to a particular distribution and smoothness of the underlying generating function. Since basis functions have local support, by adding new basis functions

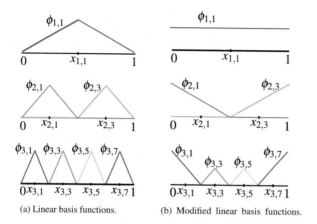

(a) Linear basis functions.   (b) Modified linear basis functions.

Fig. 1 Sparse grid piecewise linear basis functions in one-dimensional space

we are able to refine the model in the areas with high data density or high target variance, while keeping the rest of the model unchanged. This leads to higher resolution in the areas where it is really needed. The refinability gives sparse grid methods a competitive advantage over the other additive basis function methods, such as neural networks.

One can see sparse grid adaptivity as a transition from exploration and exploitation in the problem space. Exploration extends the function space broadly, even without the evidence from data supporting this extension. For example, starting with a sparse grid index-set of a certain level is exploration. Exploitation adds new basis functions to the model only if the benefit of this extension is evident from the data.

A formal definition of refinement requires the notion of hierarchical ancestry and descendancy between indices. Let Ω be a set of all possible indices

$$\Omega := \{(\mathbf{l}, \mathbf{i}) \mid (\mathbf{l}, \mathbf{i}) \in \mathbb{N}^d \times \mathbb{N}^d, \ i_t \text{ odd and } 1 \leq i_t < 2^{l_t} \text{ for each } 1 \leq t \leq d\}$$

and let desc be the function that computes the hierarchical descendants of a level-index pair $(\mathbf{l}, \mathbf{i})$ in the dimension t:

$$\text{desc} : \mathbb{N}^d \times \mathbb{N}^d \times \{1, \ldots, d\} \to \mathcal{P}(\Omega) \tag{11}$$

$$\mathbf{l}, \mathbf{i}, t \mapsto \{(\mathbf{l} + \mathbf{e}_t, \mathbf{i} + (i_t + r)\mathbf{e}_t) \mid r \in \{-1, +1\}\} .$$

Here, $\mathbf{e}_t$ is a vector containing 1 in the tth component and 0 everywhere else, and r signifies if the left or the right hierarchical descendant is created. The reverse function anc computes all hierarchical ancestors of a level-index pair $(\mathbf{l}, \mathbf{i})$:

$$\text{anc} : \mathbb{N}^d \times \mathbb{N}^d \to \mathcal{P}(\Omega) \tag{12}$$

$$\mathbf{l}, \mathbf{i} \mapsto \{(\mathbf{l} - \mathbf{e}_t, \lfloor \mathbf{i} - \tfrac{i_t - ((i_t + 1) \bmod 4)}{2} \mathbf{e}_t \rfloor) \mid 1 \leq t \leq d\}.$$

Let $G^{(k)}$ denote the sparse grid index-set after the kth refinement. The set $G^{(k)}$ is extended using the elements from the *admissible set* of candidates

$$A_{G^{(k)}} := \{(\mathbf{l}, \mathbf{i}) \mid (\mathbf{l}, \mathbf{i}) \in \mathbb{N}^d \times \mathbb{N}^d \text{ and } \text{anc}(\mathbf{l}, \mathbf{i}) \cap G^{(k)} \neq \emptyset\}. \tag{13}$$

We can either refine a point in a particular dimension:

$$G^{(k+1)} = G^{(k)} \cup (\text{desc}(\mathbf{l}, \mathbf{i}, t) \cap A_{G^{(k)}}), \tag{14}$$

or in all dimensions:

$$G^{(k+1)} = G^{(k)} \cup (\text{alldesc}(\mathbf{l}, \mathbf{i}) \cap A_{G^{(k)}}) \tag{15}$$

with

$$\text{alldesc}(\mathbf{l}, \mathbf{i}) := \bigcup_{t=1}^{d} \text{desc}(\mathbf{l}, \mathbf{i}, t). \tag{16}$$

In the first case we speak of *spatially-dimension-adaptive* refinement (Fig. 2b) and in the second case we speak of *spatially-adaptive* refinement (Fig. 2c).

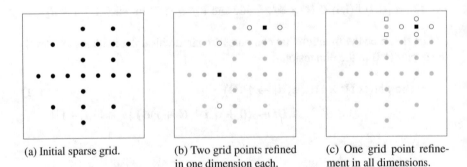

(a) Initial sparse grid.   (b) Two grid points refined   (c) One grid point refine-
                             in one dimension each.   ment in all dimensions.

Fig. 2 Illustration of *spatially-dimension-adaptive* and *spatially-adaptive* refinement schemes. The sparse grid points with indices to be refined are depicted as *black squares*, the new children points are *white circles*, the new ancestor grid points are *white squares*

Finding the optimal indices to refine can be viewed as an optimisation problem itself. The regression with adaptive sparse grids minimises the cost function (3) over a sequence of monotonically growing function spaces $V_{G^{(k)}} : V_{G^{(0)}} \subset V_{G^{(1)}} \subset \cdots \subset V$ as well as over the parameter $\mathbf{w} \in \mathbb{R}^{G^{(k)}}$.

Let $\tilde{J}(G)$ be a set function over the sparse grid index-set G that computes the minimal risk:

$$\tilde{J}(G) := \min_{\mathbf{w} \in \mathbb{R}^G} J(\mathbf{w}; G). \tag{17}$$

The refinement procedure aims to find the best n basis function indices $I^\star$ to maximise the marginal gain among all candidates in A_G:

$$I^\star = \arg\max_{\substack{I \subset A_G \\ |I|=n}} \tilde{J}(G) - \tilde{J}(G \cup I). \tag{18}$$

This formulation was first suggested by Hegland and Garcke in the context of *generalised* sparse grids and the dimension-adaptive refinement [13, 15]. They considered subsets of indices that belong together in so-called hierarchical subspaces. As a result, the authors derived a greedy method, which yields optimality bounds under certain conditions as discussed in Sect. 2.4. However, their method bears large computational costs at every step and becomes infeasible for large problems. Furthermore, the need to include complete hierarchical subspaces limits the desired ability to refine only some parts of the domain, which may be important for some regression problems.

Pflüger suggested a heuristic indicator for marginal gain and local adaptivity in order to overcome these limitations [26]. Assuming that the errors are normally distributed around 0, one can minimise $\tilde{J}(G \cup I)$ by adding new indices that correspond to the area with high local error variation. This variation is weighted by a potential influence of the basis functions. To capture this influence, Pflüger uses the absolute value of the basis functions evaluated at the data points. Altogether, he suggests to refine the indices that have the highest indicator:

$$s^* := \arg\max\{\xi(s) \mid s \in G^{(k)}, \texttt{alldesc}(s) \cap A_{G^{(k)}} \neq \emptyset\},$$

$$G^{(k+1)} := G^{(k)} \cup \texttt{alldesc}(s^*) \cup \texttt{anc}(s^*) \tag{19}$$

with

$$\xi(s) := \sum_{i=1}^{N} (y_i - f(\mathbf{x}_i; \mathbf{w}))^2 \cdot |w_s| \, \phi_s(\mathbf{x}_i). \tag{20}$$

Formula (19) ensures that all hierarchical ancestors of a new index are also in the index-set, which is a fundamental assumption for many algorithms [11, 26]. As there

---

Algorithm 1: Regression with adaptive sparse grids

---

1 start with some initial $G^{(0)}$;
2 for $k = 0, 1, \ldots$ do
3   **Fit-Step:** Compute $\mathbf{w}^{(k)}$ and $\tilde{J}(G^{(k)})$ from (17) for a given $G^{(k)}$;
4   **Refine-Step:** $G^{(k+1)} = G^{(k)} \cup I^*$ for I^* maximising the marginal gain (18);

---

is no way to tell how the new indices would contribute to the results, the refinement occurs in all directions.

The computation of this indicator is inexpensive even for large datasets. However, even though the intuition behind the indicator is clear, its optimality is difficult to analyse. It also fails to indicate the importance of the individual dimensions that will be refined. In Sect. 2.4 we come back to this idea, deriving a refinement method that combines efficiency, flexibility, and optimality guarantees.

Altogether, regression with adaptive sparse grids is performed in a succession of fitting and refinement steps, as illustrated in Algorithm 1. These steps alternate until some global convergence criterion is satisfied, for example, until the generalisation error on a validation dataset is sufficiently small, or until the computational limit is reached.

Let us consider the steps in Algorithm 1 more closely. The optimisation problem in Line 3 is often solved using the CG method, as we mentioned earlier. However, if the number of entries in a dataset is very large, the cost of even a single CG iteration becomes significant. Similarly, the spatially-adaptive refinement in Line 4 creates an unnecessarily large sparse grid, especially if the dataset has many different attributes. Finally, it is often advisable to find a suboptimal solution in the fit-step, which is sufficient to distinguish good refinement candidates from bad ones in the refine-step. The optimal parameters can then be computed only once after the last refinement step.

Hence, our new method for regression with adaptive sparse grids is based on the following principles:

Fit-Step: We use ASGD for fast parameter fitting on large datasets, which offers fast convergence to a sufficiently good solution (see Sect. 2.3 and Algorithm 2).
Refine-Step: We derive a new indicator for spatially-dimension-adaptive refinement, which is based on marginal gain maximisation (see Sect. 2.4 and Algorithm 3).

2.3 Averaged Stochastic Gradient Descent

We replace the gradient of J in (6) by its stochastic approximation – the gradient of J_{point} at a random point. Furthermore, to improve convergence, we re-shuffle the dataset every time we pass through the last point. After the last index was selected,

---

Algorithm 2: Fit(S, $G^{(k)}$, γ^0, λ) using averaged stochastic gradient descent method by Bottou [7]

---

input : dataset S, current sparse grid index-set $G^{(k)}$, step size γ^0, regularisation parameter λ
output: parameters $\bar{\mathbf{w}}^t$

1 *initialise* $\mathbf{w}^0 \leftarrow \mathbf{0}, \bar{\mathbf{w}}^0 \leftarrow \mathbf{0}, t \leftarrow 0$;
2 **repeat**
3   *compute partial derivatives at a random point* $(\mathbf{x}, y) \in S$: $\Delta \mathbf{w}^t \leftarrow \frac{\partial J_{\text{point}}}{\partial \mathbf{w}}$ *as in* (21);
4   *update learning rate* $\gamma^t \leftarrow \gamma^0 (1 + \gamma^0 \lambda t)^{-2/3}$;
5   *update parameters* $\mathbf{w}^{t+1} \leftarrow \mathbf{w}^t - \gamma^t \Delta \mathbf{w}^t$;
6   *update averaged parameters* $\bar{\mathbf{w}}^{t+1} \leftarrow (1 - \alpha)\bar{\mathbf{w}}^t + \alpha \mathbf{w}^{t+1}$ *with* $\alpha = \frac{1}{\max\{1, t - \min\{d, N\}\}}$;
7      $t \leftarrow t + 1$;
8 **until** *local convergence criteria satisfied*;

---

the random variable starts selecting from the complete dataset again. Altogether, the gradient has the form

$$\mathbf{w}^{t+1} := \mathbf{w}^t - \gamma^t \Delta \mathbf{w}^t$$

with

$$\Delta \mathbf{w}^t := \nabla_{\mathbf{w}} J_{\text{point}} \Big|_{\substack{\mathbf{x}=\mathbf{x}_{T(t)} \\ \mathbf{y}=y_{T(t)} \\ \mathbf{w}=\mathbf{w}^t \\ G=G^{(k)}}} = 2 \left(\left(\sum_{g \in G^{(k)}} w_g^t \phi_g(\mathbf{x}_{T(t)}) - y_{T(t)} \right) + \lambda \right) \mathbf{w}^t. \qquad (21)$$

We summarise the algorithm in Algorithm 2. As the individual parameter updates depend on the selected input points, it can strongly oscillate, prohibiting convergence. Therefore, we introduce an additional parameter vector $\bar{\mathbf{w}}^t$ that aggregates the values of individual parameters using exponential smoothing in Line 6 of Algorithm 2. This stabilises and accelerates the optimisation procedure in the long run, as Fig. 3 illustrates.

Unfortunately, ASGD often starts slower than the plain stochastic gradient descent and can require many steps before reaching its optimal asymptotic convergence speed. To accelerate this process, the exponential smoothing coefficient α takes the form $1 / \max\{1, t - \min\{d, N\}\}$ [7]. Suppose that d is smaller than N. Then for the first d steps the algorithm behaves as the plain stochastic gradient descent. Afterwards, the smoothing coefficient diminishes as $O(1/t)$.

The convergence of ASGD has been rigorously studied for regressors with a fixed number of basis functions [2, 28, 33]. Xu showed that, if the learning rate decreases as $\gamma^t \in \Theta(t^{-2/3})$, the objective function converges linearly [33].

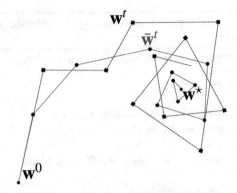

Fig. 3 Illustration of ASGD for a two-dimensional parameter space. Aggregation of coefficients $\mathbf{w}^t$ into $\bar{\mathbf{w}}^t$ leads to higher stability and faster convergence

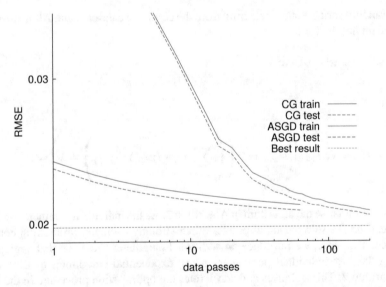

Fig. 4 Convergence comparison of ASGD and CG. The errors were computed for regression on the Sloan Digital Sky Survey dataset described in Sect. 14 and a sparse grid level 5 with same meta-parameters. Both methods started with the same initial parameter $\mathbf{w}^0 = \mathbf{0}$

We observed that ASGD exhibits a significantly faster initial convergence than CG with respect to the number of data passes. This observation is in line with observations of other researchers [6, 8]. Consider Fig. 4 that compares the decrease of the root mean square error for CG and ASGD methods over a number of data passes (a metric motivated in Sect. 2.2). ASGD shows fast initial convergence and significantly reduces the error after a single data pass. Meanwhile, CG needs approximately 20 data passes to achieve the same error.

Hence, we suggest using ASGD to solve (17) – not perfectly but close enough to compute distinguishable error indicators in Line 4 of Algorithm 1. Once the structure of the grid is established, one can compute the final optimal parameters **w** using several CG iterations if necessary.

2.4 Sparse Grid Refinement as an Optimisation Problem

In Sect. 2.2 we saw that sparse grid basis sets can be extended to befit a regression problem. Now we discuss how to select the new indices for the extension. Let $J^-(X)$ be the negative of $\tilde{J}(X)$ defined in (17):

$$J^-(X) := -\tilde{J}(X). \tag{22}$$

We begin our discussion by showing that $J^-(X)$ belongs to a special class of functions well studied in combinatorial optimisation. Then we derive the optimisation problem of finding the optimal sparse grid and explain the complexity results from combinatorial optimisation that can be directly applied to our problem. Finally, we derive and evaluate an efficient greedy algorithm for finding the optimal sparse grid structure.

Definition 1 Let Ω be a set, $\mathcal{P}(\Omega)$ its powerset, and $h : \mathcal{P}(\Omega) \to \mathbb{R}$ a set function. Let the **marginal gain** with respect to the function h be defined as

$$h(x|X) := h(X \cup \{x\}) - h(X). \tag{23}$$

Then h is said to be **submodular** if it satisfies the property of diminishing returns

$$h(x|X) \geq h(x|Y) \tag{24}$$

for all sets $X \subseteq Y \subseteq \Omega$ and all elements $x \in \Omega \setminus Y$. The submodular function h is **monotonically non-decreasing** if $h(x|X) \geq 0$ for each element $x \in \Omega \setminus X$ and each set $X \subseteq \Omega$.

In particular, if Ω is the set of all sparse grid indices and $G \subset \Omega$, it turns out that $J^-(G)$ is submodular. To show this, we first establish the following relation between projections onto subspaces.

Lemma 1 *Let V_1, V_2, and W be Euclidean vector spaces, $V_1 \subset V_2$, and let P_{V_1}, P_{V_2}, and P_W be orthogonal projections onto the corresponding spaces. Then*

$$\|(P_{V_1+W} - P_{V_1})\mathbf{y}\| \geq \|(P_{V_2+W} - P_{V_2})\mathbf{y}\|. \tag{25}$$

Proof The vector space W can be either a part of V_2 or a part of its orthogonal complement $V_2^\perp$ or consist of parts in V_2 and parts in its orthogonal complement. We consider these three cases separately.

Case 1 If $W \subseteq V_2$ then we have

$$\|(P_{V_1+W} - P_{V_1})\mathbf{y}\| \geq \|(P_{V_2+W} - P_{V_2})\mathbf{y}\| = 0.$$

Case 2 If $W \perp V_2$ then

$$\|(P_{V_1+W} - P_{V_1})\mathbf{y}\| = \|P_W \mathbf{y}\| = \|(P_{V_2+W} - P_{V_2})\mathbf{y}\|.$$

Case 3 Finally, if $W = W_1 + W_2$ such that $W_1 \subseteq V_2^\perp$ and $W_2 \subseteq V_2$ then

$$\|(P_{V_1+W_1+W_2} - P_{V_1})\mathbf{y}\| = \|(P_{V_1+W_2} + P_{W_1} - P_{V_1})\mathbf{y}\| = \|(P_{V_1+W_2} - P_{V_1})\mathbf{y} + P_{W_1}\mathbf{y}\|$$

$$= \sqrt{\|(P_{V_1+W_2} - P_{V_1})\mathbf{y}\|^2 + \|P_{W_1}\mathbf{y}\|^2} \geq \|P_{W_1}\mathbf{y}\|.$$

$$\|(P_{V_2+W_1+W_2} - P_{V_2})\mathbf{y}\| = \|(P_{V_2+W_2} + P_{W_1} - P_{V_2})\mathbf{y}\| = \|P_{W_1}\mathbf{y}\|$$

and together we have

$$\|(P_{V_1+W_1+W_2} - P_{V_1})\mathbf{y}\| \geq \|(P_{V_2+W_1+W_2} - P_{V_2})\mathbf{y}\|.$$

Let $\mathcal{I} : \{1, \ldots, |\Omega|\} \to \Omega$ be a bijective mapping that enumerates the elements of Ω. For every sparse grid index g we define a vector in $\mathbb{R}^{N+|\Omega|}$ as

$$\mathbf{v}_g := \left(\phi_g(\mathbf{x}_1), \ldots, \phi_g(\mathbf{x}_N), \sqrt{\lambda N}\delta_{\mathcal{I}(g),1}, \sqrt{\lambda N}\delta_{\mathcal{I}(g),2}, \ldots, \sqrt{\lambda N}\delta_{\mathcal{I}(g),|\Omega|} \right)^T, \tag{26}$$

with $\delta_{i,j}$ being the Kronecker delta. Then

$$V_X := \text{span}\{\mathbf{v}_g\}_{g \in X} \tag{27}$$

is a vector space embedding of X. We appropriately extend the target vector $\mathbf{y}$ to the dimensionality $N + |\Omega|$ by appending $|\Omega|$ zeros:

$$\tilde{\mathbf{y}} := \begin{bmatrix} \mathbf{y} \\ \mathbf{0} \end{bmatrix}. \tag{28}$$

Lemma 2 *Assume that V_X is defined as in (27) and $\tilde{\mathbf{y}}$ is defined as in (28). Then for the orthogonal projection P_X that satisfies*

$$P_X : \mathbb{R}^{N+|\Omega|} \to V_X$$

$$\tilde{\mathbf{y}} \mapsto \underset{\mathbf{f} \in V_X}{\arg\min} \|\mathbf{f} - \tilde{\mathbf{y}}\|^2 \tag{29}$$

we have

$$J^-(X) = -\|P_X\tilde{\mathbf{y}} - \tilde{\mathbf{y}}\|^2. \tag{30}$$

Proof For any $\mathbf{f} \in V_X$ there is a set of linear combination coefficients $\mathbf{w} := \{w_g\}_{g \in X}$ such that

$$\mathbf{f} = \sum_{g \in X} w_g \mathbf{v}_g.$$

The squared norm of the difference between $\mathbf{f}$ and $\tilde{\mathbf{y}}$ is equal to the value of the cost function $J(\mathbf{w}; X)$ defined in (3):

$$\|\mathbf{f} - \tilde{\mathbf{y}}\|^2 = \sum_{i=1}^{N} \left(\sum_{g \in X} w_g \phi_g(\mathbf{x}_i) - y_i \right)^2 + \lambda N \sum_{g \in X} (w_g - 0)^2.$$

Hence, the projection $P_X\tilde{\mathbf{y}}$ that minimises (29) corresponds to $\mathbf{w}^\star$ that solves the minimisation problem in (22). This leads directly to (30). ∎

Proposition 1 *The function J^- defined in (22) is a monotonically non-decreasing submodular function.*

Proof Suppose that $X \subset \Omega$, V_X is defined as in (27), $\tilde{\mathbf{y}}$ is defined as in (28), and P_X is defined as in (29). For simplicity of the notation we also introduce the residual vector

$$R_X := P_X\tilde{\mathbf{y}} - \tilde{\mathbf{y}}. \tag{31}$$

Let Y be another sparse grid index-set such that $X \subseteq Y \subset \Omega$, and let $s \in \Omega \setminus Y$ be a new sparse grid index. Analogously, we define the projections P_Y, P_s, $P_{X \cup \{s\}}$, $P_{Y \cup \{s\}}$ as well as the residual vectors R_Y, R_s, $R_{X \cup \{s\}}$, $R_{Y \cup \{s\}}$.

We are going to show the property of diminishing returns

$$J^-(s \mid X) \geq J^-(s \mid Y),$$

where $J^-(\cdot \mid \cdot)$ is a marginal gain function of J^- as defined in (23).

We begin by writing the relation (25) from Lemma 1 as

$$\|R_{X \cup \{s\}} - R_X\|^2 = \|(P_{X \cup \{s\}} - P_X)\tilde{\mathbf{y}}\|^2 \geq \|(P_{Y \cup \{s\}} - P_Y)\tilde{\mathbf{y}}\|^2 = \|R_{Y \cup \{s\}} - R_Y\|^2. \tag{32}$$

We now consider the norms $\|R_{X \cup \{s\}} - R_X\|^2$ and $\|R_{Y \cup \{s\}} - R_Y\|^2$ in (32). We can rewrite the inequality as

$$\|R_{X \cup \{s\}}\|^2 - 2\langle R_{X \cup \{s\}}, R_X \rangle + \|R_X\|^2 \geq \|R_{Y \cup \{s\}}\|^2 - 2\langle R_{Y \cup \{s\}}, R_Y \rangle + \|R_Y\|^2.$$

Since P_X and $P_{X \cup \{s\}}$ are orthogonal projections, we have $P_{X \cup \{s\}}(P_X \mathbf{a}) = P_X \mathbf{a}$ and $\langle R_{X \cup \{s\}}, P_{X \cup \{s\}} \mathbf{a} \rangle = 0$ for any $\mathbf{a} \in \mathbb{R}^{N+|\Omega|}$. With this in mind, we can rewrite the inner product as

$$
\begin{aligned}
\langle R_{X \cup \{s\}}, R_X \rangle &= \langle (P_{X \cup \{s\}} - I)\tilde{\mathbf{y}}, (P_X - I)\tilde{\mathbf{y}} \rangle \\
&= \langle P_{X \cup \{s\}} \tilde{\mathbf{y}}, P_X \tilde{\mathbf{y}} \rangle - \langle P_{X \cup \{s\}} \tilde{\mathbf{y}}, \tilde{\mathbf{y}} \rangle - \langle \tilde{\mathbf{y}}, P_X \tilde{\mathbf{y}} \rangle + \langle \tilde{\mathbf{y}}, \tilde{\mathbf{y}} \rangle \\
&= \langle \tilde{\mathbf{y}}, \tilde{\mathbf{y}} \rangle - \langle P_{X \cup \{s\}} \tilde{\mathbf{y}}, \tilde{\mathbf{y}} \rangle + \langle R_{X \cup \{s\}}, P_X \tilde{\mathbf{y}} \rangle \\
&= \langle \tilde{\mathbf{y}}, \tilde{\mathbf{y}} \rangle - \langle P_{X \cup \{s\}} \tilde{\mathbf{y}}, \tilde{\mathbf{y}} \rangle + \langle R_{X \cup \{s\}}, P_{X \cup \{s\}} P_X \tilde{\mathbf{y}} \rangle \\
&= \langle \tilde{\mathbf{y}}, \tilde{\mathbf{y}} \rangle - \langle P_{X \cup \{s\}} \tilde{\mathbf{y}}, \tilde{\mathbf{y}} \rangle.
\end{aligned}
\tag{33}
$$

Similarly, we obtain

$$
\|R_{X \cup \{s\}}\|^2 = \langle R_{X \cup \{s\}}, R_{X \cup \{s\}} \rangle = \langle \tilde{\mathbf{y}}, \tilde{\mathbf{y}} \rangle - \langle P_{X \cup \{s\}} \tilde{\mathbf{y}}, \tilde{\mathbf{y}} \rangle.
\tag{34}
$$

Using identities (33) and (34), we can rewrite the expression $\|R_{X \cup \{s\}} - R_X\|^2$ as

$$
\begin{aligned}
\|R_{X \cup \{s\}} - R_X\|^2 &= \|R_{X \cup \{s\}}\|^2 - 2\langle R_{X \cup \{s\}}, R_X \rangle + \|R_X\|^2 \\
&= \langle \tilde{\mathbf{y}}, \tilde{\mathbf{y}} \rangle - \langle P_{X \cup \{s\}} \tilde{\mathbf{y}}, \tilde{\mathbf{y}} \rangle - 2\langle \tilde{\mathbf{y}}, \tilde{\mathbf{y}} \rangle + 2\langle P_{X \cup \{s\}} \tilde{\mathbf{y}}, \tilde{\mathbf{y}} \rangle + \|R_X\|^2 \\
&= \|R_X\|^2 - \left(\langle \tilde{\mathbf{y}}, \tilde{\mathbf{y}} \rangle - \langle P_{X \cup \{s\}} \tilde{\mathbf{y}}, \tilde{\mathbf{y}} \rangle \right) = \|R_X\|^2 - \|R_{X \cup \{s\}}\|^2.
\end{aligned}
\tag{35}
$$

The expression $\|R_{Y \cup \{s\}} - R_Y\|^2$ can be rewritten analogously. Hence, plugging it back into (32), we get

$$
-\|R_{X \cup \{s\}}\|^2 + \|R_X\|^2 \geq -\|R_{Y \cup \{s\}}\|^2 + \|R_Y\|^2
$$

and then with (30):

$$
J^-(X \cup \{s\}) - J^-(X) \geq J^-(Y \cup \{s\}) - J^-(Y).
$$

Hence, $J^-(X)$ satisfies the diminishing returns property from Definition 1.

Finally, we show that $J^-(X)$ is monotonically non-decreasing. Suppose that $J^-(X)$ is not monotonically non-decreasing and there is $s \in \Omega \setminus Y$ such that

$$
J^-(X \cup \{s\}) - J^-(X) < 0.
$$

From (30) and (31) this would be equivalent to $\|R_X\|^2 < \|R_{X \cup \{s\}}\|^2$. In this case, there would be a vector $\mathbf{v} \in V_X$ such that $\|\mathbf{v} - \tilde{\mathbf{y}}\|^2 < \|\mathbf{v}' - \tilde{\mathbf{y}}\|^2$ for every $\mathbf{v}' \in V_{X \cup \{s\}}$. This is a contradiction, since $V_X \subset V_{X \cup \{s\}}$.

Theorem 1 *The submodular optimisation problem with cardinality constraint*

$$
\max_{G \subset \Omega, |G| \leq m} J^-(G)
\tag{36}
$$

is NP-hard. It can be solved with a greedy algorithm such that, if $\widehat{G}$ is the greedy solution and G^* is the optimal solution, we have

$$\frac{J^-(\widehat{G})}{J^-(G^*)} \geq 1 - \left(\frac{m-1}{m}\right)^m \geq \left(1 - \frac{1}{e}\right) \approx 0.632.$$

This boundary is tight unless P=NP.

Proof The result was shown in [21] for all submodular functions and, hence, can be applied to J^- as well. The tightness of the boundary follows from [12].

A greedy algorithm for (36) selects a sequence of grid index-sets $G^{(0)} \subset G^{(1)} \subset \cdots \subset G^{(k)} \subset \cdots$ such that

$$G^{(k)} := G^{(k-1)} \cup \arg\max_{s \in \Omega \backslash G^{(k-1)}} J^-(s|G^{(k-1)}). \tag{37}$$

Albeit using the marginal gain for the greedy algorithm leads to near-optimal results, the computation of the marginal gain can be expensive, since adding a new grid index to the index-set requires re-estimating all sparse grid coefficients. To make it more practical, the following lemma offers an inexpensive lower bound approximation.

Lemma 3 *The marginal gain can be bounded from below as*

$$J^-(s|G) \geq \rho^{(0)}(s,G), \quad \text{with } \rho^{(0)}(s,G) = \frac{\left(\sum_{i=1}^N r_i \phi_s(\mathbf{x}_i)\right)^2}{\sum_{i=1}^N \phi_s^2(\mathbf{x}_i) + \lambda N}. \tag{38}$$

Proof Let $\hat{\mathbf{w}}_G \in \mathbb{R}^G$ be the minimiser of (17), with $\hat{\mathbf{w}}_G = (\hat{w}_g)_{g \in G}$, and let $r_i := \sum_{g \in G} \hat{w}_g \phi_g(\mathbf{x}_i) - y_i$ be the residual at the point $(\mathbf{x}_i, y_i)$. Then, with definitions (17), (22), and (23), the marginal gain function $J^-(s|G)$ satisfies

$$J^-(s|G) = -\min_{\mathbf{w} \in \mathbb{R}^{G \cup \{s\}}} \left(\sum_{i=1}^N \left(\sum_{g \in G} w_g \phi_g(\mathbf{x}_i) - y_i + w_s \phi_s(\mathbf{x}_i) \right)^2 + \lambda N \sum_{g \in G} w_g^2 + \lambda N w_s^2 \right) +$$

$$\sum_{i=1}^N \left(\sum_{g \in G} \hat{w}_g \phi_g(\mathbf{x}_i) - y_i \right)^2 + \lambda N \sum_{g \in G} \hat{w}_g^2$$

$$\geq -\min_{w_s \in \mathbb{R}} \left(\sum_{i=1}^N (r_i + w_s \phi_s(\mathbf{x}_i))^2 + \lambda N \sum_{g \in G} \hat{w}_g^2 + \lambda N w_s^2 \right) + \sum_{i=1}^N r_i^2 + \lambda N \sum_{g \in G} \hat{w}_g^2$$

$$= - \min_{w_s \in \mathbb{R}} \sum_{i=1}^{N} \left(2w_s r_i \phi_s(\mathbf{x}_i) + w_s^2 \phi_s^2(\mathbf{x}_i)\right) + \lambda N w_s^2 =: \rho^{(0)}(s, G). \tag{39}$$

We can determine the minimising weight $\hat{w}_s$ of $\rho^{(0)}(s, G)$:

$$2 \sum_{i=1}^{N} \phi_s(\mathbf{x}_i) \left(r_i + \hat{w}_s \phi_s(\mathbf{x}_i)\right) + 2\lambda N \hat{w}_s \overset{!}{=} 0,$$

$$\sum_{i=1}^{N} \hat{w}_s \phi_s^2(\mathbf{x}_i) + \lambda N \hat{w}_s = - \sum_{i=1}^{N} r_i \phi_s(\mathbf{x}_i),$$

$$\hat{w}_s = - \frac{\sum_{i=1}^{N} r_i \phi_s(\mathbf{x}_i)}{\sum_{i=1}^{N} \phi_s^2(\mathbf{x}_i) + \lambda N}. \tag{40}$$

We then evaluate $\rho^{(0)}(s, G)$ explicitly by substituting $w_s = \hat{w}_s$ in (39):

$$\rho^{(0)}(s, G) = - \min_{w_s \in \mathbb{R}} \sum_{i=1}^{N} \left(2w_s r_i \phi_s(\mathbf{x}_i) + w_s^2 \phi_s^2(\mathbf{x}_i)\right) + \lambda N w_s^2$$

$$= - \left(2\hat{w}_s \sum_{i=1}^{N} r_i \phi_s(\mathbf{x}_i) + \hat{w}_s^2 \sum_{i=1}^{N} (\phi_s(\mathbf{x}_i))^2 + \lambda N \hat{w}_s^2\right)$$

$$= 2\frac{\sum_{i=1}^{N} r_i \phi_s(\mathbf{x}_i)}{\sum_{i=1}^{N} \phi_s^2(\mathbf{x}_i) + \lambda N} \sum_{i=1}^{N} r_i \phi_s(\mathbf{x}_i) - \frac{\left(\sum_{i=1}^{N} r_i \phi_s(\mathbf{x}_i)\right)^2}{\left(\sum_{i=1}^{N} \phi_s^2(\mathbf{x}_i) + \lambda N\right)^2} \sum_{i=1}^{N} (\phi_s(\mathbf{x}_i) + \lambda N)^2$$

$$= 2\frac{\left(\sum_{i=1}^{N} r_i \phi_s(\mathbf{x}_i)\right)^2}{\sum_{i=1}^{N} \phi_s^2(\mathbf{x}_i) + \lambda N} - \frac{\left(\sum_{i=1}^{N} r_i \phi_s(\mathbf{x}_i)\right)^2}{\sum_{i=1}^{N} \phi_s^2(\mathbf{x}_i) + \lambda N}$$

$$= \frac{\left(\sum_{i=1}^{N} r_i \phi_s(\mathbf{x}_i)\right)^2}{\sum_{i=1}^{N} \phi_s^2(\mathbf{x}_i) + \lambda N}. \tag{41}$$

This establishes Formula (38) and proves the lemma.

In our algorithm, instead of maximising the more expensive marginal gain of J^- with respect to candidate indices s, we maximise its lower bound $\rho^{(0)}(s, G)$. The computation of $\rho^{(0)}(s, G)$ is no more expensive than the computation of the refinement indicator (20). The resulting refinement procedure selects the grid index s with the highest indicator $\rho^{(0)}(s, G)$ among all indices in the candidate set.

In practice it is advisable to add multiple new grid indices in every refinement step instead of just one. However, it may be insufficient to choose the indices based on the size of the indicators. For example, taking two basis functions with a large

---

Algorithm 3: Refine($G^{(k)}$, A, n) using Fast Greedy Refinement

---

input : Old grid $G^{(k)}$, admissible index set A, number of new indices n
output: New grid $G^{(k+1)}$

1   *compute the error indicators $\rho^{(0)}(g, G^{(k)})$ for all candidates using (38)*;
2   *store the grid indices in a priority queue Q such that*
      $\rho^{(0)}(Q[0], G^{(k)}) \geq \rho^{(0)}(Q[1], G^{(k)}) \geq \ldots \geq \rho^{(0)}(Q[|A| - 1], G^{(k)})$;
3   *select the index with largest error indicator $I^{(1)} \leftarrow \{Q.pop()\}$*;
4   $j \leftarrow 2$;
5 **repeat**
6 | *select the largest element $v \leftarrow Q.pop()$*;
7 | *update the indicator refinement indicator of v using (44)*;
8 | **if** $\rho^{(j)}(v, G^{(k)})$ *is greater or equal to the refinement indicator stored for the next index in*
         the queue **then**
9 | | *add the index to the sparse grid index-set $I^{(j)} \leftarrow I^{(j-1)} \cup \{v\}$*;
10 | | *increment $j \leftarrow j + 1$*;
11 | **else**
12 | | *sort v back into the queue $Q.push(v)$*;
13 **until** $j = n$;
14 $G^{(k+1)} \leftarrow G^{(k)} \cup I^{(j)}$;

---

support overlap may be less expedient than taking functions with disjoint support. Hence, selecting a grid index may reduce the refinement indicators (potential benefit) of the other candidate indices in the admissible set.

The algorithm of our Algorithm 3 makes use of a priority queue with refinement indicators, similarly to the lazy greedy procedure for submodular optimisation [19, 20, 32]. At the step j of the refinement procedure we retrieve the largest element v from the queue and update its refinement indicator $\rho^{(j)}(v, G)$. If the updated indicator is still greater than the upper bound of the following index u in the queue, we accept it. Otherwise we put v back into the queue and repeat.

What is an efficient way to update the refinement indicators? Suppose that $I^{(j)}$ contains j grid indices already selected from the candidate set. If every $g \in I^{(j)}$ has the optimal coefficient $\hat{w}_g$ computed as in (40), the error terms become

$$r_i^{(j)} = r_i + \sum_{g \in I^{(j)}} \hat{w}_g \phi_g(\mathbf{x}_i) = r_i^{(j-1)} + \hat{w}_u \phi_u(\mathbf{x}_i), \quad \text{such that } u \in I^{(j)} \setminus I^{(j-1)}.$$

For every $v \in A \setminus I^{(j)}$ the refinement indicator is updated as

$$\rho^{(j)}(v, G) = \frac{\left(\sum_{i=1}^{N} r_i^{(j)} \phi_s(\mathbf{x}_i) \right)^2}{\sum_{i=1}^{N} \phi_s^2(\mathbf{x}_i) + \lambda N}$$

$$= \left[\frac{\sum_{i=1}^{N} r_i \phi_v(\mathbf{x}_i)}{\sqrt{\sum_{i=1}^{N} \phi_v^2(\mathbf{x}_i) + \lambda N}} + \frac{\sum_{i=1}^{N} \phi_v(\mathbf{x}_i) \sum_{g \in I^{(j)}} \hat{w}_g \phi_g(\mathbf{x}_i)}{\sqrt{\sum_{i=1}^{N} \phi_v^2(\mathbf{x}_i) + \lambda N}} \right]^2 \tag{42}$$

$$= \left[\frac{\sum_{i=1}^{N} r_i^{(j-1)} \phi_v(\mathbf{x}_i)}{\sqrt{\sum_{i=1}^{N} \phi_v^2(\mathbf{x}_i) + \lambda N}} + \frac{\sum_{i=1}^{N} \hat{w}_u \phi_u(\mathbf{x}_i) \phi_v(\mathbf{x}_i)}{\sqrt{\sum_{i=1}^{N} \phi_v^2(\mathbf{x}_i) + \lambda N}} \right]^2 , \text{ s.t. } u \in I^{(j)} \setminus I^{(j-1)}. \tag{43}$$

Equation (42) implies that the updated refinement indicator can be obtained from the original one using the formula

$$\rho^{(j)}(v, G) = \left[\sqrt{\rho^{(0)}(v, G)} + \frac{\sum_{i=1}^{N} \phi_v(\mathbf{x}_i) \sum_{g \in I^{(j)}} \hat{w}_g \phi_g(\mathbf{x}_i)}{\sqrt{\sum_{i=1}^{N} \phi_v^2(\mathbf{x}_i) + \lambda N}} \right]^2 . \tag{44}$$

Similarly, Eq. (43) implies that it can also be obtained from a previously updated indicator as

$$\rho^{(j)}(v, G) = \left[\sqrt{\rho^{(j-1)}(v, G)} + \frac{\sum_{i=1}^{N} \hat{w}_u \phi_u(\mathbf{x}_i) \phi_v(\mathbf{x}_i)}{\sqrt{\sum_{i=1}^{N} \phi_v^2(\mathbf{x}_i) + \lambda N}} \right]^2 . \tag{45}$$

What happens if more than one index is taken in the greedy procedure? While the theoretical guarantees for this case are missing, the idea was successfully applied in a number of algorithms for large sparse inverse problems using matching pursuit methods [3, 17].

To combine the greedy procedure with spatially-dimension-adaptive refinement of sparse grids, one considers not individual indices but the pairs that will be created in one dimensions. In this case, the sum of the individual refinement indicators serves as an indicator for the pair.

In this section we compare the performance of the regression method introduced in the Sects. 2 and 4 with currently used state-of-the-art techniques. The performance is assessed both with synthetic and real data. First, we focus on refinement and compare the state-of-the-art spatially-adaptive refinement procedure [27] with our greedy algorithm for spatially-dimension-adaptive refinement introduced in Sect. 2.4. After this, we consider our new regression method, which combines ASGD and spatially-dimension-adaptive refinement with a state-of-the-art technique using conjugate gradient descent and the standard spatially-adaptive refinement [26, 27].

2.5 Spatially-Dimension-Adaptive Refinement

As argued in Sect. 2.4, adding several indices in one step of spatially-dimension-adaptive refinement is preferable to adding a single index, and it comes in two variants: either taking into account the mutual influence of new basis functions or ignoring it.

The method that takes the mutual influence into account selects one best candidate pair n times in a row, as described in Algorithm 3. We call this method a spatially-dimension-adaptive *greedy* refinement. The method that ignores the mutual influence selects n pairs at once. This method may be less effective but it requires significantly less computation. We call it spatially-dimension-adaptive *block* refinement.

To test the sparse grids' ability to recognise the decision shapes of an axis-parallel square and a rhombus, we created two densely sampled datasets with strict decision boundaries, as depicted on Figs. 5a and 6a. Both datasets contain 2016 points from a Sobol sequence on a two-dimensional unit square. Points within the decision shape have the target values 1, the rest is 0.

We do a classification by regression using an adaptive sparse grid: the spatially-dimension-adaptive methods add $n = 10$ index pairs to the index-set and the spatially-adaptive method chooses 5 sparse grid indices from the index-set, adds all their descendents, and possibly the ancestors of the new indices (if those are missing from the index-set). Since the spatially-adaptive refinement method adds ca. 4 new points in every refinement, while spatially-dimension-adaptive only 2, these configurations are comparable.

We use a sparse grid with linear basis functions and 3 levels initially. The regularisation parameter λ is set to 0 in all three cases, as because of the abundant sampling and small grid sizes, there is no danger of overfitting even without regularisation.

Figures 5b–d and 6b–d illustrate the decision boundaries of the sparse grid models obtained after one or two refinement steps and have comparable sizes. These decision boundaries were estimated using Rule (8). While for the square dataset, with its axis-parallel decision boundaries, all three methods produce similar results, the spatially-adaptive refinement method has more difficulties identifying the skewed decision boundaries of the rhombus dataset. The greedy version of the spatially-dimension-adaptive refinement method is slightly better than the block version.

Besides the qualitative comparison, we show the quantitative results of the mean squared error and the number of misclassifications on Figs. 7 and 8 after 4 alternations of the fit- and refine-steps. The spatially-dimension-adaptive methods clearly outperform the spatially-adaptive refinement, showing better results with fewer grid indices. The difference in the results between the greedy and the block selection methods is not very large, such that in practice one may be willing to sacrifice the small precision advantage of the former method for the higher computational efficiency of the latter.

2.6 Spatially-Dimension-Adaptive Online Learning

We assess the overall performance of the regression method suggested in this paper using the data from the Sloan Digital Sky Survey from Data Release 5 [1]. We aim to predict the redshift of galaxies from the Main Galaxy Sample using 6 features from the dataset: dereded intensities in five broad bands (*ugriz*) together with a meta-parameter *eClass* [26]. Both training and test datasets contain 60,000 data points each.

The new regression method has the following configuration:

- **Fit-Step** is carried out using ASGD suggested in Sect. 2.3.
- **Refine-Step** is carried out using the spatially-dimension-adaptive method as follows:

    - Refinement indicators were computed as defined in (38).
    - New grid index pairs were selected all at once (block selection).

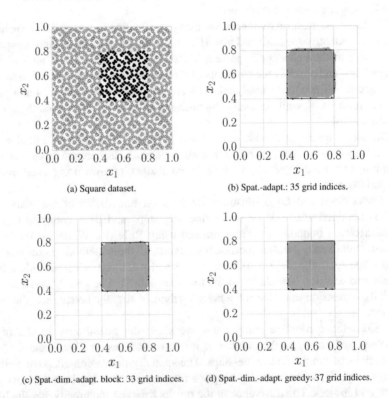

(a) Square dataset.

(b) Spat.-adapt.: 35 grid indices.

(c) Spat.-dim.-adapt. block: 33 grid indices.

(d) Spat.-dim.-adapt. greedy: 37 grid indices.

Fig. 5 Different refinement methods: *square* dataset. The *rectangle* contour indicates the true decision boundaries, the *shaded area* indicates the predicted decision boundaries, the *black points* are the misclassified entries of the train dataset

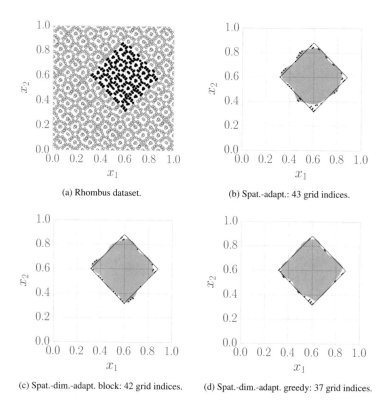

(a) Rhombus dataset.

(b) Spat.-adapt.: 43 grid indices.

(c) Spat.-dim.-adapt. block: 42 grid indices.

(d) Spat.-dim.-adapt. greedy: 37 grid indices.

Fig. 6 Different refinement methods: rhombus dataset. The *rectangle* contour indicates the true decision boundaries, the *shaded area* indicates the predicted decision boundaries, the *black points* are the misclassified entries of the train dataset

- The extension of the grid index-set was performed as in Eq. (19), adding all missing ancestors of the new grid indices to the grid.[2]

We denote this Online mode Spatially-Dimension-Adaptive method as *OSDA* for short.

We compare OSDA with the state-of-the-art method using the following configuration:

- **Fit-Step** is carried out using CG.
- **Refine-Step** is carried out using the spatially-adaptive method as follows:

  - Refinement indicators were computed as defined in (20).
  - The extension of the grid index-set was performed as in Eq. (19), adding all missing ancestors of the new grid indices to the grid.

---

[2]We noticed that this rule gives a better regularisation properties to the model with acceptable extra costs.

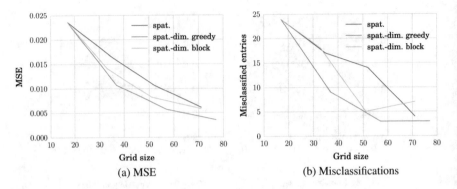

(a) MSE   (b) Misclassifications

Fig. 7 MSE and misclassifications on the training data for refinements on *square*

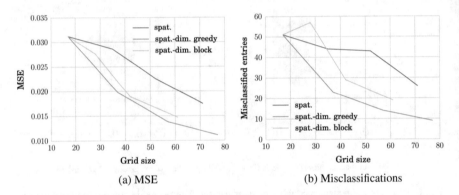

(a) MSE   (b) Misclassifications

Fig. 8 MSE and misclassifications on the training for refinements on rhombus

We denote this Batch mode Spatially-Adaptive method as *BSA* for short.

Usually, sparse grid regression utilises the normalisation technique where input variables are mapped into a unit hypercube by shifting in scaling, keeping its distributions unchanged (see Fig. 9a) [13, 26]. However, if the data are highly correlated, this can deteriorate the effectiveness of spatial adaptivity. An alternative is the Rosenblatt transformation [29], which effectively makes the data distribution uniform (see Fig. 9b). Hence, OSDA and BSA were evaluated and compared using both normalisation techniques.

The performance of the models depends on a number of parameters. The previous analysis showed that the choice of the regularisation parameter is not critical for the results, as long as it is not too large [26]. Hence, we set λ to 10^{-6} for all models. We use a sparse grid with modified linear basis functions and initial level 3 in all cases. Other model parameters were estimated in a validation procedure used in [26]. We list them in Tables 1 and 2.

Since we only want to compare the new method with the state of the art, we use a hard termination criterion: the training stops after a certain number of refinement steps (even if the global convergence has not been achieved yet).

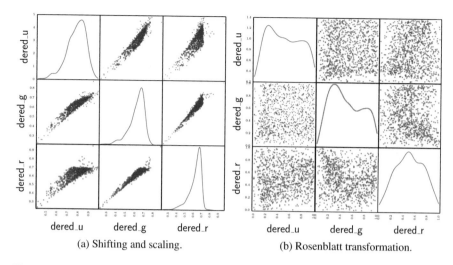

(a) Shifting and scaling.   (b) Rosenblatt transformation.

Fig. 9 Illustration of the normalisation methods for the first three attributes of the SDSS Data Release 5 dataset

Table 1 OSDA parameters

Parameter	w. Rosenblatt	w/o Rosenblatt
Training points for refinement indicator evaluation	60,000	10,000
Initial ASGD step size γ^0	$3.16 \cdot 10^{-5}$	10^{-3}
Index pairs *selected* at every refine-step	50	100
Dataset passes in ASGD (Algorithm 1 Line 3)	2	2
Iterations of outer loop (Algorithm 1 Line 2)	20	20

Table 2 BSA parameters

Parameter	w. Rosenblatt	w/o Rosenblatt
Training points for refinement indicator evaluation	60,000	60,000
Maximum CG iterations	100	250
Relative tolerance for CG	10^{-3}	10^{-2}
Grid indices *refined* at every refine-step	20	20

During the validation we observed that some parameters are more important than others. One or two passes through the data in the ASGD optimisation procedure were seen to be sufficient for proper pre-training before refinement. The results are significantly worse if only a small portion of lately seen data points in OSDA were used for refinement indicator computation. However, using between 10,000 and all 60,000 training points shows uniformly good results.

We also noticed that the algorithms favour different normalisation methods, as illustrated in Fig. 10. While the baseline BSA produces better results if we only shift and scale the data to the unit square, the spatially-dimension-adaptive refinement

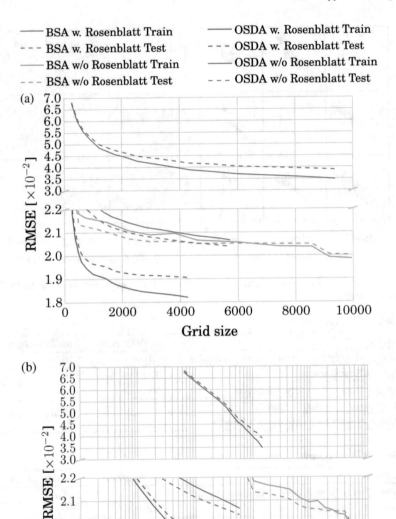

Fig. 10 Performance comparison between OSDA and BSA. The new method clearly shows the ability to create smaller sparse grids with the same representation power (**a**) using less operations (**b**)

Table 3 Final results of comparison between OSDA and BSA

	OSDA		BSA	
	Rosenblatt	w/o Rosenblatt	Rosenblatt	w/o Rosenblatt
Train RMSE [$\times 10^{-2}$]	1.82/**1.71**	2.06/1.99	3.48	1.99
Test RMSE [$\times 10^{-2}$]	**1.91**/1.93	2.04/2.00	3.88	2.00
Grid size	**4247**	5709	9366	9940
Data passes	**408**	570	1446	2224

Note: For the OSDA method we provide the RMSE before (*left*) and after (*right*) the final CG iterations

of OSDA benefits from the uniform distribution of the input data. There is a big gap between the performance of BSA with the Rosenblatt transformation and the performance of the rest, so that we had to break the continuity of the ordinate to compact the representation. The ordinate axis starts at $1.8 \cdot 10^{-2}$ instead of 0. Prediction of SDSS DR5 is well studied in the literature, and $1.8 \cdot 10^{-2}$ seems to be the limit that nonlinear models do not surpass [26, Tab. 6.8].[3]

As Fig. 10a shows, the best OSDA model has RMSE of $1.82 \cdot 10^{-2}$ on the test data with 4247 grid indices. This is better than $2.00 \cdot 10^{-2}$ with 9940 grid indices of the best BSA model. Figure 10b illustrates the convergence of the methods with increasing number of passes through the data.[4] Comparing the best results, the difference between OSDA and BSA is over 250 %: 408 data passes for OSDA vs. 1446 for BSA.

Table 3 summarises the final results. We performed several CG steps at the end of the OSDA procedure. For both OSDA results we give the errors before and after the finalising CG iterations. While the CG finalisation leads to a slight degradation of the generalisation performance for the OSDA model trained on the data with Rosenblatt transformation, for the OSDA model without Rosenblatt transformation CG yields exactly the same performance and that of the BSA model, but with fewer grid indices and data passes.

Altogether, the results show that the new method can build smaller sparse grid models which have the same predictive power as the state of the art. It also requires just a fraction of the number of data passes to build and train a new model.

---

[3] In this experiment we focused on the comparison between OSDA and BSA. Hence, we terminated the training of OSDA with Rosenblatt transformation prematurely. However, Fig. 10 suggests that further improvement may be possible.

[4] For OSDA we counted 2 passes through data in online optimisation loop and one for computing the refinement indicators.

3 Conclusions

We presented a new method for solving regression problems using spatially-dimension-adaptive sparse grids and ASGD. Our experiments on synthetic and real data show that the new method exhibits a competitive advantage over the state of the art. This advantage is twofold. On the one hand, the proposed spatially-dimension-adaptive refinement leads to models with lower complexity and similar representation power. On the other hand, the method reduces the number of required numerical operations between refinements and hence trains the models faster.

There are still questions to be answered. As the performance of ASGD crucially depends on the initial learning rate, a fast way to determine the optimal initial learning rate is paramount. The problem of determining the initial learning rate, however, is not specific to sparse grid models and its solution would benefit other machine learning methods as well. Furthermore, the spatially-dimension-adaptive refinement procedure derived in Sect. 2.4, while very effective at minimising the training error, may overfit a model faster than other refinement procedures. To prevent this overfitting one may need to develop new regularisation methods.

References

1. J.K. Adelman-McCarthy et al., The Fifth Data Release of the Sloan Digital Sky Survey. Astrophys. J. Suppl. Ser. **172**(2), 634–644 (2007)
2. F. Bach, E. Moulines, Non-strongly-convex smooth stochastic approximation with convergence rate o(1/n), in *Advances in Neural Information Processing Systems 26*, ed. by C. Burges, L. Bottou, M. Welling, Z. Ghahramani, K. Weinberger (Curran Associates, Inc., Red Hook, 2013), pp. 773–781
3. T. Blumensath, M.E. Davies, Stagewise weak gradient pursuits. IEEE Trans. Signal Process. **57**(11), 4333–4346 (2009)
4. L. Bottou, Stochastic learning, in *Advanced Lectures on Machine Learning* (Springer, Berlin/Heidelberg, 2004), pp. 146–168
5. L. Bottou, Online algorithms and stochastic approximations, in *Online Learning and Neural Networks*, ed. by D. Saad (Cambridge University Press, Cambridge, 1998), pp. 9–42. Revised (2012)
6. L. Bottou, Large-scale machine learning with stochastic gradient descent, in *Proceedings of the 19th International Conference on Computational Statistics (COMPSTAT'2010)*, Paris, ed. by Y. Lechevallier, G. Saporta (Physica-Verlag, Heidelberg, 2010), pp. 177–187. ISBN:978-3-7908-2603-6
7. L. Bottou, Stochastic gradient tricks, in *Neural Networks, Tricks of the Trade, Reloaded*, ed. by G. Montavon, G.B. Orr, K.-R. Müller. Lecture Notes in Computer Science (LNCS 7700) (Springer, 2012), pp. 430–445
8. L. Bottou, Y. LeCun, Large scale online learning, in *Advances in Neural Information Processing Systems 16*, ed. by S. Thrun, L. Saul, B. Schölkopf (MIT, Cambridge MA, 2004), pp. 217–224
9. H.-J. Bungartz, M. Griebel, Sparse grids. Acta Numer. **13**, 147–269 (2004)

10. H.-J. Bungartz, D. Pflüger, S. Zimmer, Adaptive sparse grid techniques for data mining, in *Modeling, Simulation and Optimization of Complex Processes*, ed. by H.G. Bock, E. Kostina, H.X. Phu, R. Rannacher (Springer, Berlin/Heidelberg, 2008), pp. 121–130. ISBN:978-3-540-79408-0

11. G. Buse, *Exploiting Many-Core Architectures for Dimensionally Adaptive Sparse Grids*. Dissertation, Institut für Informatik, Technische Universität München, München, 2015

12. U. Feige, A threshold of ln n for approximating set cover. J. ACM **45**(4), 634–652 (1998)

13. J. Garcke, *Maschinelles Lernen durch Funktionsrekonstruktion mit verallgemeinerten dünnen Gittern*. Doktorarbeit, Institut für Numerische Simulation, Universität Bonn, 2004

14. J. Garcke, M. Griebel, M. Thess, Data mining with sparse grids. Computing **67**(3), 225–253 (2001)

15. M. Hegland, Adaptive sparse grids, in *Proceedings of 10th Computational Techniques and Applications Conference CTAC-2001*, Brisbane, vol. 44, ed. by K. Burrage, R.B. Sidje (2003), pp. C335–C353

16. A. Heinecke, D. Pflüger, Multi- and many-core data mining with adaptive sparse grids, in *Proceedings of the 8th ACM International Conference on Computing Frontiers*, New York, May 2011 (ACM, 2011), pp. 29:1–29:10

17. P. Kambadur, A.C. Lozano, A parallel, block greedy method for sparse inverse covariance estimation for ultra-high dimensions, in *Proceedings of the Sixteenth International Conference on Artificial Intelligence and Statistics*, Scottsdale (2013), pp. 351–359

18. V. Khakhutskyy, D. Pflüger, M. Hegland, Scalability and fault tolerance of the alternating direction method of multipliers for sparse grids, in *Parallel Computing: Accelerating Computational Science and Engineering (CSE)*, Amsterdam, 2014, ed. by M. Bader, H.-J. Bungartz, A. Bode, M. Gerndt, G.R. Joubert. Volume 25 of Advances in Parallel Computing (IOS, 2014), pp. 603–612

19. J. Leskovec, A. Krause, C. Guestrin, C. Faloutsos, J. VanBriesen, N. Glance, Cost-effective outbreak detection in networks, in *Proceedings of the 13th ACM SIGKDD International Conference on Knowledge Discovery and Data Mining*, KDD '07, New York (ACM, 2007), pp. 420–429

20. M. Minoux, Accelerated greedy algorithms for maximizing submodular set functions, in *Optimization Techniques, Lecture Notes in Control and Information Sciences* 7:234–243, (1978)

21. G.L. Nemhauser, L.a. Wolsey, M.L. Fisher, An analysis of approximations for maximizing submodular set functions-I. Math. Program. **14**, 265–294 (1978)

22. J. Nocedal, S.J. Wright, *Numerical Optimization*. Springer Series in Operations Research and Financial Engineering, 2nd edn. (Springer, New York, 2006)

23. B. Peherstorfer, *Model Order Reduction of Parametrized Systems with Sparse Grid Learning Techniques*. Dissertation, Department of Informatics, Technische Universität München, Oct. 2013

24. B. Peherstorfer, D. Pflüger, H.-J. Bungartz, A Sparse-grid-based out-of-sample extension for dimensionality reduction and clustering with laplacian eigenmaps, in *AI 2011: Advances in Artificial Intelligence*, ed. by D. Wang, M. Reynolds (Springer, Berlin/Heidelberg, 2011), pp. 112–121

25. B. Peherstorfer, F. Franzelin, D. Pflüger, H.-J. Bungartz, Classification with probability density estimation on sparse grids, in *Sparse Grids and Applications – Munich 2012*, ed. by J. Garcke, D. Pflüger. Volume 97 of Lecture Notes in Computational Science and Engineering, pp. 255–270 (Springer, Cham/New York, 2014)

26. D. Pflüger, *Spatially Adaptive Sparse Grids for High-Dimensional Problems* (Verlag Dr. Hut, München, 2010).

27. D. Pflüger, Spatially adaptive refinement, in *Sparse Grids and Applications*, ed. by J. Garcke, M. Griebel. Lecture Notes in Computational Science and Engineering (Springer, Berlin/Heidelberg, 2012), pp. 243–262

28. B. Polyak, A. Juditsky, Acceleration of stochastic approximation by averaging. SIAM J. Control Optim. **30**(4), 838–855 (1992)

29. M. Rosenblatt, Remarks on a multivariate transformation. Ann. Math. Stat. **23**(3), 470–472 (1952)
30. T. Schaul, S. Zhang, Y. LeCun, No More Pesky learning rates. J. Mach. Learn. Res. **28**, 343–351 (2013)
31. D. Strätling, *Concept drift with adaptive sparse grids*. Bachelor Thesis, Technische Universität München, 2015
32. K. Wei, J. Bilmes, R.U.W. Edu, B.U.W. Edu, Fast multi-stage submodular maximization, in *International Conference on Machine Learning*, Beijing, 2014
33. W. Xu, Towards optimal one pass large scale learning with averaged stochastic gradient descent. Arxiv preprint arXiv:1107.2490. (2011), pp. 1–19

Sparse Grids for the Vlasov–Poisson Equation

Katharina Kormann and Eric Sonnendrücker

Abstract The Vlasov–Poisson equation models the evolution of a plasma in an external or self-consistent electric field. The model consists of an advection equation in six dimensional phase space coupled to Poisson's equation. Due to the high dimensionality and the development of small structures the numerical solution is quite challenging. For two or four dimensional Vlasov problems, semi-Lagrangian solvers have been successfully applied. Introducing a sparse grid, the number of grid points can be reduced in higher dimensions. In this paper, we present a semi-Lagrangian Vlasov–Poisson solver on a tensor product of two sparse grids. In order to defeat the problem of poor representation of Gaussians on the sparse grid, we introduce a multiplicative delta-f method and separate a Gaussian part that is then handled analytically. In the semi-Lagrangian setting, we have to evaluate the hierarchical surplus on each mesh point. This interpolation step is quite expensive on a sparse grid due to the global nature of the basis functions. In our method, we use an operator splitting so that the advection steps boil down to a number of one dimensional interpolation problems. With this structure in mind we devise an evaluation algorithm with constant instead of logarithmic complexity per grid point. Results are shown for standard test cases and in four dimensional phase space the results are compared to a full-grid solution and a solution on the four dimensional sparse grid.

1 Introduction

In order to build fusion energy devices, it is necessary to understand the behavior of plasmas in external and self-consistent electromagnetic fields. Within the kinetic theory, the plasma is described by its distribution function in phase-space and its

K. Kormann (✉) • E. Sonnendrücker
Technische Universität München, Zentrum Mathematik, Boltzmannstr. 3, 85747 Garching, Germany

Max-Planck-Institut für Plasmaphysik, Boltzmannstr. 2, 85748 Garching, Germany
e-mail: katharina.kormann@tum.de; katharina.kormann@ipp.mpg.de; eric.sonnendruecker@ipp.mpg.de

© Springer International Publishing Switzerland 2016
J. Garcke, D. Pflüger (eds.), *Sparse Grids and Applications – Stuttgart 2014*,
Lecture Notes in Computational Science and Engineering 109,
DOI 10.1007/978-3-319-28262-6_7

163

motion is described by the Vlasov–Maxwell or, for low-frequency phenomena, by the Vlasov–Poisson equation. Since analytic solutions are not known, numerical simulations are inevitable to understand the behavior of plasmas in a fusion device. Because the distribution is a function in phase-space, the problem is six-dimensional. Several approximative models exist with the aim of reducing the dimensionality of the problem. However, only a simulation of the full 6D problem can reveal the full structure of the problem and can help to understand the validity of lower dimensional models. Three types of methods are widely used in the simulation of the Vlasov equation: Particle-In-Cell (PIC) methods, semi-Lagrangian and Eulerian methods. Since the latter two are grid-based, they suffer from the curse of dimensionality. For this reason PIC methods are most common for high-dimensional Vlasov simulations. On the other hand, the semi-Lagrangian method [21] has proven to be an accurate alternative in simulations of the Vlasov equation in two and four dimensional phase space and there are deterministic numerical methods that are specially designed to solve high-dimensional problems efficiently.

The sparse grid method [6] is targeted at moderately high-dimensional problems with solutions of bounded mixed derivatives. The subject of the present paper is to introduce an efficient semi-Lagrangian method to solve the Vlasov–Poisson equation on a sparse grid. In [3] sparse grids are combined with the semi-Lagrangian method to solve the Hamilton–Jacobi–Bellman equation.

The evolution of small filaments in the plasma distribution function yields large mixed derivatives which causes a major problem for the performance of sparse grids. However, if the filaments evolve mainly along the coordinate axes, a tensor product of separate spatial and velocity sparse grids can yield good compression as we will demonstrate.

Sparse grids have previously been introduced to plasma physics in the context of solving an eigenvalue problem in linear gyrokinetics with the sparse grid combination technique [13]. For this application the sparse grid method is particularly suited since field-aligned coordinates are used and nonlinear effects are not considered.

The outline of the paper is as follows. First we will briefly define a semi-Lagrangian Vlasov solver in Sect. 2 and discuss typical features of the solution to the Vlasov equation. The sparse grid method is introduced in Sect. 3 before we describe the main components of our combined semi-Lagrangian solver on a sparse grid in Sect. 4. Section 5 explains how to use the structure of the problem in order to improve on the implementation and the accuracy of the problem. In Sect. 6, we explain how to improve the interpolation of Gaussians on a sparse grid and introduce our multiplicative δf method. A brief discussion on conservation and stability properties of our method is given in Sect. 7 before we show the numerical performance for benchmark problems in Sect. 8. Finally, conclusions are drawn and further research directions indicated in Sect. 9.

2 The Vlasov–Poisson Equation

The Vlasov–Poisson equation describes the motion of a plasma in its self-consistent electric field for low-frequency phenomena. In this paper, we will consider the dimensionless Vlasov–Poisson equation for electrons with a neutralizing ion background,

$$\partial_t f(\mathbf{x}, \mathbf{v}, t) + \mathbf{v} \cdot \nabla_{\mathbf{x}} f(\mathbf{x}, \mathbf{v}, t) - \mathbf{E}(\mathbf{x}, t) \cdot \nabla_{\mathbf{v}} f(\mathbf{x}, \mathbf{v}, t) = 0,$$

$$- \Delta\phi(\mathbf{x}, t) = 1 - \rho(\mathbf{x}, t), \quad \mathbf{E}(\mathbf{x}, t) = -\nabla\phi(\mathbf{x}, t), \quad \rho(\mathbf{x}, t) = \int f(\mathbf{x}, \mathbf{v}, t)\, d\mathbf{v}.$$

Here, f denotes the probability density of finding a particle at position $\mathbf{x} \in D \subset \mathbb{R}^3$ and velocity $\mathbf{v} \in \mathbb{R}^3$, $\mathbf{E}$ denotes the electric field, ϕ the electric potential, and ρ the charge density. As for any scalar hyperbolic conservation law with divergence-free advection field, the value of f is constant along the characteristics defined by

$$\frac{d\mathbf{X}}{dt} = \mathbf{V}, \quad \frac{d\mathbf{V}}{dt} = -\mathbf{E}(\mathbf{X}, t). \tag{1}$$

Let us denote by $\mathbf{X}(t; \mathbf{x}, \mathbf{v}, s), \mathbf{V}(t; \mathbf{x}, \mathbf{v}, s)$ the solution of the characteristic equations (1) at time t with initial conditions $\mathbf{X}(s) = \mathbf{x}$ and $\mathbf{V}(s) = \mathbf{v}$. Given an initial distribution f_0 at time 0, the solution at time $s > 0$ is given by

$$f(\mathbf{x}, \mathbf{v}, s) = f_0(\mathbf{X}(0; \mathbf{x}, \mathbf{v}, s), \mathbf{V}(0; \mathbf{x}, \mathbf{v}, s)), \tag{2}$$

i.e. we can find the solution at any time s by solving the characteristic equations backwards in time. However, we cannot solve (1) analytically since the right-hand side depends on f through the Poisson equation. Existence and uniqueness of the solution are shown in [9, Ch. 4]. The regularity of the solution depends on the initial condition [14] and typical features will be discussed in the following subsection. In Sect. 2.2, we will then discuss how to use the characteristic equations to find a numerical solution of the Vlasov equation.

2.1 Some Linear Analysis of the Solution to the Vlasov–Poisson Equation

The typical equilibrium solution to the Vlasov–Poisson problem is a Maxwellian of the form

$$f_0(\mathbf{x}, \mathbf{v}) = \frac{1}{(2\pi v_{th})^{3/2}} \exp\left(-\frac{\|\mathbf{v}\|_2^2}{2v_{th}^2}\right). \tag{3}$$

In this case, the electric field is zero. Also linear combinations of several Maxwellians are possible. This can be verified by inserting the Maxwellian into the Vlasov equation. The dynamics are typically initiated by small perturbations in space from equilibrium. For example the Landau test case refers to an initial condition of the type

$$f_0(\mathbf{x}, \mathbf{v}) = \frac{1}{(2\pi v_{th})^{3/2}} \exp\left(-\frac{\|\mathbf{v}\|_2^2}{2v_{th}^2}\right)\left(1 + \varepsilon \sum_{i=1}^{3} \cos(kx_i)\right), \tag{4}$$

where ε is a small parameter. While the solution stays close to equilibrium for small values of ε, a filamentation in phase space depending on ε develops over time. In Fig. 1a, b, the projection of $f(\mathbf{x}, \mathbf{v}, t) - f_0(\mathbf{x}, \mathbf{v})$ to the (x_1, v_1) plane is visualized at time 5 and 15 for a four dimensional simulation of the Landau problem with $\varepsilon = 0.01$ and $k = 0.5$. Comparing the phase-space distribution one can clearly see the filamentation at time 15. This filamentation is typical for solutions to the Vlasov–Poisson system (cf. [7]). Other configurations give rise to instabilities where the solution does not only show filamentation but also new structures form. In this case, the solution does not stay close to equilibrium and nonlinear effects eventually become dominant. A typical unstable problem for the two dimensional Vlasov–Poisson problem is the two stream instability defined by the initial condition

$$f_0(x, v) = \frac{1}{2\sqrt{2\pi}} \left(1 + 0.001 \cos(0.2x)\right) \left(e^{-0.5(v-2.4)^2} + e^{-0.5(v+2.4)^2}\right). \tag{5}$$

Figure 1d, e shows the phase-space distribution for a simulation of the two stream instability at time 5—where the distribution is close to the initial distribution— and time 30 where a hole structure has formed. A good approximation of the dynamics can often be obtained by linearizing the Vlasov–Poisson equations around the equilibrium (cf. e.g. [4, Ch. 7]). This yields a description to the first order in ε. For instance for the Landau damping problem with initial value (4), linear analysis tells us that

$$\phi^{lin}(\mathbf{x}, t) \propto \varepsilon e^{-\gamma t} \cos(\beta t - \phi) \sum_{i=1}^{3} \cos(kx_i), \tag{6}$$

with $\gamma, \beta, \phi \in \mathbb{R}$ depending on k. The (oscillating) electric field is damped by the rate γ. For weak Landau damping—that is if ε is small—linear theory gives a good description of the actual phenomenon. Due to the fact that the initial perturbation in (4) is separable, the dispersion relation is the sum of three two-dimensional cases. Note that the form (6) of the electric potential implies that $E_i^{lin}(\mathbf{x}, t) = E_i^{lin}(x_i, t)$ for $i = 1, 2, 3$, and the characteristic equations separate into three independent subproblems for the tuples $(x_i, v_i), i = 1, 2, 3$. Therefore, the filamentation is limited to the (x_i, v_i)-planes. In Fig. 1c, we have plotted the projection of $f(\mathbf{x}, \mathbf{v}, 15) - f_0(\mathbf{x}, \mathbf{v})$

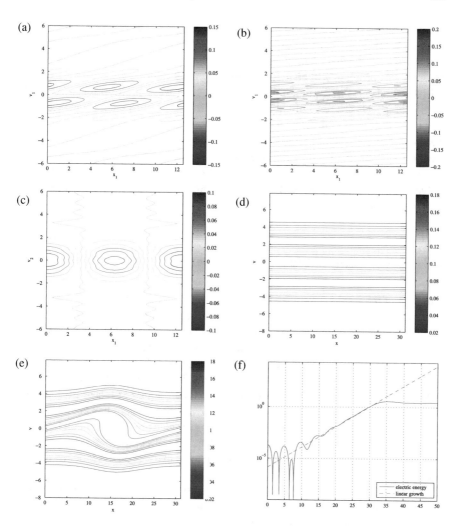

Fig. 1 Phase space distributions and electrical energy for Landau and two stream instability test cases. (**a**) Landau damping: (x_1, v_1)-projection of $f(\mathbf{x}, \mathbf{v}, 5) - f_0(\mathbf{x}, \mathbf{v})$. (**b**) Landau damping: (x_1, v_1)-projection of $f(\mathbf{x}, \mathbf{v}, 15) - f_0(\mathbf{x}, \mathbf{v})$. (**c**) Landau damping: (x_1, v_2)-projection of $f(\mathbf{x}, \mathbf{v}, 15) - f_0(\mathbf{x}, \mathbf{v})$. (**d**) Two stream instability: $f(x, v, 5)$. (**e**) Two stream instability: $f(x, v, 30)$. (**f**) Two stream instability: electric energy

to the (x_1, v_2)-plane. As opposed to the situation in (x_1, v_1)-plane, there is no filamentation.

For the two-stream instability, a linear perturbation analysis is also possible. For the given parameters, the electrical field is growing in the linear description. Hence, the perturbation is increasing over time. Figure 1f shows the time evolution of the simulated electric energy, $\frac{1}{2}\|\mathbf{E}\|_{L_2}$, together with the straight line with slope equal to the predicted growth rate of the electrical energy by the linear theory. The numerical solution of the two stream instability indeed shows that the electrical energy does not follow the linear behavior over long times but the electrical energy stops growing at a certain point where nonlinear effects become dominant. For the following two reasons, this problem will be more challenging for a solution on a sparse grid as we will discuss later: Firstly, the solution is no longer close to equilibrium and secondly nonlinear effects are not limited to couplings in the (x_i, v_i)-planes. Numerically, it has been observed that there occur considerable couplings between the different coordinate directions for a similar problem in [12] where the solution of the Vlasov–Poisson system has been represented in tensor train format. In the tensor train format, a function on a tensor product grid it compressed by higher order singular value decompositions. This can roughly be considered as an adaptive procedure to sparsify the grid.

2.2 A Semi-Lagrangian Vlasov Solver

The idea of the semi-Lagrangian method is to introduce a grid G in phase space and to solve the characteristics successively on small time intervals Δt. Given the values f^n of the solution at the grid points for some time t_n, the values of f^{n+1} at time $t_{n+1} = t_n + \Delta t$ can be found by numerically solving the characteristics equation starting at the grid points and then interpolating the value at the origins for time t_n from the values f^n. The advantage of the semi-Lagrangian method over Eulerian solvers is the fact that usually no CFL conditions are required for stability. Stability of the semi-Lagrangian method on a bounded velocity domain has been shown in [2]. Since the Vlasov problem is posed on an unbounded domain, artificial boundary conditions have to be added to close the system. The effect of boundary conditions has not been considered in [2]. A reasonable choice of Δt is on the order of the grid spacing or slightly below.

The solution of the characteristic equations can be found using a numerical ODE solver. For the Vlasov–Poisson system, one can instead split the $\mathbf{x}$ and $\mathbf{v}$ advection steps applying a Strang splitting, solving the two problems

$$\partial_t f - \mathbf{E}(\mathbf{x}) \cdot \nabla_{\mathbf{v}} f = 0, \quad \partial_t f + \mathbf{v} \cdot \nabla_{\mathbf{x}} f = 0 \tag{7}$$

separately. Note that the $\mathbf{v}$-advection equation is a constant-coefficient advection for each $\mathbf{x}$ and the characteristics are given as

$$\mathbf{X}(t; \mathbf{x}, \mathbf{v}, s) = \mathbf{x}, \quad \mathbf{V}(t; \mathbf{x}, \mathbf{v}, s) = \mathbf{v} - (t - s)\mathbf{E}(\mathbf{x}). \tag{8}$$

Note that the density ρ and hence the field $\mathbf{E}$ do not change in the $\mathbf{v}$-advection step. Also the $\mathbf{x}$-advection equation has constant coefficients for each $\mathbf{v}$ and the solution for the characteristics are

$$\mathbf{X}(t; x, v, s) = \mathbf{x} + (t - s)\mathbf{v}, \quad \mathbf{V}(t; x, v, s) = \mathbf{v}. \tag{9}$$

This idea yields the following split semi-Lagrangian scheme originally introduced by Cheng and Knorr [7]

1. Solve $\mathbf{v}$-advection on half time step: $f^{(n,*)}(\mathbf{x_i}, \mathbf{v_j}) = I_{f^{(n)}}(\mathbf{x_i}, \mathbf{v_j} + \mathbf{E}^{(n)}(\mathbf{x_i})\frac{\Delta t}{2})$.
2. Solve $\mathbf{x}$-advection on full time step: $f^{(n,**)}(\mathbf{x_i}, \mathbf{v_j}) = I_{f^{(n,*)}}(\mathbf{x_i} - \mathbf{v_j}\Delta t, \mathbf{v_j})$.
3. Compute $\rho(\mathbf{x_i}, \mathbf{v_i})$ from $f^{(n,**)}(\mathbf{x_i}, \mathbf{v_j})$ and solve the Poisson equation for $\mathbf{E}^{(n+1)}$.
4. Solve $\mathbf{v}$-advection on half time step: $f^{(n+1)}(\mathbf{x_i}, \mathbf{v_j}) = I_{f^{(n,**)}}(\mathbf{x_i}, \mathbf{v_j} + \mathbf{E}^{(n+1)}(\mathbf{x_i})\frac{\Delta t}{2})$.

For a given function g at all grid points, I_g evaluated at any point $(\mathbf{X}, \mathbf{V})$, generally not a grid point, denotes the interpolated value at $(\mathbf{X}, \mathbf{V})$ from the values of g on the grid points. In the split step method, the interpolation along the three $\mathbf{x}$ and $\mathbf{v}$ directions, respectively, is split again into three successive one-dimensional interpolations. For instance, we have the one dimensional interpolation problem

$$g^*(x_{1,i_1}, x_{2,i_2}, x_{3,i_3}, \mathbf{v_j}) = I_g(x_{1,i_1} - \Delta t v_{1,j_1}, x_{2,i_2}, x_{3,i_3}, \mathbf{v_j}) \quad \text{for all } (\mathbf{x_i}, \mathbf{v_j}) \in G.$$
(10)

Defining a one dimensional stripe of varying x_1 component from the grid by $S_{y_2, y_3, \mathbf{w}} = \{(\mathbf{x}, \mathbf{v}) \in G | x_2 = y_2, x_3 = y_3, \mathbf{v} = \mathbf{w}\}$, the interpolation problem (10) splits into a one-dimensional interpolation problem for each such stripe: All the points x_{1,i_1} are shifted by the constant displacement $-\Delta t v_{1,j_1}$.
The building blocks of the split semi-Lagrangian scheme are hence

1. Interpolation along one-dimensional stripes to propagate.
2. Integration over velocity dimension.
3. Solution of the three dimensional Poisson problem.

Using linear interpolation for the one-dimensional interpolation problems is too diffusive [7] and high-order interpolation needs to be used. The use of a cubic spline interpolator is common in Vlasov codes since cubic splines are a good compromise between accuracy and complexity (cf. [20, Ch. 2]). As mentioned in [7], trigonometric interpolation gives very good results for periodic problems, however, this is very specific and does not easily generalize to more complex geometries or mesh adaptivity.

3 The Sparse Grid Method

In this section, we give a brief introduction to the concept of sparse grids and introduce the notation used throughout the rest of the paper. For the definition of a sparse grid, we restrict ourself to the domain $[0, 1]^d$. However, a scaling of the domain is straightforward. Moreover, we will concentrate on piecewise linear functions first.
On the interval $[0, 1]$, we define for each $\ell \in \mathbb{N}$ the grid points

$$x_{\ell,k} = \frac{k}{2^\ell}, \quad k = 0, \dots, 2^\ell.$$
(11)

Associated to each grid point, we have the nodal hat function

$$\varphi_{\ell,k}(x) = \begin{cases} 1 - 2^{\ell}\mathrm{abs}(x - x_{\ell,k}) & \text{for } x \in [x_{\ell,k} - \frac{1}{2^{\ell}}, x_{\ell,k} + \frac{1}{2^{\ell}}] \\ 0 & \text{elsewhere.} \end{cases} \tag{12}$$

with $\varphi_{\ell,k}(x_{\ell,j}) = \delta_{k,j}$ and support of size $\frac{1}{2^{\ell-1}}$ centered around $x_{\ell,k}$. In d dimensions, we define the grid of level $\boldsymbol{\ell} \in \mathbb{N}^d$ by

$$\Omega_{\boldsymbol{\ell}} := \left\{ \mathbf{x}_{\boldsymbol{\ell},k} = (x_{\ell_1,k_1}, \ldots, x_{\ell_d,k_d}) | k_i = 0, \ldots, 2^{\ell_i} \right\} \tag{13}$$

and, associated to $\Omega_{\boldsymbol{\ell}}$, the space spanned by the piecewise d-linear nodal functions

$$V_{\boldsymbol{\ell}} := \mathrm{span} \left\{ \varphi_{\boldsymbol{\ell},k}(\mathbf{x}) = \prod_{i=1}^{d} \varphi_{\ell_i,k_i}(x_i) | k_i = 0, \ldots, 2^{\ell_i} \right\}. \tag{14}$$

From the definition of Ω_{ℓ} in one dimension, we can see that $\Omega_{\ell-1} \subset \Omega_{\ell}$ contains all the points from Ω_{ℓ} with even index, that is Ω_{ℓ} is the disjoint union of $\Omega_{\ell-1}$ and

$$\Omega_{\ell}^{\mathrm{odd}} := \{x_{\ell,k} \in \Omega_{\ell}, k \text{ odd}\}. \tag{15}$$

This leads us to the definition of the hierarchical increment

$$W_{\ell} := \mathrm{span}\left\{\varphi_{\ell,k} | k = 1, 3, \ldots, 2^{\ell-1} - 1\right\}. \tag{16}$$

The space $V_{\mathcal{L}}$ can be decomposed into the direct sum of hierarchical increments with level indices smaller equal $\mathcal{L}$, i.e.

$$V_{\mathcal{L}} = \bigoplus_{\ell \leq \mathcal{L}} W_{\ell}. \tag{17}$$

Instead of expanding a function $f \in V_{\mathcal{L}}$ in the nodal basis, we can represent it by its hierarchical surplus $v_{\ell,k}$ as

$$f(\mathbf{x}) = \sum_{|\boldsymbol{\ell}|_{\ell_{\infty}} \leq \mathcal{L}} \sum_{k \in \Omega_{\ell}^{\mathrm{odd}}} v_{\boldsymbol{\ell},k} \varphi_{\boldsymbol{\ell},k}(\mathbf{x}). \tag{18}$$

There are two important differences between the nodal and the hierarchical representation: Firstly, the number of functions different from zero at a point $\mathbf{x} \in [0, 1]^d$ increases. Except for the points on the grid, one function per hierarchical-increment basis is different from zero. On the other hand, the hierarchical surpluses express the additional information on the corresponding hierarchical increments compared to the representation of the solution on the space spanned by the hierarchical increments with smaller indices.

The sparse grid method is based on the possibility to leave out hierarchical increments where the hierarchical surpluses are smaller. In a standard sparse grid this is not done adaptively but a priori according with the aim of optimizing the cost-benefit ratio: The ℓ_1 norm of the index vector is restricted yielding

$$V_{\mathcal{L}}^s = \bigoplus_{|\ell|_{\ell_1} \leq \mathcal{L}} W_\ell. \tag{19}$$

This reduces the number of hierarchical increments to

$$\sum_{i=0}^{\mathcal{L}} \binom{d-1+i}{d-1}. \tag{20}$$

Note that the number of increments is still of the order $\frac{\mathcal{L}^d}{d!}$. On the other hand, the hierarchical increments that are left out are the ones with more points. The number of points for $V_{\mathcal{L}}^s$ is $O(2^{\mathcal{L}} \mathcal{L}^{d-1})$. This means the exponential growth in the dimension is only for the basis $\mathcal{L} = \ln(N)$, while on the other hand the accuracy is only decreased to

$$O(2^{-\mathcal{L}(p+1)} \mathcal{L}^{d-1}) \tag{21}$$

for functions of bounded mixed derivatives, i.e. again by the dth power of the logarithm. Here, the parameter p denotes the degree of the basis functions. So far, we have only defined linear basis functions. A construction of higher order basis functions for sparse grids has been proposed by Bungartz [5]. For each $x_{\ell,k} \in \Omega_\ell^{\text{odd}}$, the basis function $\varphi_{\ell,k}^{(p)}$ of degree p is defined by the $p + 1$ conditions

$$\varphi_{\ell,k}^{(p)}(x_{\ell,k}) = 1, \quad \varphi_{\ell,k}^{(p)}(x_j) = 0, \tag{22}$$

where x_j are the two neighbors $x_{\ell,k} - \frac{1}{2^\ell}$ and $x_{\ell,k} + \frac{1}{2^\ell}$ as well as the $p - 2$ next hierarchical ancestors of $x_{\ell,k}$. The support of $\varphi_{\ell,k}^{(p)}$ is restricted to $[x_{\ell,k} - \frac{1}{2^\ell}, x_{\ell,k} + \frac{1}{2^\ell}]$. In the following, let us denote by $M_{\mathcal{L}}$ the number of points of $V_{\mathcal{L}}^s$.

Now that we have exploited the benefits of a hierarchical representation, we need to discuss the downside that the support of the bulk of the basis function is increased. If we want to evaluate a representation in the nodal piecewise d-linear basis, the number of basis functions different from zero are mostly 2^d, whilst we have one function per hierarchical increment, i.e. $\mathcal{L}^d$ functions on a sparse grid of maximum level $\mathcal{L}$, in a hierarchical representation. Hence, the function evaluation has a complexity that is exponential in d for the basis $\mathcal{L}$ (cf. also [17, Sec. 2]). Also forming the hierarchical surplus in a naive implementation suffers from logarithmic scaling. However, for the piecewise d-linear basis, there is a simple relation between the hierarchical surplus and the function values that can be evaluated in linear complexity. A similar relation applies between the hierarchical surplus for a basis of order p and order $p + 1$ when they are constructed as described by Bungartz [5]. Hence, we can hierarchize and also dehierarchize the representation of a function on a sparse grid in linear complexity. Note, however, that this algorithm for hierarchization is specific to the basis functions constructed in [5] which is why using another basis is generally not advisable on sparse grids. One other basis that can efficiently be used on sparse grids is a hierarchical Fourier basis since a fast Fourier transform on sparse grids based on the so-called unidirectional principle [5] can be devised [10].

4 A Semi-Lagrangian Solver on a Sparse Grid

In the previous two sections, we have introduced the central ingredients of our novel Vlasov solver. Next, we will discuss various variants of introducing a sparse grid to the phase space. Once we have introduced the sparse grid, we discuss the design of the three building blocks of the split semi-Lagrangian method on the sparse grid.

4.1 Representation on a Sparse Grid

The basic ansatz would be to represent the distribution function f on a 6D sparse grid. However, since there is a natural splitting of $\mathbf{x}$ and $\mathbf{v}$ coordinates, a tensor product of a sparse grid in $\mathbf{x}$ and a sparse grid in $\mathbf{v}$ is also an interesting variant. This is actually a very special case of a dimension adaptive [11] 6D sparse grid. Using dimension or also spatial adaptivity [18] intermediate variants can be designed. However, we concentrate on non-adaptive variants in this paper where we have a simple construction principle whose structure can be exploited when implementing the building blocks of our solver.

Comparing the full 6D sparse grid (SGxv) to the tensor product of two 3D sparse grids (SGxSGv), it becomes immediately clear that the number of points will grow quadratically in $N = 2^{\mathcal{L}}$ (up to a logarithmic scaling $\mathcal{L}^d$) for the SGxSGv variant as opposed to the linear growth for the SGxv variant since we have the product of two sparse grids. On the other hand, the structure of the interpolation becomes simpler and parallelization is much simpler for the SGxSGv variant as will be discussed in Sect. 5. The strongest argument in favour of the tensor product variant SGxSGv lays, however, in the structure of the problem: Since the sparse grid prefers hierarchical increments with anisotropic refinement, functions that can be well approximated by the sum of univariate functions are also well approximated on a sparse grid. When we consider the Vlasov–Poisson system, we know that filaments evolve in phase-space which cause large mixed derivatives. Hence, a sparse grid representation of the solution to the Vlasov equation can be problematic. However, the filamentation is limited to the (x_i, v_i)-planes, $i = 1, 2, 3$, as long as linear effects dominate the dynamics and the initial perturbation is aligned with the coordinate axes (cf. the discussion in Sect. 2.1). Since the (x_i, v_i)-plane is not sparsified in the SGxSGv sparse grid, a good representation of the distribution function is expected for problems with field-aligned coordinate axes and where no instabilities arise. Also note that the SGxSGv is a structured special case of a dimension-adaptive sparse grid. Since not only the pairs (x_i, v_i) are fully resolved but also pairs with different indices, we expect that a dimension adaptive algorithm will further increase compression. However, this will be at the price of less structure, making efficient implementation more difficult.

Finally, let us discuss the boundary conditions. We note that in our definition of the sparse grid, level 0 is the level of the boundary points. Due to the constraint on the ℓ_1 norm, the points of the sparse grid are most dense on the boundary. Since the domain is periodic in **x** in our model, choosing periodic boundary conditions along the **x**-coordinates is natural. In this case, only the left boundary is included along each dimension. Along the velocity dimensions, we have an unbounded domain. On the other hand, our solutions have a Gaussian shape as discussed in Sect. 2.1 and therefore decay fast towards infinity. We therefore truncate the computational domain where the value of the distribution function is negligible and we have to set artificial boundary conditions. Since the solution is very small, not much information is kept at the boundary points which is a bit problematic in light of the fact that the sparse grid has most points at the boundary. A simple solution would be to set the solution equal to zero at the boundary and to skip level zero of the sparse grid. However, the Vlasov equation is a first order equation that does not allow for outflow boundary conditions which lead to unstable discretizations. Zero inflow boundary conditions are more suitable but require many boundary points. Another mathematically correct boundary closure are periodic boundaries. This is, of course, unphysical but since the solution is small at the boundary, the effect of the boundary condition is small and the advantage on a sparse grid is that we only need half the boundary points. In our experiments, we have chosen periodic boundary conditions.

In order to reduce the fraction of boundary points, we can modify the definition of the sparse grid by requiring the following two conditions on the level vector of the sparse grid

$$|\boldsymbol{\ell}|_{\ell_\infty} \leq \mathcal{L}, \quad |\boldsymbol{\ell}|_{\ell_1} \leq \mathcal{L} + d - 1. \tag{23}$$

This reduces the number of points on the boundary compared to a sparse grid defined by $|\boldsymbol{\ell}|_{\ell_1} \leq \mathcal{L}+d-1$ only. On the other hand, we have experimentally seen accuracies of the same order for both grids for the velocity sparse grid. Alternatively, we could have placed boundary and mid points on the same level which would have a similar—but not exactly the same—effect. We have chosen the former modification due to its ease of implementation. Note that the modification (23) yields worse results for the spatial sparse grid.

4.2 Interpolation

The first building block of the split-semi-Lagrangian algorithm, is one-dimensional interpolation. Interpolation on a sparse grid can be realized by computing and evaluating the hierarchical surplus. The major problem is the complexity of this operation. As long as we use piecewise d-linear basis functions or the pth order functions constructed in [5], computing the hierarchical surplus is comparably cheap as explained in Sect. 3. As a next step, we have to evaluate the function on the origin of each characteristic, i.e. at $M_{\mathcal{L}}$ points. As mentioned in Sect. 3, each evaluation of the sparse grid function has a complexity of $p\mathcal{L}^d$. In Sect. 5, we will design an algorithm that exploits the structure of this problem which reduces the complexity of the whole interpolation step to $O(pM_{\mathcal{L}})$ for the SGxSGv variant and to $O(p\mathcal{L}^3 M_{\mathcal{L}})$ for the SGxv variant.

Another question is about a suitable interpolation method. On the full grid, using a cubic spline interpolator is common as mentioned in Sect. 2.2. However, computing the hierarchical surplus for splines is generally more expensive [22, Sec. 4.1]. On sparse grids, the piecewise d-linear basis is most-frequently used. Even though linear interpolation has its advantages on sparse grids when it comes to conservation properties (cf. Sect. 7), linear interpolation is too diffusive yielding much worse results compared to higher order interpolation [7]. Hence, using sparse grids with third or fourth order polynomial basis functions constructed as in [5] are a good compromise between accuracy and computational complexity. However, the main interpolation task is only one dimensional and we will see in Sect. 5.4 how we can efficiently combine spline interpolation along the advection direction with higher order sparse grid interpolation along all other dimensions to improve accuracy.

4.3 Integration Over Velocity Coordinates

In order to compute the particle density ρ we need to integrate over the velocity dimensions. If we have a tensor product of a sparse grid in space and velocity, this is a simple sparse grid integration [8, 16]: For each point in the x-sparse grid, we have a v-sparse grid over which we have to integrate.

If we have a full sparse grid (SGxv), the velocity coordinates of all the points on the sparse grid with one particular spatial coordinate form a three dimensional sparse grid and we can perform a three dimensional sparse grid integration for each point in space. Note that the representation of the distribution function needs to be in the hierarchical surplus along the spatial dimensions before performing the integration on the three dimensional sparse grids. If we want to have the value of the density at each point, we have to dehierarchize over the spatial dimensions in the end.

In our numerical experiments, we use the sparse grid trapezoidal rule to compute the integrals.

4.4 Solution of the Poisson Problem

Once we have computed ρ and hence the right-hand side of the Poisson problem, we need to solve a three dimensional Poisson problem. In this paper, we focus on problems posed on a periodic domain in $\mathbf{x}$ and use a pseudospectral Poisson solver based on the sparse grid fast Fourier transform [10].

5 Improved Efficiency and Accuracy

As mentioned in Sect. 3 the advantage of the hierarchical basis is that we can leave out points without loosing too much in accuracy. On the other hand, the computational complexity of the algorithms increases. Efficient algorithms exist for some tasks but the evaluation of a sparse grid interpolant is the task in our algorithm with highest complexity, namely $O(p\mathcal{L}^d M_\mathcal{L})$. However, our interpolation problem exhibits a certain structure: We only have a displacement along one direction and the displacement only depends on some of the dimensions. In this section, we are going to explain how to exploit this structure to improve on the efficiency. Finally, in the last subsection we will exploit the same idea also in order to improve on the accuracy of our method.

5.1 Mixed Nodal-Hierarchical Representation

A key ingredient to our implementation is the observation that we can represent a sparse grid interpolant by its semi-hierarchical surplus, i.e. by values that are hierarchized along all but one dimension. In the derivation, we will only consider a two dimensional function and a hierarchization along dimension one for the ease of notation. Replacing the one hierarchized dimension by several ones or interchanging the indices is straightforward.

Let us consider the sparse grid interpolant I_f of a function f. It can be expressed as

$$
\begin{aligned}
I_f(x_1, x_2) &= \sum_{\ell_1=0}^{\mathcal{L}} \sum_{\ell_2=0}^{\mathcal{L}-\ell_1} \sum_{k_1 \in \Omega_{\ell_1}^{\mathrm{odd}}} \sum_{k_2 \in \Omega_{\ell_2}^{\mathrm{odd}}} v_{\ell,k} \varphi_{\ell,k}(x_1, x_2) \\
&= \sum_{\ell_1=0}^{\mathcal{L}} \sum_{k_1 \in \Omega_{\ell_1}^{\mathrm{odd}}} \varphi_{\ell_1,k_1}(x_1) \left[\sum_{\ell_2=0}^{\mathcal{L}-\ell_1} \sum_{k_2 \in \Omega_{\ell_2}^{\mathrm{odd}}} v_{\ell,k} \varphi_{\ell_2,k_2}(x_2) \right].
\end{aligned}
\tag{24}
$$

In the second step, we have reordered the sum such that the expression in brackets is a one-dimensional hierarchical sum of basis functions for all index pairs (ℓ_1, k_1).

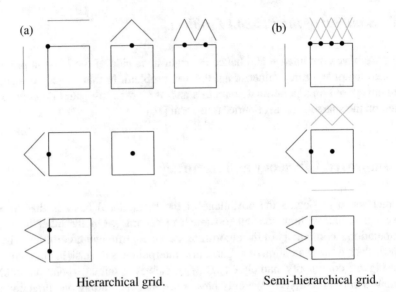

Hierarchical grid.   Semi-hierarchical grid.

Fig. 2 Sparse grid with maximum level $\mathcal{L} = 2$ and periodic boundary conditions. Part (**a**) shows the hierarchical increments with hierarchical basis functions. Part (**b**) shows the semi-hierarchical version where the horizontal basis functions are nodal with different level for each horizontal stripe. The hierarchical basis functions are drawn in *blue* and the nodal ones in *green*

Since the nodal basis is equivalent to the hierarchical, we can dehierarchize $v_{\ell,k}$ along dimension 2 and use a nodal representation along dimension 2. This gives

$$I_f(x_1, x_2) = \sum_{\ell_1=0}^{\mathcal{L}} \sum_{k_1 \in \Omega_{\ell_1}^{\text{odd}}} \varphi_{\ell_1,k_1}(x_1) \left[\sum_{k_2 \in \Omega_{\mathcal{L}-\ell_1}} v_{\ell_1,k_1}(x_{\ell_2,k_2}) \varphi_{\mathcal{L}-\ell_1,k_2}(x_2) \right] \tag{25}$$

In Fig. 2, we have visualized the hierarchical representation (24) and the semi-hierarchical representation (25) for a sparse grid with maximum level $\mathcal{L} = 2$ and periodic boundaries.

If we want to evaluate the representation (25), we only have a hierarchical representation along dimension 1 and a nodal representation along dimension 2. Hence, the cost of one evaluation goes down from order $\mathcal{L}^2$ to $\mathcal{L}$, or in the d-dimensional case from $\mathcal{L}^d$ to $\mathcal{L}^{d-1}$. Of course this requires that we have the semi-hierarchical surplus available. If we start from the function values of the grid points, we can compute this by just hierarchizing along $d - 1$ dimensions in the same way, but even slightly cheaper than the full hierarchical surplus. If we start of from the hierarchical surplus, we have to apply dehierarchization along one dimension with a complexity of order $M_{\mathcal{L}}$. So in this case, this way of evaluating a sparse grid function is only worthwhile if we want to evaluate the functions about $M_{\mathcal{L}}$ times. But this is the case in our method. The total complexity of the interpolation step will hence reduce to $O(p\mathcal{L}^{d-1}M_\mathcal{L})$.

5.2 Constant Displacement

As a next step, we want to exploit the fact that we have a one-dimensional interpolation and that the displacement is constant along some dimensions. First, we consider the case where the displacement is fully constant. In the variant SGxSGv, all interpolations fall into this category. Even though the coefficients are not constant, the coefficient of the **x**-advections only depend on **v** and vice versa and the sparse grids are only including the **x** or **v** directions separately. In our analysis, we also allow for displacements that are dependent on the dimension along which we have the displacement, even though this is not the case for the interpolations in the SGxSGv method.

Let $f(x_1, x_2)$ be the original function and let I_f be its sparse grid interpolant. Now, we want to find a sparse grid representation of a function $g(x_1, x_2)$ that satisfies

$$I_g(x_1, x_2) = I_f(x_1, x_2 + c(x_2)) \text{ for all } (x_1, x_2) \in S, \tag{26}$$

where S is the set of points representing the sparse grid and $c(x_2)$ is the displacement along x_2 that is a function of x_2, i.e. constant along x_1. Note that we can again interchange the indices or add more hierarchical dimensions without displacement.

Now, we choose a semi-hierarchical representation that is nodal along dimension 2. For a point $(x_{L_1,K_1}, x_{L_2,K_2}) \in S$, this representation reads

$$I_f(x_{L_1,K_1}, x_{L_2,K_2} + c(x_{L_2,K_2})) = \sum_{\ell_1=0}^{\mathcal{L}} \sum_{k_1 \in \Omega_{\ell_1}^{\mathrm{odd}}} \varphi_{\ell_1,k_1}(x_{L_1,K_1}) \cdot$$

$$\left[\sum_{k_2 \in \Omega_{\mathcal{L}-\ell_1}} v_{\ell_1,k_1}^f(x_{L_2,K_2}) \varphi_{\mathcal{L}-\ell_1,k_2}(x_{L_2,K_2} + c(x_{L_2,K_2})) \right]$$

$$I_g(x_{L_1,K_1}, x_{L_2,K_2}) = \sum_{\ell_1=0}^{\mathcal{L}} \sum_{k_1 \in \Omega_{\ell_1}^{\mathrm{odd}}} \varphi_{\ell_1,k_1}(x_{L_1,K_1}) v_{\ell_1,k_1}^g(x_{L_2,K_2}).$$

$$\tag{27}$$

If we now set

$$v_{\ell_1,k_1}^g(y_{L_2,K_2}) = \sum_{k_2 \in \Omega_{\mathcal{L}-\ell_1}} v_{\ell_1,k_1}^f(x_{L_2,K_2}) \varphi_{\mathcal{L}-\ell_1,k_2}(x_{L_2,K_2} + c(x_{L_2,K_2})) \tag{28}$$

for all combinations of (ℓ_1, k_1) and (L_2, K_2) on the sparse grid, Eq. (26) is satisfied. Now, we found a representation of the semi-hierarchical surplus of I_g from which we can either compute the hierarchical surplus by hierarchization along dimension 2 or the function values by dehierarchization along dimension 1. Evaluating equation (28) is independent of $\mathcal{L}$ since we have a nodal representation. In this way, we can find the representation of the interpolant of the shifted function by a combination of algorithms with linear complexity.

Following this implementation the complexity of each advection step reduces to $O(pM_{\mathcal{L}})$ for a sparse grid interpolation on the SGxSGv variant.

5.3   Interpolation with Coefficients Constant Along Some Dimensions

While we have found a very efficient implementation of the interpolation steps for the SGxSGv variant, the situation is more complicated for the SGxv variant since now the displacement is still dependent on some of the dimensions in the sparse grid. In order to still be able to somewhat reduce the complexity, we would like to partly apply the algorithm derived in the previous section. In order to understand the algorithm for a simple example, we consider the case of a three dimensional sparse grid with displacement along x_3 only depending on x_2, x_3. This means, we consider a function $f(x_1, x_2, x_3)$ with known interpolant I_f on the sparse grid $\mathcal{S}$ and want to compute a sparse grid representation of the function $g(x_1, x_2, x_3)$ such that

$$I_g(x_1, x_2, x_3) = I_f(x_1, x_2, x_3 + c(x_2, x_3)) \text{ for all } (x_1, x_2, x_3) \in \mathcal{S}. \tag{29}$$

Let us consider the point $(x_{L_1,K_1}, x_{L_2,K_2}, x_{L_3,K_3}) \in \mathcal{S}$ and use a representation of I_f that is nodal along x_3 and a representation of I_g that is nodal in x_2, x_3. We then have

$$I_f(x_{L_1,K_1}, x_{L_2,K_2}, x_{L_3,K_3} + c(x_{L_2,K_2}, x_{L_3,K_3})) =$$

$$\sum_{\ell_1=0}^{\mathcal{L}} \sum_{k_1 \in \Omega_{\ell_1}^{odd}} \varphi_{\ell_1,k_1}(x_{L_1,K_1}) \sum_{\ell_2=0}^{\mathcal{L}-\ell_1} \sum_{k_2 \in \Omega_{\ell_2}^{odd}} \varphi_{\ell_2,k_2}(x_{L_2,K_2}) \cdot$$

$$\left[\sum_{k_3 \in \Omega_{\mathcal{L}-\ell_1-\ell_2}} v_{\ell_1,k_1,\ell_2,k_2}^f(x_{L_3,K_3}) \varphi_{\mathcal{L}-\ell_1-\ell_2,k_3}(x_{L_3,K_3} + c(x_{L_2,K_2}, x_{L_3,K_3})) \right]$$

$$I_g(x_{L_1,K_1}, x_{L_2,K_2}, x_{L_3,K_3}) = \sum_{\ell_1=0}^{\mathcal{L}} \sum_{k_1 \in \Omega_{\ell_1}^{odd}} \varphi_{\ell_1,k_1}(x_{L_1,K_1}) v_{\ell_1,k_1}^g(x_{L_2,K_2}, x_{L_3,K_3}).$$

$$\tag{30}$$

From this representation, we can see that (29) is satisfied for all $(x_{L_1,K_1}, x_{L_2,K_2}, x_{L_3,K_3})$ $\in S$ if we set for all combinations of (ℓ_1, k_1), (L_2, K_2) and (L_3, K_3)

$$
v_{\ell_1,k_1}^g (x_{L_2,K_2}, x_{L_3,K_3}) = \sum_{\ell_2=0}^{\mathcal{L}-\ell_1} \sum_{k_2 \in \Omega_{\ell_2}^{odd}} \varphi_{\ell_2,k_2}(x_{L_2,K_2}) \cdot
$$

$$
\sum_{k_3 \in \Omega_{\mathcal{L}-\ell_1-\ell_2}^{odd}} v_{\ell_1,k_1,\ell_2,k_2}^f (x_{L_3,K_3}) \varphi_{\mathcal{L}-\ell_1-\ell_2,k_3}(x_{L_2,K_2} + c(x_{L_2,K_2}, x_{L_3,K_3})).
$$

(31)

This can now no longer be computed within linear complexity since we have one sum over hierarchical basis functions. Hence, we get a logarithmic scaling by $\mathcal{L}$. For the general case of a d dimensional sparse grid and a displacement depending on γ directions (different from the direction of the displacement), the complexity of the evaluation step is $O(p\mathcal{L}^\gamma M_\mathcal{L})$. In particular, each velocity advection step on a six-dimensional sparse grid will have complexity of $O(p\mathcal{L}^3 M_\mathcal{L})$ and each spatial advection step $O(p\mathcal{L} M_\mathcal{L})$.

5.4 Mixed Interpolation

When we solve the interpolation problem based on the efficient evaluation formulas discussed above, we always need the semi-hierarchical surplus that is hierarchical along all directions except for the one along which the displacement appears. In this section, we discuss how to exploit this fact to improve on the accuracy. As mentioned in Sect. 2.2, spline interpolation has proven very efficient for semi-Lagrangian methods on the full grid but computing the hierarchical surplus for a spline basis on a sparse grid has too high complexity. However, we do not need to hierarchize along the dimension along which we displace the points. This gives us the freedom to use another basis (i.e. another interpolator) along that dimension. To illustrate this, let us revisit the example of a function $f(x_1, x_2)$ of two variables where we have a displacement $c_1(x_1, x_2)$ along x_1 and a displacement $c_2(x_1, x_2)$ along x_2. For a given one-dimensional set of equidistant points Ω_ℓ as defined in (13), let us consider an arbitrary set of basis functions $(\phi_{\ell,k})_{k=1,\dots,2^\ell}$ associated with Ω_ℓ. Then, we represent f on the sparse grid by

$$
I_f^{(1)}(x_1, x_2) = \sum_{\ell_1=0}^{\mathcal{L}} \sum_{k_1 \in \Omega_{\ell_1}^{odd}} \varphi_{\ell_1,k_1}(x_1) \left[\sum_{k_2 \in \Omega_{\mathcal{L}-\ell_1}} v_{\ell_1,k_1}^{(1)}(x_{\ell_2,k_2}) \phi_{\mathcal{L}-\ell_1,k_2}(x_2) \right]
$$

(32)

for the interpolation problem along x_2. In order to solve the interpolation problem, we can thus succeed in two steps:

1. Compute the semi-hierarchical surplus $v_{\ell_1,k_1}^{(1)}(x_{\ell_2,k_2})$.

2. For each one-dimensional stripe $S_{x_{L_1,K_1}} = \{(x_1,x_2) \in S_{\mathcal{L}} | x_1 = x_{L_1,K_1}\}$ defined by the x_1-coordinate x_{L_1,K_1} appearing on the sparse grid $S_{\mathcal{L}}$, solve the one-dimensional interpolation problem in the basis $(\phi_{\mathcal{L}-\ell_1,k})_k$.

For the interpolation along x_1, we will then exchange the roles of the coordinates and use the following interpolant

$$I_f^{(2)}(x_1,x_2) = \sum_{\ell_2=0}^{\mathcal{L}} \sum_{k_2 \in \Omega_{\ell_2}^{odd}} \varphi_{\ell_2,k_2}(x_2) \left[\sum_{k_1 \in \Omega_{\mathcal{L}-\ell_2}} v_{\ell_2,k_2}^{(2)}(x_{\ell_1,k_1})\phi_{\mathcal{L}-\ell_2,k_1}(x_1) \right]. \qquad (33)$$

Note that unless we choose the basis $(\phi_{\mathcal{L}-\ell_1,k})_k$ equivalent to the basis used on the sparse grid, the representations (32) and (33) will not be equivalent, i.e. we use different representations for each one-dimensional interpolation problem. This also means (32) and (33) are—other than (25)—not equivalent to the representation with the fully hierarchical sparse grid interpolation. Generalization to higher dimensions is straight-forward. In our experiments, we will use cubic spline interpolations along the one-dimensional stripes combined with sparse grids with cubic basis functions in order to improve the accuracy (cf. Sect. 8.2).

5.5 Parallelization

Even though the number of grid points is reduced when using a sparse grid, for six-dimensional problems with reasonable resolution the number of points and the number of arithmetic operations can become so large that parallel computations become necessary. Due to the non-locality of the hierarchical basis, parallelizing sparse grid routines is not trivial.

On the other hand, the split-step semi-Lagrangian method provides trivial parallelism: Each one-dimensional interpolation step reduces to operations on one-dimensional stripes if a nodal basis is used. This is exploited in the parallelization strategy of the semi-Lagrangian library SeLaLib [1] which is the basis for our implementation: When we are computing the advection step along one dimension, the domain is decomposed along one or several of the other dimensions and distributed between the processors. Then each of the processors can work on its one-dimensional stripes independently of the other. The only thing that has to be done is a redistribution of the data once we turn to an advection step along a direction over which the data was distributed.

One of the advantages of the SGxSGv variant is the fact that we can apply this parallelization strategy. When computing the x-advection steps, the problem is nodal along the v directions and vice versa. We can therefore distribute the distribution function over the points of the velocity sparse grid when performing the x-advection steps, and along the points of the spatial sparse grid when performing the v-advection steps.

6 Multiplicative δf Method

It is well-known that sparse grids and, especially higher order polynomials on sparse grids, are badly suited to interpolate Gaussians [18, Sec. 4.2]. The major problem is that a quite large number of points is necessary before the interpolation starts to converge. Since the initial value of the distribution function is a Gaussian, the quality of the representation along the velocity dimensions needs a considerable resolution. On the other hand, we know that the solution often stays close to the equilibrium distribution (cf. Sect. 2.1). For this reason, we could only simulate the difference of the solution from the Gaussian equilibrium. Since the perturbation typically follows the same Gaussian decay, we consider a multiplicative splitting which we call multiplicative δf method.

In order to keep the presentation simple, we will explain our multiplicative splitting for the two-dimensional case. The main idea is to split the distribution function into a time-dependent part and a time-constant Gaussian part,

$$f(x, v, t) = g(x, v, t)h(v), \qquad (34)$$

and to represent the function g on the sparse grid only while h is known analytically. If our solution is close to equilibrium, $g(x, v, t)$ will be close to one. For the Landau damping problem introduced in Sect. 2.1, the initial splitting would be

$$g(x, v, 0) = 1 + \varepsilon \cos(kx), \quad h(v) = \frac{1}{2\pi} \exp(-0.5v^2). \qquad (35)$$

Since h is independent of x, this part of the distribution function will not change for the x-advection. For a v-advection, on the other hand, we have

$$f(x, v + E(x)\Delta t, t) = g(x, v + E(x)\Delta t)h(v + E(x)\Delta t)$$
$$= g(x, v + E(x)\Delta t)\frac{h(v + E(x)\Delta t)}{h(v)}h(v). \qquad (36)$$

Hence, we have to perform a usual sparse grid integration for g followed by a scaling by $\frac{h(v+E(x)\Delta t)}{h(v)}$. Finally, when integrating over the velocity dimension, we analytically compute the weighted integrals

$$w_{\ell,k} := \int h(v)\varphi_{\ell,k}(v)\,dv, \qquad (37)$$

as quadrature weights for the hierarchical surpluses $v_{\ell,k}$. In three dimensions, on the SGxSGv grid, the density ρ at grid point $\mathbf{x}$ is thus computed as

$$\rho(\mathbf{x}) = \sum_{(\ell,k)\in G} w_{\ell_1,k_1} w_{\ell_2,k_2} w_{\ell_3,k_3} v_{\ell,k}(\mathbf{x}), \qquad (38)$$

where G denotes the index set defining the sparse grid along $\mathbf{v}$ direction. Of course, this would result in a considerable computational overhead since evaluating (37) requires evaluations of the error function. However, the weights are not depending on time and can be precomputed once.

In Sect. 8.3 we will show that a considerable improvement of the solution can be obtained for the Landau damping problem when the multiplicative δf method is applied. On the other hand, this procedure fails to improve accuracy when, for instance, an instability occurs, at least as long as $h(v)$ is not updated when the instability occurs.

7   A Note on Stability and Conservation Properties

A major disadvantage of a sparse grid solver is the fact that stability of sparse grid algorithms is not very well-understood or not guaranteed. A lack of L_2 stability was discussed in [6]. Also Bokanowski et al. [3] point out that they cannot provide stability estimates for their semi-Lagrangian solver.

Stability for a one-dimensional semi-Lagrangian Vlasov solver on a uniform grid was analyzed in [2]. However, effects from domain truncation are not considered. Due to the fact that we have stripes with very coarse refinement in a sparse grid, the sphere of influence of the boundary points is increased which is a potential source of numerical instability.

Indeed, if the solution is not well-resolved, unstable results have been obtained in our numerical experiments (cf. Sect. 8.4). From our experience, instabilities arise first when the solution is severly underresolved and mixed derivatives become large. As shown in Sect. 8.4, the method can be stabilized by switching to linear interpolation or alternating between linear and higher-order interpolation once the resolution becomes too bad. This introduces diffusion to the system and we observe that energy is dissipated depending on the amount of diffusion added.

Alternatively, we could add a small diffusive term on the scale of the smallest grid size as it was discussed in [19] for gyrokinetic simulations. However, a smallest grid size is not well-defined on a sparse grid. We have seen that diffusion on an increasing scale must be introduced when increasing simulation time. A better theoretical understanding of the stability of the method would be necessary to develop a robust stabilization method that does not unnecessarily deteriorate accuracy. Also the influence of the time step and the boundary conditions needs to be better understood.

Moreover, the sparse grid method does not mimic the conservation properties of the continuous model. In the continuous model, we have conservation of mass, momentum, energy and all L^p-norms. Only when using a linear interpolator and trapezoidal sparse grid integration, mass conservation is assured. We assume that improving on the conservation properties might increase the stability of the method as well.

8 Numerical Results

We have implemented the sparse grid method as part of the Fortran library SeLaLib [1]. In this section, we study the performance of our method. We will often compare the sparse grid solution to a full grid solution. The full grid solution is computed with the split-step semi-Lagrangian scheme based on a cubic spline interpolator. In all simulations on the SGxSGv grid, the spatial sparse grid is a sparse grid of the form (19) with ℓ_1 bound and periodic boundary conditions and the velocity sparse grid is a modified sparse grid with ℓ_1 and ℓ_∞ bound on the level vector according to (23) and with periodic boundary conditions. The given maximal level $\mathcal{L}_x$ refers to the upper bound in the ℓ_1 norm of the level vector and $\mathcal{L}_v$ to the upper bound in the ℓ_∞ norm of the level vector. Except for the comparative study in Sect. 8.2 a sparse grid with cubic basis functions is used together with cubic spline interpolation along the one dimensional stripes with displacement (cf. Sect. 5.4). The interpolation steps are implemented with the efficient algorithms devised in Sects. 5.1, 5.2 and 5.3. The multiplicative δf method is only applied if this is explicitly noted.

In Sects. 8.1, 8.2 and 8.3, we will consider the weak Landau problem in d dimensions ($d = 2, 3$) with initial value

$$f(\mathbf{x}, \mathbf{v}) = \left(1 + 0.01 \sum_{i=1}^{d} \cos(0.5x_i)\right) \frac{1}{(2\pi)^{d/2}} \exp(-0.5\|\mathbf{v}\|_2^2). \tag{39}$$

In order to assess the quality of the solution, we will compare the time evolution of the electric energy (cf. e.g. Fig. 3) to the decay rate predicted by linear theory

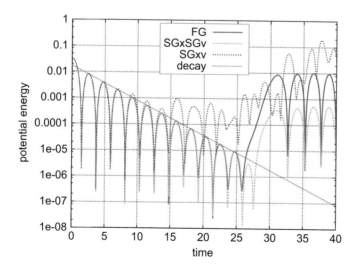

Fig. 3 Weak Landau damping in 4D. Electric energy for simulations on full grid (FG), SGxSGv and SGxv

(cf. Sect. 2.1). For the given parameter of $\varepsilon = 0.01$ the electric energy should show this decay over long times. However, in numerical solutions on a grid, one typically observes an approximative recurrence of the initial state after a certain time that is proportional to the reciprocal of the velocity grid spacing, $\frac{1}{\Delta v}$ (see [15]). Of course, the simulated solution is incorrect as soon as this artificial recurrence appears.

8.1 Comparison of Full Grid, SGxSGv, and SGxv

Let us consider the weak Landau damping problem in four dimensions. We compare the two variants of the sparse grid with a full grid solution. The time evolution of the potential energy is shown in Fig. 3. In this experiment, we use a sparse grid with cubic polynomial basis functions and cubic spline interpolation along the dimension with displacement. On the full grid, we use 32 points along each dimension. The number of levels on the SGxSGv grid is chosen such that the damping rate is recovered for at least the same time interval as on the full grid. For this we need $\mathcal{L}_x = 5$ and $\mathcal{L}_v = 7$. This means we have a grid of $M_x \times M_v = 112 \times 1024$ grid points. Compared to the full grid, we can considerably reduce the number of grid points along the **x** directions. However, there is no reduction along the **v** directions. Also using the multiplicative δf method, we cannot recover the damping rate with a maximum level less than $\mathcal{L}_v = 7$. Therefore, we can conclude that the use of a sparse grid only reduces the number of points in the **x** grid for this two-dimensional simulation. Note that it is clear from the structure of the initial value that a representation of f on a sparse grid is more difficult in velocity space where we have a function depending on v_1 multiplied by a function depending on v_2 while the perturbations along x_1 and x_2 are additive. Note that we observe an artificial recurrence on the sparse grid as well which is, however, damped and appears earlier than it would for a full grid with the same resolution of the finest level present in the sparse grid.

In Fig. 3, we also show the results obtained with a full sparse grid SGxv of maximum level $\mathcal{L} = 10$ (ℓ_1 bound) and periodic boundary conditions. In this case, the number of grid points is 66,304, i.e. only 6 % compared to the full and 58 % compared to the SGxSGv variant. On the other hand, we have to bear in mind that the complexity of the **v**-advection steps is of the order $O(p\mathcal{L}^2)$ per grid point, i.e. about a factor $\mathcal{L}^2 = 100$ higher than for the FG and SGxSGv variants (cf. Sect. 5). Hence, the computing time is expected to be highest for our SGxv variant. Since the solution of the SGxv method is the worst, the numerical experiments are in line with our theoretical considerations in Sect. 4.1 that a complete sparse grid will not be very successful in representing the distribution function governed by the Vlasov–Poisson equation.

In order to numerically verify our estimates on the complexity of our algorithms from Sect. 5, the CPU times for the three experiments are reported in Table 1. For each run, 2000 time steps have been simulated. The simulations were performed in serial on an Intel Ivy Bridge notebook processor at 3.0 GHz and the SeLaLib

Table 1 Comparison of CPU times for FG, SGxSGv, and SGxv

Method	No. of grid points	CPU time [s]	CPU time/grid point [s]
FG	1,048,576	448	$4.3 \cdot 10^{-4}$
SGxSGv	114,688	216	$2.3 \cdot 10^{-3}$
SGxv	66,304	8270	$1.2 \cdot 10^{-1}$

library was compiled with the GNU Fortran compiler 4.8 and optimization level -O3. Normalizing the CPU times by the number of grid points, we see that the time per grid point is increased by a factor 5 for SGxSGv and by a factor 292 for SGxv, respectively, compared to the FG solution. Considering the fact that the complexity is increased by a factor $\mathcal{L}^2$ for the SGxv grid, this means the complexity constant is increased by a factor 5 or 2.9 for SGxSGv and SGxv, respectively, in our prototype implementation. Given the fact that the sparse grid algorithms described in Sect. 5 require (de)hierarchization steps, the constants were expected to be larger than on the full grid. Note however that the codes are not completely optimized for the complexity constant so that these values should rather be used as qualitative estimates.

8.2 Comparison of Various Interpolators

In this section, we again consider the four dimensional Landau problem on a $M_x \times M_v = 112 \times 1024$ SGxSGv grid and compare various interpolators. The resulting potential energy plots are shown in Fig. 4. It can be clearly seen that linear interpolation is too dissipative. This is not an effect of the sparse grid but can likewise be observed on the full grid. Comparing the cubic sparse grid interpolator with the cubic sparse grid with cubic splines on the one-dimensional stripes, we can clearly see that applying mixed interpolation helps in increasing the accuracy of the interpolation.

8.3 Effects of the Multiplicative δf Method

So far, we have only looked at the Landau damping in four dimensions. If we use the same values for the maximum level $\mathcal{L}_x = 5$ and $\mathcal{L}_v = 7$ also in six-dimensions, the damping rate can only be recovered over a time interval of about 10 (see Fig. 5). However, if we apply the multiplicative δf method as described in Sect. 6, we can again recover the damping rate until time 25 as in the four dimensional case. The sparse grid only contains $M_x \times M_v = 272 \times 7808$ mesh points which only amounts to 0.2 % of the 32^6 points on the full grid with similar accuracy. This shows the potential of the multiplicative δf method. Note that the multiplicative δf method

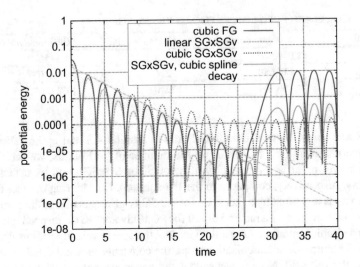

Fig. 4 Weak Landau damping in 4D. Electric energy for simulations on SGxSGv grid with 112 × 1024 points and various interpolators

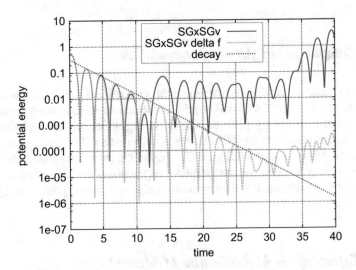

Fig. 5 Weak Landau damping in 6D. Electric energy for simulations with SGxSGv grid with $M_x \times M_v = 272 \times 7808$ points with and without multiplicative δf modeling

did not noticeably improve the results of our four dimensional experiments and is therefore not used in the four dimensional experiments reported here.

8.4 The Two Stream Instability: Effects of Instabilities

In this section, we want to consider a second test case: the two stream instability
with initial value

$$f(\mathbf{x}, \mathbf{v}) = \frac{1}{4\pi} \left(1 + 0.001 \sum_{j=1}^{2} \cos(0.2x_j) \right) \left(e^{-0.5(v_1 - 2.4)^2} + e^{-0.5(v_1 + 2.4)^2} \right) e^{-0.5v_2^2}.$$

(40)

The perturbation along x_1 will yield an instability as discussed in Sect. 2.1. During
a first linear phase the energy grows and a hole structure evolves in (x_1, v_1) space.
Around time 35, the energy stops to grow and nonlinear effects take over. During the
nonlinear phase, particles are trapped in the hole structure and smaller and smaller
filaments evolve. In the light of the discussion on the structure of the problem in
Sect. 4.1, this problem is more difficult to represent on a sparse grid than the Landau
damping problem. We can see from the results in Fig. 6 that a SGxSGv sparse grid
of 262,144 points in total ($\mathcal{L}_x = 6, \mathcal{L}_v = 7$) can nicely recover the linear phase.
In the nonlinear phase, the electric energy only keeps an oscillating structure on the
right level but the error in the solution is rather large.

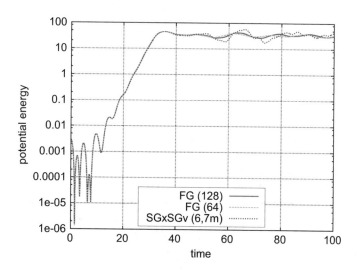

Fig. 6 Two stream instability in 4D. Electric energy for simulation on full grid and SGxSGv

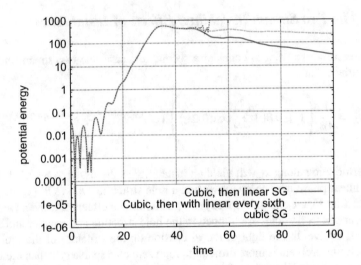

Fig. 7 Two stream instability in 6D. Electric energy for simulation on SGxSGv with cubic interpolation, cubic interpolation until time 45 and linear afterwards, and linear interpolation every 6th step from time 40 onwards

When solving the extension of the problem to six dimensions

$$f(\mathbf{x}, \mathbf{v}) = \frac{0.5}{(2\pi)^{1.5}} \left(1 + 0.001 \sum_{j=1}^{3} \cos(0.2x_j) \right) \left(e^{-0.5(v_1-2.4)^2} + e^{-0.5(v_1+2.4)^2} \right)$$

$$e^{-0.5(v_2^2+v_3^2)}.$$

(41)

we have observed a numerical instability at the same resolution (see Fig. 7). For this simulation, we have applied the multiplicative δf method along v_2 and v_3. We can see that the simulation can be stabilized when switching to a linear interpolator at time 45. At this point in time the resolution is poor and dissipation needs to be added to keep the simulation stable. However, we can see that the propagation method becomes too diffusive and energy is dissipated. In this case, we have added too much diffusion. If we instead use a linear interpolation each 6th step from time 40, the solution still remains stable but is much less diffusive. From our experiments, we have seen that more and more diffusion needs to be added as time evolves. In order to design a stable algorithm that adds as little diffusion as possible, one would need to analyze stability of the sparse grid method.

9 Conclusions

We have introduced a semi-Lagrangian Vlasov solver on a sparse grid. From both theoretical considerations and numerical performance, we have seen that a tensor product of a sparse grid in spatial and a sparse grid in velocity coordinates is better suited than a full six-dimensional sparse grid. We have introduced an efficient implementation that improves on the efficiency by exploiting the special structure of the problem. Moreover, we have devised a multiplicative δf method to defeat the problem of poor representation of Gaussians on a sparse grid.

From the results, we can conclude that good compression and reduced computational complexity can be obtained in six dimensions for problems close to equilibrium or when the filaments are aligned with the coordinate directions. For this reason, sparse grids might be interesting in a hybrid method where the bulk of the domain is resolved by a sparse grid while small structures are additionally resolved using localized full grids or particles.

References

1. SeLaLib, http://selalib.gforge.inria.fr/. Accessed 2015
2. N. Besse, M. Mehrenberger, Convergence of classes of high-order semi-Lagrangian schemes for the Vlasov–Poisson system. Math. Comput. **77**(261), 93–123 (2008)
3. O. Bokanowski, J. Garcke, M. Griebel, I. Klompmaker, An adaptive sparse grid semi-Lagrangian scheme for first order Hamilton–Jacobi Bellman equations. J. Sci. Comput. **55**(3), 575–605 (2013)
4. T. Boyd, J. Sanderson, *The Physics of Plasmas* (Cambridge University Press, Cambridge, 2003)
5. H.-J. Bungartz, Finite elements of higher order on sparse grids, Habilitationsschrift, 1998
6. H.-J. Bungartz, M. Griebel, Sparse grids. Acta Numer. **13**, 147–269 (2004)
7. C.Z. Cheng, G. Knorr, The integration of the Vlasov equation in configuration space. J. Comput. Phys. **22**(3), 330–351 (1976)
8. T. Gerstner, M. Griebel, Numerical integration using sparse grids. Numer. Algorithms **18**(3–4), 209–232 (1998)
9. R.T. Glassey, *The Cauchy Problem in Kinetic Theory* (SIAM, Philadelphia, 1996)
10. K. Hallatschek, Fouriertransformation auf dünnen Gittern mit hierarchischen Basen. Numer. Math. **63**, 83–97 (1992)
11. M. Hegland, Adaptive sparse grids. ANZIAM J. **44**, C335–C353 (2003)
12. K. Kormann, A semi-Lagrangian Vlasov solver in tensor train format. SIAM J. Sci. Comput. **37**(4), B613–B632 (2015)
13. C. Kowitz, D. Pflüger, F. Jenko, M. Hegland, The combination technique for the initial value problem in linear gyrokinetics, in *Sparse Grids and Applications*, ed. by J. Garcke, M. Griebel. Volume 88 of Lecture Notes in Computational Science and Engineering (Springer, Berlin/Heidelberg, 2013), pp. 205–222
14. P. Lions, B. Perthame, Propagation of moments and regularity for the 3-dimensional Vlasov–Poisson system. Invent. Math. **105**, 415–430 (1991)
15. G. Manfredi, Long-time behavior of nonlinear Landau damping. Phys. Rev. Lett. **79**, 2815–2818 (1997)
16. E. Novak, K. Ritter, High dimensional integration of smooth functions over cubes. Numer. Math. **75**(1), 79–97 (1996)

17. B. Peherstorfer, S. Zimmer, H.-J. Bungartz, Model reduction with the reduced basis method and sparse grids, in *Sparse Grids and Applications*, ed. by J. Garcke, M. Griebel. Volume 88 of Lecture Notes in Computational Science and Engineering (Springer, Berlin/Heidelberg, 2013), pp. 223–242
18. D. Pflüger, *Spatially Adaptive Sparse Grids for High-Dimensional Problems* (Verlag Dr. Hut, München, 2010)
19. M. Pueschel, T. Dannert, F. Jenko, On the role of numerical dissipation in gyrokinetic Vlasov simulations of plasma microturbulence. Comput. Phys. Commun. **181**(8), 1428–1437 (2010)
20. E. Sonnendrücker, in *Numerical Methods for the Vlasov Equations*. Lecture notes from Technische Universität München 2013, http://www-m16.ma.tum.de/foswiki/pub/M16/Allgemeines/NumMethVlasov/Num-Meth-Vlasov-Notes.pdf
21. E. Sonnendrücker, J. Roche, P. Bertrand, A. Ghizzo, The semi-Lagrangian method for the numerical resolution of the Vlasov equation. J. Comput. Phys. **149**(2), 201–220 (1999)
22. J. Valentin, D. Pflüger, Hierarchical gradient-based optimization with B-splines on sparse grids, in *Sparse Grid and Applications – Stuttgart 2014*, ed. by P. Garcke (Springer, Cham, 2015, to appear)

An Adaptive Sparse Grid Algorithm for Elliptic PDEs with Lognormal Diffusion Coefficient

Fabio Nobile, Lorenzo Tamellini, Francesco Tesei, and Raúl Tempone

Abstract In this work we build on the classical adaptive sparse grid algorithm (*T. Gerstner and M. Griebel, Dimension-adaptive tensor-product quadrature*), obtaining an enhanced version capable of using non-nested collocation points, and supporting quadrature and interpolation on unbounded sets. We also consider several profit indicators that are suitable to drive the adaptation process. We then use such algorithm to solve an important test case in Uncertainty Quantification problem, namely the Darcy equation with lognormal permeability random field, and compare the results with those obtained with the quasi-optimal sparse grids based on profit estimates, which we have proposed in our previous works (cf. e.g. *Convergence of quasi-optimal sparse grids approximation of Hilbert-valued functions: application to random elliptic PDEs*). To treat the case of rough permeability fields, in which a sparse grid approach may not be suitable, we propose to use the adaptive sparse grid quadrature as a control variate in a Monte Carlo simulation. Numerical results show that the adaptive sparse grids have performances similar to those of the quasi-optimal sparse grids and are very effective in the case of smooth permeability fields. Moreover, their use as control variate in a Monte Carlo simulation allows to tackle efficiently also problems with rough coefficients, significantly improving the performances of a standard Monte Carlo scheme.

1 Introduction

In this work we consider the problem of building a sparse grid approximation of a multivariate function $f(\mathbf{y}) : \Gamma \to V$ with global polynomials, where Γ is an N-dimensional hypercube $\Gamma = \Gamma_1 \times \Gamma_2 \times \ldots \times \Gamma_N$ (with $\Gamma_n \subseteq \mathbb{R}, n = 1, \ldots, N$), and

F. Nobile • L. Tamellini (✉) • F. Tesei
SB-MATHICSE-CSQI-EPFL, Station 8, CH-1015, Lausanne, Switzerland
e-mail: fabio.nobile@epfl.ch; lorenzo.tamellini@epfl.ch; francesco.tesei@epfl.ch

R. Tempone
SRI Center for Uncertainty Quantification in Computational Science and Engineering, KAUST, Thuwal, Saudi Arabia
e-mail: raul.tempone@kaust.edu.sa

© Springer International Publishing Switzerland 2016   191
J. Garcke, D. Pflüger (eds.), *Sparse Grids and Applications – Stuttgart 2014*,
Lecture Notes in Computational Science and Engineering 109,
DOI 10.1007/978-3-319-28262-6_8

V is a Hilbert space [1, 3, 6, 24, 29]. We also assume that each Γ_n is endowed with a probability measure $\varrho_n(y_n)dy_n$, so that $\varrho(\mathbf{y})d\mathbf{y} = \prod_{n=1}^{N} \varrho_n(y_n)dy_n$ is a probability measure on Γ. This setting is common in many optimization and Uncertainty Quantification problems, where sparse grids have been increasingly used to perform tasks such as quadrature, interpolation and surrogate modeling, since they allow for trivial parallelization and maximal reuse of legacy codes, with little or no expertise required by the end-user. While very effective for moderate dimensions (say $N \approx 10$), the basic sparse grid algorithms show a significant performance degradation when N increases (the so-called "curse of dimensionality" effect). The search for advanced sparse grid implementations, ideally immune to this effect, has thus become a very relevant research topic.

A general consensus has been reached on the fact that the "curse of dimensionality" should be tackled by exploiting the anisotropy of f, i.e. by assessing the amount of variability of f due to each parameter y_i and enriching the sparse grid approximation accordingly. Two broad classes of algorithms can be individuated to this end: those that discover the anisotropy structure "a-posteriori", i.e. at run-time, based on suitable indicators, and those based on "a-priori" theoretical estimates, possibly aided by some preliminary computations (we refer to the latter as "a-priori/a-posteriori" methods). A-priori algorithms based on a sharp theoretical analysis save the cost of the exploration of the anisotropy structure, while a-posteriori approaches are to a certain extent more flexible and robust. Focusing on the field of Uncertainty Quantification, examples of a-priori/a-posteriori algorithms can be found e.g. in [4, 24, 26], while the classical a-posteriori algorithm originally proposed in [17] has been further considered e.g. in [8, 28, 32].

A-posteriori sparse grid algorithms have always been used in the literature in combination with nested univariate quadrature rules, since this choice eases the computation of the anisotropy indicators, cf. [17]. In Uncertainty Quantification it is quite natural to choose univariate quadrature points according to the probability measures $\rho_n(y_n)dy_n$, see e.g. [1]: hence, one is left with the problem of computing good univariate nested quadrature rules for the probability measures at hand. While the case of the uniform measure has been thoroughly investigated and several choices of appropriate nested quadratures are available, like Leja, Gauss–Patterson or Clenshaw–Curtis points (see e.g. [24, 25] and references therein), non-uniform measures have been less explored. In the very relevant case of normal probability distribution a common choice is represented by Genz-Keister quadrature rules [16]; however, the cardinality of such quadrature rules increases very quickly when moving from one quadrature level to the following one, hence leading to an heavy computational burden when tensorized in a high-dimensional setting. The very recent work [23] develops instead generalized Leja quadrature rules for arbitrary measures on unbounded intervals: the main advantage of such quadrature rules over the Genz-Keister points is that two consecutive quadrature rules differ by one point only, rendering the Leja points more suitable for sparse grids construction. In this work we will approach the problem from a different perspective and propose a slight generalization of the classical a-posteriori adaptive algorithm that allows to use non-nested quadrature rules: this immediately permits to build adaptive sparse grids using gaussian-type quadrature nodes, which are readily available for practically every common probability measure. We will also consider different profit indicators and compare the performances of the corresponding adaptive schemes.

We will then test our version of the adaptive algorithm on a classical Uncertainty Quantification test problem, i.e. an elliptic PDE describing a Darcy flow in a porous medium, whose diffusion coefficient is modeled as a lognormal random field [5, 7, 9, 14, 18] and discretized by a Karhunen–Loève expansion. The covariance structure of the random field will be described by a tensor Matérn covariance model [11], which is a family of covariance structures parametrized by a scalar value ν that governs the smoothness of each realization of the random field and includes the Gaussian and the Exponential covariance structure as particular cases ($\nu = \infty$ and $\nu = 0.5$, respectively); more specifically, we will first consider the case $\nu = 2.5$, that results in fairly smooth random field realizations, and then move to the rough case $\nu = 0.5$, which leads to continuous but not differentiable field realizations. In both cases we will compare the performance of the adaptive sparse grid procedure with the "a-priori/a-posteriori" quasi-optimal sparse grid proposed in [5] for the same problem.

In the case $\nu = 2.5$, the lognormal random field can be very accurately described by including a moderate number of random variables in the Karhunen–Loève expansion, and a sparse grid approach to solve the Darcy problem is quite effective. Note however that we will not fix a-priori the number of random variables to be considered, but rather propose a version of the adaptive algorithm that progressively adds dimensions to the search space, thus formally working with $N = \infty$ random variables. Yet, even such dimension adaptive sparse grids (as well as the quasi-optimal ones) may suffer from a deterioration of the performance when the lognormal random field gets rougher. In particular, in the case $\nu = 0.5$, numerical experience seems to indicate that their performance might be asymptotically not better than a standard Monte Carlo method. Thus, in this case we will actually compare the performances of the adaptive and quasi-optimal sparse grids in the framework proposed in [27], in which they will be applied to a smoothed version of the problem (where a sparse grid approach can be effective), and the results used as control variates in a Monte Carlo approach.

The rest of the paper is organized as follows. We start by introducing the general construction of sparse grids in Sect. 2. Then, we discuss in detail the construction of the quasi-optimal and adaptive sparse grids in Sect. 3: in particular, we will setup a common framework for the two methods in the context of the resolution of discrete optimization problems, and specify the details of the two algorithms in Sects. 3.1 and 3.2 respectively. The details of the Darcy problem are presented in Sect. 4, and in particular we will describe the dimension adaptive algorithm in Sect. 4.1 and the Monte Carlo Control Variate approach in Sect. 4.2. The numerical results are shown in Sect. 5, while Sect. 6 presents the conclusions of this work.

In what follows, $\mathbb{N}$ will denote the set of integer numbers including 0, and $\mathbb{N}_+$ that of integer numbers excluding 0. Given two vectors $\mathbf{v}, \mathbf{w} \in \mathbb{N}^N$, $|\mathbf{v}|_0, |\mathbf{v}|_1, |\mathbf{v}|_2$ denote respectively the number of non-zero entries of $\mathbf{v}$, the sum of their absolute values and the euclidean norm of $\mathbf{v}$, and we write $\mathbf{v} \leq \mathbf{w}$ if and only if $v_j \leq w_j$

for every $1 \leq j \leq N$. Moreover, $\mathbf{0}$ will denote the vector $(0, 0, \ldots, 0) \in \mathbb{N}^N$, $\mathbf{1}$ the vector $(1, 1, \ldots, 1) \in \mathbb{N}^N$, and $\mathbf{e}_j$ the j-th canonical vector in $\mathbb{R}^N$, i.e. a vector whose components are all zero but the j-th, whose value is one. To close our introduction, we recall the definition of some functional spaces that will be useful in the following. In particular, we will need the weighted L^p spaces

$$L_\varrho^p(\Gamma; V) = \left\{ f : \Gamma \to V \text{ s.t. } \int_\Gamma \|f(\mathbf{y})\|_V^p \varrho(\mathbf{y}) d\mathbf{y} < \infty \right\}, \quad \forall p \in (0, \infty),$$

and the space of continuous functions with weighted maximum norm

$$C_\pi^0(\Gamma; V) = \left\{ f : \Gamma \to V \text{ s.t. } f \text{ is continuous and } \max_\Gamma \|f(\mathbf{y})\|_V \pi(\mathbf{y}) < \infty \right\},$$

where $\pi = \prod_{n=1}^N \pi_n(y_n)$, $\pi_n : \Gamma_n \to \mathbb{R}$, is a positive and smooth function. The reasons for introducing two different weight functions ϱ and π will be clearer later on. Observe in particular that since V and $L_\varrho^2(\Gamma)$ are Hilbert spaces, $L_\varrho^2(\Gamma; V)$ is isomorphic to the tensor space $V \otimes L_\varrho^2(\Gamma)$, and is itself an Hilbert space.

2   Sparse Grid Approximation of Multivariate Functions

As already mentioned in the introduction, we consider the problem of constructing a sparse grid approximation with global polynomials of the V-valued multivariate function f, defined over the hypercube Γ with associated probability measure $\varrho(\mathbf{y})d\mathbf{y} = \prod_{n=1}^N \varrho_n(y_n)dy_n$. More precisely, we will consider functions f that are continuous with respect to $\mathbf{y}$ and with finite variance, i.e. belonging to $L_\varrho^2(\Gamma; V) \cap C_\pi^0(\Gamma; V)$ for some suitable weight π (which can be often taken equal to ϱ, but not always, as indeed in certain instances of the stochastic Darcy problem we will consider in the numerical part of this paper, see e.g. [1, 19]). Observe that approximating f with global polynomials is a sound approach if f is not just continuous, but actually a smooth function of $\mathbf{y}$, see [1, 24]. Sparse grids based on piecewise polynomial approximations, which are suitable for non-smooth or even discontinuous functions, have been developed e.g. in [15, 20].

To begin with the construction of the sparse grid, we consider a sequence $\{\mathcal{U}_n^{m(i_n)}\}_{i_n \in \mathbb{N}}$ of univariate Lagrangian interpolant operators along each dimension Γ_n of the hypercube Γ,

$$\mathcal{U}_n^{m(i_n)} : C_{\pi_n}^0(\Gamma_n) \to \mathbb{P}_{m(i_n)-1}(\Gamma_n),$$

where $m(i_n)$ denotes the number of collocation points used by the i_n-th interpolant, and $\mathbb{P}_q(\Gamma_n)$ is the set of polynomials in y_n of degree at most q. The function $m : \mathbb{N} \to \mathbb{N}$ is called "level-to-nodes function" and is a strictly increasing function, with

$m(0) = 0$ and $m(1) = 1$; consistently, we set $\mathcal{U}_n^0[f] = 0$. Next, for any $\mathbf{i} \in \mathbb{N}_+^N$ we define the tensor interpolant operator

$$\mathcal{T}_{\mathbf{i}}^m[f](\mathbf{y}) = \bigotimes_{n=1}^{N} \mathcal{U}_n^{m(i_n)}[f](\mathbf{y}), \tag{1}$$

and the *hierarchical surplus* operator

$$\Delta^{m(\mathbf{i})} = \bigotimes_{n=1}^{N} \left(\mathcal{U}_n^{m(i_n)} - \mathcal{U}_n^{m(i_n-1)} \right), \tag{2}$$

where with a slight abuse of notation we have denoted with $m(\mathbf{i})$ the vector $[m(i_1)m(i_2)\ldots m(i_N)]$. A sparse grid approximation is built as a sum of *hierarchical surplus* operators; more specifically, we consider a sequence of index sets $\mathbf{I}(w) \subset \mathbb{N}_+^N$ such that $\mathbf{I}(w) \subset \mathbf{I}(w+1)$, $\mathbf{I}(0) = \{\mathbf{1}\}$ and $\cup_{w \in \mathbb{N}} \mathbf{I}(w) = \mathbb{N}_+^N$, and we define the sparse grid approximation of $f(\mathbf{y})$ at level $w \in \mathbb{N}$ as

$$S_{\mathbf{I}(w)}^m : L_\varrho^2(\Gamma; V) \cap C_\pi^0(\Gamma; V) \to L_\varrho^2(\Gamma; V), \quad S_{\mathbf{I}(w)}^m[f](\mathbf{y}) = \sum_{\mathbf{i} \in \mathbf{I}(w)} \Delta^{m(\mathbf{i})}[f](\mathbf{y}). \tag{3}$$

To ensure good approximation properties to the sparse approximation, the sum (3) must be telescopic, cf. [17]: to this end we require that

$$\forall \mathbf{i} \in \mathbf{I}, \quad \mathbf{i} - \mathbf{e}_j \in \mathbf{I} \text{ for } 1 \le j \le N \text{ such that } i_j > 1.$$

A set $\mathbf{I}$ satisfying the above property is said to be a *lower set* or a *downward closed set*, see e.g. [10]. The choice of the set $\mathbf{I}(w)$ plays a crucial role in devising effective sparse grid schemes: the next section will be entirely devoted to the discussion of two possible strategies to this end, namely the a-posteriori adaptive and the "a-priori/a-posteriori" quasi-optimal procedures that have been mentioned in the introduction.

Further insight into the structure of sparse grid operators can be obtained by rewriting (3) as a linear combination of tensor interpolant operators (1), see e.g. [31]. Assuming that $\mathbf{I}(w)$ is downward closed, we get indeed

$$S_{\mathbf{I}(w)}^m[f](\mathbf{y}) = \sum_{\mathbf{i} \in \mathbf{I}(w)} c_{\mathbf{i}} \mathcal{T}_{\mathbf{i}}^m[f](\mathbf{y}), \qquad c_{\mathbf{i}} = \sum_{\substack{\mathbf{j} \in \{0,1\}^N \\ (\mathbf{i}+\mathbf{j}) \in \mathbf{I}(w)}} (-1)^{|\mathbf{j}|}. \tag{4}$$

Observe that many of the coefficients $c_{\mathbf{i}}$ in (4) may be zero: in particular $c_{\mathbf{i}}$ is zero whenever $\mathbf{i} + \mathbf{j} \in \mathbf{I}(w) \ \forall \mathbf{j} \in \{0,1\}^N$. The set of all collocation points needed by (4) is actually called a *sparse grid*, and we denote its cardinality by $W_{\mathbf{I}(w),m}$. It is useful to introduce the operator $\text{pts}(S)$ that returns the set of points associated to a tensor/sparse grid operator, and the operator card(S) that returns the cardinality of pts(S):

$$\text{card}(\mathcal{T}_{\mathbf{i}}^m) = \prod_{n=1}^{N} m(i_n), \qquad \text{card}(\mathcal{S}_{\mathbf{I}(w)}^m) = W_{\mathbf{I}(w),m}. \tag{5}$$

Finally, consider a sequence of univariate quadrature operators built over the same set of points of $\{\mathcal{U}_n^{m(i_n)}\}_{i_n \in \mathbb{N}}$; it is then relatively straightforward to derive a sparse grid quadrature scheme $\mathcal{Q}_{\mathbf{I}(w)}^m[\cdot]$ starting from (4):

$$\int_{\Gamma} f(\mathbf{y})\varrho(\mathbf{y})d\mathbf{y} \approx \int_{\Gamma} \mathcal{S}_{\mathbf{I}(w)}^m[f]\varrho(\mathbf{y})d\mathbf{y} = \sum_{j=1}^{W_{\mathbf{I}(w)}^m} f(\mathbf{y}_j)\beta_j = \mathcal{Q}_{\mathbf{I}(w)}^m[f], \tag{6}$$

for suitable quadrature weights $\beta_j \in \mathbb{R}$.

Coming to the choice of the univariate collocation points used to build $\mathcal{U}_n^{m(i_n)}$, as mentioned in the introduction they should be chosen according to the probability measure $\varrho_n(y_n)dy_n$ on Γ_n. Although the use of nested points seems to be particularly indicated for the hierarchical construction (3), as the $\mathbf{\Delta}^{m(i)}$ operator would entail evaluations only on the new points added going from the tensor grid $\mathcal{T}_{\mathbf{i}-1}^m$ to $\mathcal{T}_{\mathbf{i}}^m$, at this point any choice of univariate collocation points is allowed (see Table 1), and in particular Gauss interpolation/quadrature points, associated to the underlying probability density functions $\varrho_n(y_n)$, have been widely used, cf. e.g. [2, 12, 13, 26]. Note however that non-nested interpolatory rules have not been used in the adaptive context, for reasons that will be clearer in a moment; the aim of this work is to extend the adaptive algorithm to non-nested quadrature rules.

Table 1 Common choices of univariate collocation points for sparse grids

Collocation points			
	Measure	Nested	m(i)
Gauss–Legendre	Uniform	No	i
Clenshaw–Curtis	Uniform	Yes	$2^{i-1} + 1$
Gauss–Patterson	Uniform	Yes	$2^i - 1$
Leja	Uniform	Yes	$m(i) = i$ or $m(i) = 2i - 1$
Gauss–Hermite	Gaussian	No	i
Genz–Keister	Gaussian	Yes	Tabulated: $m(i) = 1, 3, 9, 19, 35$
Generalized Leja	Gaussian	Yes	i

3 On the Choice of I(w)

In this section we detail two possible strategies to design the sequence of sets $\mathbf{I}(w)$. To simplify the notation, let us assume that $V = \mathbb{R}$, i.e. f is a real-valued N-variate function over Γ, and that we are measuring the sparse grid approximation error by some non-negative sublinear functional[1] $\mathcal{E}[\cdot]$, e.g. a semi-norm on $L_\varrho^p(\Gamma)$ (we will give three such examples in the following). Furthermore, assume that we can formally write $f = S^m_{\mathbb{N}^N_+}[f] = \sum_{i\in\mathbb{N}^N_+} \mathbf{\Delta}^{m(i)}[f]$. Then, we have

$$\mathcal{E}\left[f - S^m_{\mathbf{I}(w)}[f]\right] = \mathcal{E}\left[\sum_{i\notin\mathbf{I}(w)} \mathbf{\Delta}^{m(i)}[f]\right] \leq \sum_{i\notin\mathbf{I}(w)} \mathcal{E}\left[\mathbf{\Delta}^{m(i)}[f]\right]. \tag{7}$$

Since the exact value of $\mathcal{E}\left[\mathbf{\Delta}^{m(i)}[f]\right]$ may not be at disposal, we further define the *error contribution operator* $\Delta E(\mathbf{i})$ as any computable (and hopefully tight) approximation of $\mathcal{E}\left[\mathbf{\Delta}^{m(i)}[f]\right]$, namely $\Delta E(\mathbf{i}) \approx \mathcal{E}\left[\mathbf{\Delta}^{m(i)}[f]\right]$. Moreover, we also introduce the *work contribution* $\Delta W(\mathbf{i})$, i.e. the number of evaluations of f implied by the addition of the hierarchical surplus operator $\mathbf{\Delta}^{m(i)}[f]$ to the sparse grid approximation. Observe that this is actually a quite delicate issue when using non-nested points as discussed later on.

Upon having assigned an error and a work contribution to each hierarchical surplus operator, the selection of the sequence of sets $\mathbf{I}(w)$ can be rewritten as a "binary knapsack problem" [6, 22],

$$\max \sum_{i\in\mathbb{N}^N_+} \Delta E(\mathbf{i})x_{\mathbf{i}} \quad \text{s.t.} \quad \sum_{i\in\mathbb{N}^N_+} \Delta W(\mathbf{i})x_{\mathbf{i}} \leq W_{max}(w) \text{ and } x_{\mathbf{i}} \in \{0,1\},$$

where $W_{\max}(w)$ is the maximum computational work allowed for the approximation level w. Note that we are not explicitly enforcing that the resulting sets $\mathbf{I}(w)$ be downward closed (which will have to be verified a-posteriori).

While the binary knapsack problem is known to be computationally intractable (NP-hard) its *linear programming relaxation*, in which fractional values of $x_{\mathbf{i}}$ are allowed, can be solved analytically by the so-called Dantzig algorithm [22]:

1. Assign a "profit" to each multi-index $\mathbf{i}$,

$$P(\mathbf{i}) = \frac{\Delta E(\mathbf{i})}{\Delta W(\mathbf{i})}; \tag{8}$$

---

[1]A sublinear functional over a vector space X is a function $\Theta : X \to \mathbb{R}$ such that

- $\Theta(\alpha x) = \alpha\Theta(x), \forall \alpha > 0$ and $x \in X$;
- $\Theta(x + y) \leq \Theta(x) + \Theta(y), \forall x, y \in X$.

2. sort multi-indices by decreasing profit;
3. set $x_\mathbf{i} = 1$, i.e. add $\mathbf{i}$ to $\mathbf{I}(w)$, until the constraint on the maximum work is fulfilled. In particular, whenever the multi-index $\mathbf{1} + \mathbf{e}_n$ enters the set $\mathbf{I}(w)$ we say that the random variable y_n is *activated*.

Note that only the last multi-index included in the selection is possibly taken not entirely (i.e. with $x_\mathbf{i} < 1$), whereas all the previous ones are taken entirely (i.e. with $x_\mathbf{i} = 1$). However, if this is the case, we assume that we could slightly adjust the computational budget, so that all $x_\mathbf{i}$ have integer values; observe that such integer solution is also the solution of the original binary knapsack problem with modified work constraint.

Both the quasi-optimal and the a-posteriori adaptive sparse grids strategies fit in this general framework. What changes between the two schemes are just the choice of the error indicator $\mathcal{E}[\cdot]$ and the way $\Delta W(\mathbf{i})$ and $\Delta E(\mathbf{i})$ are computed.

3.1 Quasi-Optimal Sparse Grids

In this section we briefly summarize the quasi-optimal sparse grids construction, see [24] for a thorough discussion. In this case, the error indicator $\mathcal{E}[\cdot]$ is the L_ϱ^2-norm, so that (7) becomes

$$\left\| f - S_{\mathbf{I}(w)}^m [f] \right\|_{L_\varrho^2} \leq \sum_{\mathbf{i} \notin \mathbf{I}(w)} \left\| \Delta^{m(\mathbf{i})} [f] \right\|_{L_\varrho^2},$$

and we need to provide a computable approximation $\left\| \Delta^{m(\mathbf{i})} [f] \right\|_{L_\varrho^2} \approx \Delta E(\mathbf{i})$. Following [4, 5, 24], this can be obtained by further introducing the spectral expansion of f over a N-variate $\tilde{\rho}$-orthonormal polynomial basis $\varphi_\mathbf{q}(\mathbf{y})$,[2] with $\tilde{\rho}$ not necessarily equal to ϱ; for example, in the case where y_n are uniform random variables, $\rho_n(y_n) = 1/|\Gamma_n|$, one is allowed to expand f on tensorized Chebyshev polynomials, which are orthonormal with respect to $\tilde{\rho} = \prod_{n=1}^N \widetilde{\rho}_n$, with $\widetilde{\rho}_n(y_n) = 1/\sqrt{1 - y_n^2}$. Next, let us denote by $f_\mathbf{q}$ the $\mathbf{q}$-th coefficient of the $\tilde{\varrho}$-expansion of f and by $\mathbb{M}_n^{m(i_n)}$ the "$C_\pi^0 \to L_\varrho^2$ Lebesgue constant" of the univariate interpolant operators $\mathcal{U}_n^{m(i_n)}$ for a suitable weight π, i.e.

$$\mathbb{M}_n^{m(i_n)} = \sup_{\|f\|_{C_\pi^0(\Gamma_n)} = 1} \left\| \mathcal{U}_n^{m(i_n)} [f] \right\|_{L_\varrho^2(\Gamma_n)}.$$

Then, assuming that the coefficients $f_\mathbf{q}$ are at least exponentially decreasing in each y_n, $|f_\mathbf{q}| \leq C \prod_n \exp(-g_n q_n)$, and that $\|\varphi_\mathbf{q}\|_{C_\pi^0} \leq C^{|\mathbf{q}|_0}$, following [24] we have that

---

[2]Here the n-th component of $\mathbf{q}$ denotes the polynomial degree with respect to y_n.

for a suitable constant C there holds

$$\left\| \mathbf{\Delta}^{m(\mathbf{i})}[f] \right\|_{L^2_\varrho} \leq \Delta E(\mathbf{i}) = C(N) \left| f_{m(\mathbf{i}-\mathbf{1})} \right| \prod_{n=1}^{N} \mathbb{M}_n^{m(i_n)}, \tag{9}$$

where $m(\mathbf{i}-\mathbf{1})$ indicates the vector $[m(i_1-1), m(i_2-1), \dots, m(i_N-1)]$. Observe that in practical cases, the constant $\mathbb{M}_n^{m(i_n)}$ can be estimated numerically, and computable ansatzes for $f_{m(\mathbf{i}-\mathbf{1})}$ can be derived, so that it is possible to obtain numerical estimates of the quantities $\Delta E(\mathbf{i})$. Such computable ansatzes depend on the exponential coefficients $g_1, \dots, g_N$, that can be conveniently precomputed with a numerical procedure that requires $O(N)$ evaluations of f. We will return on this matter in the next sections, proposing an ansatz for the Darcy problem, as well as giving details on the numerical procedure needed to estimate $g_1, \dots, g_N$.

Concerning the work contributions $\Delta W(\mathbf{i})$, the definitions are different depending on whether the family of nodes considered is nested or non-nested (see [24] for details). In the former case, we can set

$$\Delta W(\mathbf{i}) = \prod_{n=1}^{N} \left(m(i_n) - m(i_n - 1) \right), \tag{10}$$

and there holds

$$W_{\mathbf{I}(w),m} = \sum_{\mathbf{i} \in \mathbf{I}(w)} \Delta W(\mathbf{i}),$$

i.e. the cardinality of the sparse grid is equal to the sum of the work contributions. On the contrary, when considering non-nested points the number of new evaluations of f needed by the addition of $\mathbf{\Delta}^{m(\mathbf{i})}$ will depend in general on the set $\mathbf{I}$ to which $\mathbf{i}$ is added to, i.e. if $\mathbf{I}, \mathbf{I}'$ are two index sets such that both $\mathbf{I} \cup \{\mathbf{j}\}$ and $\mathbf{I}' \cup \{\mathbf{j}\}$ are downward closed, it can happen that

$$\mathrm{card}(S^m_{\mathbf{I} \cup \{\mathbf{j}\}}) \neq \mathrm{card}(S^m_{\mathbf{I}' \cup \{\mathbf{j}\}}), \tag{11}$$

and nodes that are present in the sparse grid built over $\mathbf{I}$ are not necessarily present in the one built over $\mathbf{I} \cup \{\mathbf{j}\}$, i.e.

$$\mathrm{pts}(S^m_{\mathbf{I}}) \not\subset \mathrm{pts}(S^m_{\mathbf{I} \cup \{\mathbf{j}\}}). \tag{12}$$

Therefore, we have to use the pessimistic estimate

$$\Delta W(\mathbf{i}) = \prod_{n=1}^{N} m(i_n) = \mathrm{card}(\mathcal{T}^m_{\mathbf{i}}[f]), \tag{13}$$

i.e. the cardinality of the entire tensor grid associated to $\mathbf{i}$, which ensures

$$W_{\mathbf{I}(w),m} \leq \sum_{\mathbf{i} \in \mathbf{I}(w)} \Delta W(\mathbf{i}).$$

Once the numerical values of $\Delta E(\mathbf{i})$ and $\Delta W(\mathbf{i})$ are available, the profits (8) and the sequence of optimal sets $\mathbf{I}(w)$ can be computed right-away, and the sparse grid construction can proceed. Thus, this algorithm is said to be "a-priori"/"a-posteriori" since it relies on a-priori estimates whose constants $g_1 \ldots, g_N$ need however to be tuned numerically.

3.2 An Extended Adaptive Sparse Grid Algorithm

We now describe the adaptive sparse grid construction algorithm [8, 17, 28], and its extension to non-nested points and unbounded intervals. To this end, we introduce the concepts of *margin* and *reduced margin* of a multi-index set $\mathbf{I}$, and the concept of *neighbors* of a multi-index. The margin of $\mathbf{I}$, which we denote by $\mathbf{M_I}$, contains all the multi-indices $\mathbf{i}$ that can be reached within "one-step forward" from $\mathbf{I}$, i.e.

$$\mathbf{M_I} = \{\mathbf{i} \in \mathbb{N}_+^N \setminus \mathbf{I} : \exists \mathbf{j} \in \mathbf{I} : |\mathbf{i} - \mathbf{j}|_1 = 1\}.$$

The reduced margin of $\mathbf{I}$, denoted by $\mathbf{R_I}$, is the subset of the margin of $\mathbf{I}$ containing only those indices $\mathbf{i}$ such that "one-step backward" in any direction takes into $\mathbf{I}$, i.e.

$$\mathbf{R_I} = \{\mathbf{i} \in \mathbb{N}_+^N \setminus \mathbf{I} : \mathbf{i} - \mathbf{e}_j \in \mathbf{I}, \ \forall j = 1, \ldots, N : i_j > 1\} \subset \mathbf{M_I}.$$

This means that the reduced margin of $\mathbf{I}$ contains all indices $\mathbf{i}$ such that $\mathbf{I} \cup \{\mathbf{i}\}$ is downward closed, provided that $\mathbf{I}$ itself is downward closed. Furthermore, given an index $\mathbf{i}$ on the boundary of $\mathbf{I}$, we call *neighbors* of $\mathbf{i}$ with respect to $\mathbf{I}$, neigh($\mathbf{i}, \mathbf{I}$), the indices $\mathbf{j}$ not included in $\mathbf{I}$ that can be reached with "one step forward" from $\mathbf{i}$, so that $\mathbf{M_I} = \bigcup_{\mathbf{i} \in \mathbf{I}} \text{neigh}(\mathbf{i}, \mathbf{I})$.

Instead of computing the profits and the sets $\mathbf{I}(w)$ beforehand as in the quasi-optimal algorithm, the idea of the adaptive algorithm is to compute the profits and the sets $\mathbf{I}(w)$ at run-time, proceeding iteratively in a greedy way. More specifically, given a multi-index set $\mathbf{I}$ and its reduced margin $\mathbf{R_I}$, the adaptive algorithm operates as follows:

1. the profits of $\mathbf{i} \in \mathbf{R_I}$ are computed;
2. the index $\mathbf{i}$ with the highest profit is moved from $\mathbf{R_I}$ to $\mathbf{I}$;
3. the reduced margin is updated and the algorithm moves to the next iteration, until some stopping criterion is met (usually, a check on the number of evaluations of f or on the values of the profits or error contributions of the multi-indices in $\mathbf{R_I}$).

Note that the profits of the indices $\mathbf{i} \in \mathbf{R_I}$ are computed by actually adding the hierarchical surpluses to the sparse grid operator (as will be clearer in a moment), hence the definition of "a-posteriori"; therefore, the outcome of the algorithm at each iteration is the sparse grid approximation built on $\mathbf{I} \cup \mathbf{R_I}$ and not on $\mathbf{I}$ only.

In this work, we consider two different error indicators $\mathcal{E}[\cdot]$, namely the absolute value of the expectation of $f - S^m_{\mathbf{I}(w)}[f]$, which is a semi-norm on $L^1_\varrho(\Gamma)$, and the weighted $C^0_\pi(\Gamma)$ norm, so that (7) becomes

$$\left| \mathbb{E}[f] - \mathbb{E}\left[S^m_{\mathbf{I}}[f]\right] \right| \leq \sum_{\mathbf{i} \notin \mathbf{I}} \left| \mathbb{E}\left[\Delta^{m(\mathbf{i})}[f]\right] \right|,$$

$$\left\| f - S^m_{\mathbf{I}}[f] \right\|_{C^0_\pi(\Gamma)} \leq \sum_{\mathbf{i} \notin \mathbf{I}} \left\| \Delta^{m(\mathbf{i})}[f] \right\|_{C^0_\pi(\Gamma)}.$$

To derive the error indicator $\Delta E(\mathbf{i})$ for the quantity $\mathcal{E}\left[\Delta^{m(\mathbf{i})}[f]\right]$, let us consider an arbitrary set $\mathbf{I}$, and let $\mathbf{J} = \mathbf{I} \cup \{\mathbf{i}\}$, with both $\mathbf{I}, \mathbf{J}$ downward closed index sets and observe that $\Delta^{m(\mathbf{i})}[f] = S^m_{\mathbf{J}}[f] - S^m_{\mathbf{I}}[f]$. For the $L^1_\varrho(\Gamma)$ seminorm, we immediately have

$$\mathbb{E}\left[\Delta^{m(\mathbf{i})}[f]\right] = \mathbb{E}\left[S^m_{\mathbf{J}}[f] - S^m_{\mathbf{I}}[f] \right] = Q^m_{\mathbf{J}}[f] - Q^m_{\mathbf{I}}[f],$$

and therefore we define $\Delta E(\mathbf{i})$ as

$$\Delta E(\mathbf{i}) = \left| \mathbb{E}\left[\Delta^{m(\mathbf{i})}[f]\right] \right| = \left| Q^m_{\mathbf{J}}[f] - Q^m_{\mathbf{I}}[f] \right|. \tag{14}$$

In the case of the $C^0_\pi(\Gamma)$ norm, the computation is different for nested and non-nested points. For nested points the sparse grid operator is interpolatory (see e.g. [3, Prop. 6]), hence $\Delta^{m(\mathbf{i})}[f](\mathbf{y}) = S^m_{\mathbf{J}}[f](\mathbf{y}) - S^m_{\mathbf{I}}[f](\mathbf{y}) = 0$ for any $\mathbf{y} \in \text{pts}(S^m_{\mathbf{I}})$ and $S^m_{\mathbf{J}}[f](\mathbf{y}) = f(\mathbf{y}) \neq S^m_{\mathbf{I}}[f](\mathbf{y})$ for any $\mathbf{y} \in \mathcal{N}\text{ew} = \text{pts}(S^m_{\mathbf{J}}) \setminus \text{pts}(S^m_{\mathbf{I}})$, cf. Eq. (5). Therefore we can estimate the $C^0_\pi(\Gamma)$ norm of $\Delta^{m(\mathbf{i})}[f]$ by looking only at the values on $\mathbf{y} \in \mathcal{N}\text{ew}$. Thus we approximate the norm $\left\| \Delta^{m(\mathbf{i})}[f] \right\|_{C^0_\pi(\Gamma)}$ by

$$\left\| \Delta^{m(\mathbf{i})}[f] \right\|_{C^0_\pi(\Gamma)} \approx \max_{\mathbf{y} \in \mathcal{N}\text{ew}} \left| \Delta^{m(\mathbf{i})}[f](\mathbf{y})\pi(\mathbf{y}) \right|$$

and define

$$\Delta E(\mathbf{i}) = \max_{\mathbf{y} \in \mathcal{N}\text{ew}} \left| \Delta^{m(\mathbf{i})}[f](\mathbf{y})\pi(\mathbf{y}) \right| = \max_{\mathbf{y} \in \mathcal{N}\text{ew}} \left| \left(f(\mathbf{y}) - S^m_{\mathbf{I}}[f](\mathbf{y})\right)\pi(\mathbf{y}) \right|. \tag{15}$$

On the other hand, sparse grids built with non-nested points are not interpolatory, and the set of points added to a sparse grid by $\Delta^{m(\mathbf{i})}$ is not unique, cf. Eq. (11), as it depends on the current index set $\mathbf{I}$ to which $\mathbf{i}$ is added. Thus, we define $\mathcal{N}ew = \mathrm{pts}(\mathcal{T}_{\mathbf{i}}^m)$ and approximate the $C_{\pi}^0(\Gamma)$ norm as

$$\Delta E(\mathbf{i}) = \max_{\mathbf{y} \in \mathcal{N}ew} \left| \Delta^{m(\mathbf{i})}[f](\mathbf{y})\pi(\mathbf{y}) \right| = \max_{\mathbf{y} \in \mathcal{N}ew} \left| \left(S_{\mathbf{J}}^m[f](\mathbf{y}) - S_{\mathbf{I}}^m[f](\mathbf{y}) \right) \pi(\mathbf{y}) \right|. \quad (16)$$

The right hand side in (14), (15) and (16), with suitably chosen $\mathbf{I}$ and $\mathbf{J}$, are the actual formulas we have used to compute $\Delta E(\mathbf{i})$ (see Algorithm 1 and 2 for further details). However, we remark that these values do not depend on the set $\mathbf{I}$ chosen for evaluation, since the quantities $\Delta E(\mathbf{i})$ are defined starting from the hierarchical surplus operator $\Delta^{m(\mathbf{i})}$. This means that we can consider the indices of the reduced margin $\mathbf{R_I}$ in any order, and that the values of $\Delta E(\mathbf{i})$ do not need to be recomputed at each iteration.

As for the work contribution $\Delta W(\mathbf{i})$, we consider the same indicators defined in the quasi-optimal case, i.e. (10) for nested points and (13) for non-nested points, which is equivalent to setting the work contributions equal to the cardinality of the sets $\mathcal{N}ew$ introduced above. A third option is to consider $\Delta W(\mathbf{i}) = 1$, i.e. driving the adaptivity only by the error contributions. This is the choice considered e.g. in [8, 28], while [17, 21] combine $\Delta E(\mathbf{i})$ and $\Delta W(\mathbf{i})$ in a different way. To summarize, we will drive the adaptive algorithm with any of the four profit definitions listed next, whose formulas differ depending on whether nested or non-nested points are used:

- **"deltaint"**: set $\Delta E(\mathbf{i})$ as in (14) and $\Delta W(\mathbf{i}) = 1$;
- **"deltaint/new points"** combine $\Delta E(\mathbf{i})$ as in (14) with $\Delta W(\mathbf{i})$ in (10) for nested points and in (13) for non-nested points;
- **"weighted Linf"** set $\Delta E(\mathbf{i})$ as in (15) and $\Delta W(\mathbf{i}) = 1$ for nested points, and $\Delta E(\mathbf{i})$ as in (16) and $\Delta W(\mathbf{i}) = 1$ for non-nested points;
- **"weighted Linf/new points"** combine $\Delta E(\mathbf{i})$ in (15) with $\Delta W(\mathbf{i})$ in (10) for nested points and $\Delta E(\mathbf{i})$ in (16) with $\Delta W(\mathbf{i})$ in (13) for non-nested points.

The pseudo-code of the algorithm is listed in Algorithm 1. Since nodes that are present in a given sparse grid are not necessarily present in the following ones when using non-nested points, cf. Eq. (12), the full work count in this case is not simply $\mathrm{pts}(S)$ (as it would be for nested points), but should rather include all the points "visited" to reach that grid in the adaptive algorithm, which motivates lines L1–L2 in Algorithm 1. Observe however that all Gaussian quadrature rules associated to a symmetric weight (or probability density) are in a sense "partially nested", meaning that rules with odd number of points place a quadrature node in the midpoint of the interval, implying that a non-negligible number of points can still be in common between two grids (e.g., the grid with 3×5 Gauss–Legendre points shares 5 of its 15 points with the grid 1×5).

Algorithm 1: Adaptive sparse grids algorithm

Adaptive sparse grids(*MaxPts, ProfTol,* π, *<ProfitName>*)

   $I = \{1\}, G = \{1\}, R_I = \varnothing, i = 1$;

   $S_{old} = S_I^m[f], Q_{old} = Q_I^m[f]$;

   $\mathcal{H} = \text{pts}(S_{old})$, NbPts=card($S_{old}$), ProfStop=$\infty$;

   while *NbPts < MaxPts* **and** *ProfStop > ProfTol* **do**

      $Ng = \text{neigh}(i, I)$

      for $j \in Ng$ **and** $I \cup \{j\}$ *downward closed* **do**

         $G = G \cup \{j\}$;   *at the end of the for loop,* $G = I \cup R_I$

         $S = S_G^m[f]$;   *j must be added to S to evaluate its profit.*

         $Q = Q_G^m[f]$;

         if *using nested points* **then**

            $New = \text{pts}(S) \setminus \text{pts}(S_{old})$;   *i.e. the points added by j to S*

            NbPts = NbPts + card(New) ;

            $\mathbf{v}$ = evaluations of f on each $\mathbf{y} \in New$;   *cf. Eq.* (15)

         else

            $New = \text{pts}(\mathcal{T}_i^m)$

L1 $\mathcal{H} = \mathcal{H} \cup \text{pts}(S)$;   *add points of S to $\mathcal{H}$ (no repetitions)*

L2   NbPts = card($\mathcal{H}$) ;   *for non-nested points, card($\mathcal{H}$)>card(S)*

            $\mathbf{v}$ = evaluations of S on each $\mathbf{y} \in New$;   *cf. Eq.* (16)

         $\mathbf{v}_{old}$ = evaluations of S_{old} on each $\mathbf{y} \in New$;

         π = evaluations of π on each $\mathbf{y} \in New$;

         $P(j) = \text{Compute\_profit}(New, \mathbf{v}, \mathbf{v}_{old}, \pi, Q, Q_{old}, <\text{ProfitName}>)$

         $R_I = R_I \cup \{j\}$

         $S_{old} = S, Q_{old} = Q$;

      choose the i from R_I with highest profit;

      $I = I \cup \{i\}, R_I = R_I \setminus \{i\}$

      update ProfStop with a suitable criterion based on the values of P

   return S, Q

Compute_profit(*New,* $\mathbf{v}$, $\mathbf{v}_{old}$, π, Q, Q_{old}, *<ProfitName>*)

   switch *ProfitName* **do**

      case *deltaint* **do**

         $\text{profit}(i) = |Q - Q_{old}|$;

      case *deltaint/new points* **do**

         $\text{profit}(i) = \dfrac{|Q - Q_{old}|}{\text{card}(New)}$;

      case *Weighted Linf* **do**

         $\text{profit}(i) = \max\{|\mathbf{v} - \mathbf{v}_{old}| \odot \pi\}$; $\odot$ *denotes element-wise multiplication*

      case *Weighted Linf/new points* **do**

         $\text{profit}(i) = \dfrac{\max\{|\mathbf{v} - \mathbf{v}_{old}| \odot \pi\}}{\text{card}(New)}$;

   return $\text{profit}(i)$

---

4 Darcy Problem

As mentioned in the introduction, in this work we are concerned with the application of the adaptive sparse grid algorithm in the Uncertainty Quantification context. In particular, we focus on the numerical approximation of the solution of the stochastic version of the Darcy problem [5, 7, 14, 18] in which an unknown Darcy pressure p is obtained as solution of an elliptic PDE having a lognormal random field a as diffusion coefficient; a models the permeability of the medium in which the flow takes place and, since it is a quantity that often can not be properly estimated, it is modeled as a random field over a suitable probability space $(\Omega, \mathcal{F}_\Omega, \mathbb{P})$, where Ω is the set of possible outcomes ω, $\mathcal{F}_\Omega$ a σ-algebra and $\mathbb{P} : \mathcal{F}_\Omega \to [0, 1]$ a probability measure. The mathematical formulation of the problem is the following:

Problem 1 *Given $D \in \mathbb{R}^d$, find a real-valued function $p : \overline{D} \times \Omega \to \mathbb{R}$, such that $\mathbb{P}$-almost surely (a.s) there holds:*

$$\begin{cases} -\operatorname{div}(a(\mathbf{x}, \omega)\nabla p(\mathbf{x}, \omega)) = f(\mathbf{x}) & \mathbf{x} \in D, \\ p(\mathbf{x}, \omega) = g(\mathbf{x}) & \mathbf{x} \in \partial D_j^D, \ j = 1, \dots, k_D, \\ \nabla p(\mathbf{x}, \omega) \cdot \mathbf{n} = 0 & \mathbf{x} \in \partial D_j^N, \ j = 1, \dots, k_N, \end{cases}$$

where the operators div *and* ∇ *imply differentiation with respect to the physical coordinates only, $a : \overline{D} \times \Omega \to \mathbb{R}$ is a given random field, $\mathbf{n}$ is the outward normal to the boundary, $\partial D_D = \cup_{j=1}^{k_D} \partial D_j^D$ denotes the Dirichlet boundary, $\partial D_N = \cup_{j=1}^{k_N} \partial D_j^N$ denotes the Neumann boundary and $\overline{\partial D_D} \cup \overline{\partial D_N} = \partial D$, $\mathring{\partial D}_D \cap \mathring{\partial D}_N = \emptyset$.*

More specifically, we set $a(\mathbf{x}, \omega) = e^{\gamma(\mathbf{x}, \omega)}$, γ being a mean-free stationary Gaussian random field having a tensor covariance function belonging to the so-called Matérn family [11], namely:

$$\operatorname{cov}_\nu(\mathbf{x}, \mathbf{x}') = \sigma^2 \prod_{i=1}^d \frac{\left(\sqrt{2\nu} \frac{|x_i - x_i'|}{L_c}\right)^\nu K_\nu \left(\sqrt{2\nu} \frac{|x_i - x_i'|}{L_c}\right)}{\Gamma(\nu) 2^{\nu-1}}, \quad \nu \geq 0.5, \tag{17}$$

where σ^2 is the pointwise variance, L_c is a correlation length, Γ is the gamma function, K_ν is the modified Bessel function of the second kind and ν is a parameter that governs the regularity of the covariance function and, in turn, of the realizations of the random field. In particular, for $\nu = 1/2$ we obtain a tensor Exponential covariance function $\operatorname{cov}_\nu(\mathbf{x}, \mathbf{x}') = \sigma^2 exp\{-|\mathbf{x} - \mathbf{x}'|_1/L_c\}$ which is only Lipschitz continuous and produces realizations of the random field $a(\mathbf{x}, \omega_i)$, $\omega_i \in \Omega$, that are a.s. Hölder continuous $C^{0,s}(\overline{D})$ with parameter $s < 1/2^3$; on the other hand when $\nu \to \infty$ we obtain a Gaussian covariance function $\operatorname{cov}_\nu(\mathbf{x}, \mathbf{x}') = \sigma^2 exp\{-|\mathbf{x} - \mathbf{x}'|_2^2/L_c^2\}$ which is analytic and generates infinitely differentiable realizations; in between, depending on ν, we have all the possible regularities; in general realizations with $\nu = n + \alpha$ with $n \in \mathbb{N}$ and $\alpha \in (0, 1]$, are

---

[3]A function $f : \overline{D} \subset \mathbb{R}^d \to \mathbb{R}$ is said to be Hölder continuous with parameter $s \in (0, 1]$, $f \in C^{0,s}(\overline{D})$, if there exist non-negative real constants C and s such that

$$|f(\mathbf{x}) - f(\mathbf{y})| \leq C|\mathbf{x} - \mathbf{y}|_2^s \quad \forall \mathbf{x}, \mathbf{y} \in \overline{D}.$$

n times a.s. differentiable and have all the n-th derivatives a.s. Hölder continuous $C^{0,s}(\overline{D})$ with parameter $s < \alpha$ (see e.g. [27, Lemma C.2]). The well-posedness of Problem 1 has been studied e.g. in [7, 18]. The choice of the diffusion coefficient a just detailed guarantees that Problem 1 has a unique solution $p \in L^2_{\mathbb{P}}(\Omega, V)$, $V = H^1(D)$, see e.g. [7, 18] for details, under standard regularity assumptions on f, g, that will be fulfilled by the test case that we will detail later on.

To make Problem 1 suitable for the sparse grid methodology developed in the previous sections, we consider a truncated Karhunen-Loève (KL) expansion of the Gaussian random field $\gamma(\mathbf{x}, \omega)$ with N i.i.d. standard normal random variables $\{y_i\}_{i=1}^N$ and approximate $a(\mathbf{x}, \omega) = e^{\gamma(\mathbf{x}, \omega)}$ accordingly, namely

$$\gamma(\mathbf{x}, \omega) = \sum_{n=1}^{\infty} \sqrt{\lambda_n} \psi_n(\mathbf{x}) y_n(\omega) \approx \sum_{n=1}^{N} \sqrt{\lambda_n} \psi_n(\mathbf{x}) y_n(\omega) = \gamma(\mathbf{x}, \mathbf{y}(\omega)),$$

$$a(\mathbf{x}, \omega) = e^{\gamma(\mathbf{x}, \omega)} \approx e^{\gamma(\mathbf{x}, \mathbf{y}(\omega))} = a(\mathbf{x}, \mathbf{y}(\omega)); \tag{18}$$

where the functions $\psi_n(\mathbf{x}) : D \to \mathbb{R}$, $n = 1, 2, 3 \ldots$, and the positive coefficients $\{\lambda_n\}_{n=1}^{\infty}$ are the solutions of the eigenvalue problem

$$\int_D \text{cov}_\nu(\mathbf{x}, \mathbf{x}') \psi(\mathbf{x}) d\mathbf{x} = \lambda \psi(\mathbf{x}').$$

Once the random field has been (approximately) parametrized with a random vector $\mathbf{y} = (y_1, \ldots, y_N)$ belonging to the probability space $(\Gamma, \mathcal{F}_\Gamma, \varrho(\mathbf{y}) d\mathbf{y})$, where $\Gamma = \mathbb{R}^N$ is the image of $\mathbf{y}$, $\mathcal{F}_\Gamma$ is the Borel σ-algebra and $\varrho(\mathbf{y}) = (2\pi)^{-\frac{N}{2}} \exp\left(-\frac{|\mathbf{y}|_2^2}{2}\right)$ is the probability density function of $\mathbf{y}$, we can approximate Problem 1 with the following finite dimensional parametric problem:

Problem 2 *Find a real-valued function* $p : \overline{D} \times \Gamma \to \mathbb{R}$, *such that* $\varrho(\mathbf{y}) d\mathbf{y}$-*almost everywhere there holds:*

$$\begin{cases} -\text{div}(a(\mathbf{x}, \mathbf{y}) \nabla p(\mathbf{x}, \mathbf{y})) = f(\mathbf{x}) & \mathbf{x} \in D, \\ p(\mathbf{x}, \mathbf{y}) = g(\mathbf{x}) & \mathbf{x} \in \partial D_j^D, j = 1, \ldots, k_D, \\ \nabla p(\mathbf{x}, \mathbf{y}) \cdot \mathbf{n} = 0 & \mathbf{x} \in \partial D_j^N, j = 1, \ldots, k_N. \end{cases}$$

Consistently with what we said about the infinite dimensional case, Problem 2 admits a unique solution $p \in L^2_\varrho(\Gamma, V)$, and it is now ready to be solved numerically. In particular, in our analysis we will be interested in computing the expectation of some quantity of interest (QoI) related to the solution p of the Darcy problem, defined as $u(\omega) = L(p(\cdot, \omega))$, where L is a functional $L : V \to \mathbb{R}$ that we will detail later on. At this point it is also crucial to remark that solving the stochastic Darcy problem with sparse grids is a sound approach since it can be shown that the dependence of p on the random parameters $\mathbf{y}$ is smooth, and more precisely analytic, as shown in [1, 14]; moreover it can be shown that $p \in C^0_\pi(\Gamma, V)$ with

$\pi(\mathbf{y}) = \prod_{n=1}^{N} exp(-|y_n|\sqrt{\lambda_n}\|\psi_n\|_{L^\infty(D)})$, see [19]. Nonetheless, we will choose $\varrho = \tilde{\rho} = \pi$ in the computations, cf. Eqs. (9), (15), (16). In particular, this means that we can use Hermite polynomials $\varphi_\mathbf{q}$ in the quasi-optimal approach, for which indeed $\|\varphi_\mathbf{q}\|_{C_\pi^0} \leq C$, and we will use the following ansatz for the Hermite coefficient of u:

$$|u_\mathbf{q}| \approx C \prod_{n=1}^{N} \frac{e^{-g_n q_n}}{\sqrt{q_n!}}, \tag{19}$$

cf. [5]. Concerning the truncation of the Karhunen–Loève expansion of γ, it is desirable to select the number of random variables N such that essentially the entire spatial variability is taken into account (say more than 99.9 %), in order to obtain a negligible distance between the exact solution of Problem 1 $p(\mathbf{x}, \omega)$ and the exact solution of Problem 2 $p(\mathbf{x}, \mathbf{y})$, and, in turn, between $u(\omega)$ and $u(\mathbf{y})$. As a consequence, the problem will depend on a number of random variables N ranging from a few tens (for choices of ν that yield smooth realizations of a) to several hundreds (for $\nu \to 1/2$). This will require some adaptations of the adaptive algorithm introduced earlier, that will be detailed in the following sections.

4.1 Dimension-Adaptive Sparse Grid Algorithm

When considering a large number of random variables, generating and exploring the reduced margin of $\mathbf{I}$ might be computationally intensive. Thus, in the following we present a modified version of the adaptive sparse grid algorithm that starts by working over a parameter space $\widetilde{\Gamma}$ with a moderate dimensionality $\widetilde{N}$ and progressively increases $\widetilde{N}$. Crucially, such strategy actually relieves us from fixing a-priori a truncation for the Karhunen–Loève expansion, i.e. it allows to work with $N = \infty$ random variables.

To this end, we start by assuming that the importance of the random variables in the approximation of the QoI u follows to a good extent the Karhunen–Loève ordering: in other words, y_n may contribute less than y_{n+1} to the variability of the QoI but there is a certain "dimensional buffer" N_b such that y_n is guaranteed to be more important than y_{n+N_b}. Then, the adaptive algorithm starts by considering $\widetilde{N} = N_b$ random variables only, and whenever a variable y_n with $n < \widetilde{N}$ is activated (cf. Sect. 3.2), the random variable $y_{\widetilde{N}+1}$ enters the approximation (i.e. the multi-index $\mathbf{i} = \mathbf{1} + \mathbf{e}_{\widetilde{N}+1}$ is included in the reduced margin) and the counter $\widetilde{N}$ is increased by one, so that there is always a buffer of N_b non-activated directions. This strategy is detailed in Algorithm 2.

Algorithm 2: Dimension adaptive algorithm

Note: To avoid ambiguities we write $\mathbf{v}^N$ *to make clear that the vector* $\mathbf{v}$ *has N components; analogously* $\mathbf{I}^N$ *indicates that the multi-index set* $\mathbf{I}$ *is composed of N-dimensional vectors.*

Dimension adaptive sparse grids(*MaxPts, ProfTol,* π, *<ProfitName>,N_b*)

$\widetilde{N} = N_b, \mathbf{A}^{\widetilde{N}} = \mathbf{0}^{\widetilde{N}}$;   *A is a Boolean vector indicating which variables are active*

$\mathbf{I}^{\widetilde{N}} = \{\mathbf{1}^{\widetilde{N}}\}, \mathbf{G}^{\widetilde{N}} = \{\mathbf{1}^{\widetilde{N}}\}, \mathbf{R}_\mathbf{I}^{\widetilde{N}} = \varnothing, \mathbf{i}^{\widetilde{N}} = \mathbf{1}^{\widetilde{N}}, S_{old} = S_{\underset{\mathbf{I}^{\widetilde{N}}}{m}}[f], Q_{old} = Q_{\underset{\mathbf{I}^{\widetilde{N}}}{m}}[f]$;

$\mathcal{H} = \text{pts}(S_{old}), \text{NbPts=card}(S_{old}), \text{ProfStop}=\infty$;

while *NbPts < MaxPts* **and** *ProfStop > ProfTol* **do**

$\quad \mathcal{N}g = \text{neigh}(\mathbf{i}^{\widetilde{N}}, \mathbf{I}^{\widetilde{N}})$;

$\quad$ **for** $\mathbf{j} \in \mathcal{N}g$ **and** $\mathbf{I}^{\widetilde{N}} \cup \{\mathbf{j}\}$ *is downward closed* **do**

$\qquad \mathbf{G}^{\widetilde{N}} = \mathbf{G}^{\widetilde{N}} \cup \{\mathbf{j}\}$;

$\qquad S = S_{\underset{\mathbf{G}^{\widetilde{N}}}{m}}[f]$;

$\qquad Q = Q_{\underset{\mathbf{G}^{\widetilde{N}}}{m}}[f]$;

$\qquad$ **if** *using nested points* **then**

$\qquad\quad \mathcal{N}ew = \text{pts}(S) \setminus \text{pts}(S_{old}), \text{NbPts} = \text{NbPts} + \text{card}(\mathcal{N}ew)$;

$\qquad\quad \mathbf{v} = \text{evaluations of } f \text{ on each } \mathbf{y} \in \mathcal{N}ew$;   *cf. Eq.*(15)

$\qquad$ **else**

$\qquad\quad \mathcal{N}ew = \text{pts}(T_\mathbf{i}^m), \mathcal{H} = \mathcal{H} \cup \text{pts}(S), \text{NbPts} = \text{card}(\mathcal{H})$;

$\qquad\quad \mathbf{v} = \text{evaluations of } S \text{ on each } \mathbf{y} \in \mathcal{N}ew$;   *cf. Eq.*(16)

$\qquad \mathbf{v}_{old} = \text{evaluations of } S_{old} \text{ on each } \mathbf{y} \in \mathcal{N}ew$;

$\qquad \pi = \text{evaluations of } \pi \text{ on each } \mathbf{y} \in \mathcal{N}ew$;

$\qquad P(\mathbf{j}) = \text{Compute\_profit}(\mathcal{N}ew, \mathbf{v}, \mathbf{v}_{old}, \pi, Q, Q_{old}, \text{<ProfitName>})$

$\qquad \mathbf{R}_\mathbf{I}^{\widetilde{N}} = \mathbf{R}_\mathbf{I}^{\widetilde{N}} \cup \{\mathbf{j}\}$

$\qquad S_{old} = S, Q_{old} = Q$;

$\quad$ choose $\mathbf{k}^{\widetilde{N}}$ from $\mathbf{R}_\mathbf{I}^{\widetilde{N}}$ with highest profit; $\mathbf{i}^{\widetilde{N}} = \mathbf{k}^{\widetilde{N}}$;

$\quad$ **if** $\exists n = 1, \ldots, \widetilde{N}$ *s.t.* $A_n = 0$ **and** $k_n > 1$ **then**

$\qquad A_n = 1, \widetilde{N} = \widetilde{N} + 1$;   *activate n-th variable and update* $\widetilde{N}$

$\qquad$ extend the containers $\mathbf{I}, \mathbf{R}_\mathbf{I}, \mathbf{G}, \mathbf{k}, \mathbf{A}$ by adding the new direction.

$\qquad \mathbf{G}^{\widetilde{N}} = \mathbf{G}^{\widetilde{N}} \cup \{\mathbf{1}^{\widetilde{N}} + \mathbf{e}_{\widetilde{N}}^{\widetilde{N}}\}; S = S_{\underset{\mathbf{G}^{\widetilde{N}}}{m}}[f]; Q = Q_{\underset{\mathbf{G}^{\widetilde{N}}}{m}}[f]$;

$\qquad$ **if** *using nested points* **then**

$\qquad\quad \mathcal{N}ew = \text{pts}(S) \setminus \text{pts}(S_{old}), \text{NbPts} = \text{NbPts} + \text{card}(\mathcal{N}ew)$;

$\qquad\quad \mathbf{v} = \text{evaluations of } f \text{ on each } \mathbf{y} \in \mathcal{N}ew$;   *cf. Eq.*(15)

$\qquad$ **else**

$\qquad\quad \mathcal{N}ew = \text{pts}(T_\mathbf{i}^m), \mathcal{H} = \mathcal{H} \cup \text{pts}(S), \text{NbPts} = \text{card}(\mathcal{H})$;

$\qquad\quad \mathbf{v} = \text{evaluations of } S \text{ on each } \mathbf{y} \in \mathcal{N}ew$;   *cf. Eq.*(16)

$\qquad \mathbf{v}_{old} = \text{evaluations of } S_{old} \text{ on each } \mathbf{y} \in \mathcal{N}ew$;

$\qquad \pi = \text{evaluations of } \pi \text{ on each } \mathbf{y} \in \mathcal{N}ew$;

$\qquad P(\mathbf{1}^{\widetilde{N}} + \mathbf{e}_{\widetilde{N}}^{\widetilde{N}}) = \text{Compute\_profit}(\mathcal{N}ew, \mathbf{v}, \mathbf{v}_{old}, \pi, Q, Q_{old}, \text{<ProfitName>})$

$\qquad \mathbf{R}_\mathbf{I}^{\widetilde{N}} = \mathbf{R}_\mathbf{I}^{\widetilde{N}} \cup \{\mathbf{1}^{\widetilde{N}} + \mathbf{e}_{\widetilde{N}}^{\widetilde{N}}\}$,

$\qquad \mathbf{i}^{\widetilde{N}} = \text{argmax}(\max(P(\mathbf{1}^{\widetilde{N}} + \mathbf{e}_{\widetilde{N}}^{\widetilde{N}}), P(\mathbf{k}^{\widetilde{N}})))$;   *select* $\mathbf{i}^{\widetilde{N}}$ *with highest profit*

$\quad \mathbf{I}^{\widetilde{N}} = \mathbf{I}^{\widetilde{N}} \cup \{\mathbf{i}^{\widetilde{N}}\}, \mathbf{R}_\mathbf{I}^{\widetilde{N}} = \mathbf{R}_\mathbf{I}^{\widetilde{N}} \setminus \{\mathbf{i}^{\widetilde{N}}\}$

$\quad$ update ProfStop with a suitable criterion based on the values of $P(\mathbf{j})$

return S, Q

4.2 Monte Carlo Method with Control Variate (MCCV)

For values of ν close to $1/2$, the decay of the eigenvalues of the KL expansion of γ is so slow that a very large number of random variables will equally contribute to the variability of the QoI; therefore, even the dimension-adaptive sparse grid algorithm detailed in the previous section may not be effective. In such a case we propose to combine the sparse grid approximation with a Monte Carlo sampling following the ideas proposed in [27]. More precisely, we will introduce an auxiliary problem having a smoothed coefficient a^ϵ as random permeability, whose solution u^ϵ can be effectively approximated by a quasi-optimal or an adaptive sparse grid scheme. Then we will use u^ϵ as control variate in order to define a new QoI, namely u^{CV}, upon which we build a MC estimator.

The first step in order to apply this strategy is to define a proper smoothed random field. Thus, let $\gamma(x, \omega)$ and $\gamma^\epsilon(x, \omega)$ be two random fields obtained respectively by considering a covariance function of the Matérn family and the convolution of $\gamma(x, \omega)$ with a smooth kernel (e.g. Gaussian),

$$\gamma^\epsilon(\cdot, \omega) = \gamma(\cdot, \omega) * \phi_\epsilon(\cdot) \quad \text{where} \quad \phi_\epsilon(\mathbf{x}) = e^{-\frac{|\mathbf{x}|^2}{2\epsilon^2}} / (2\pi\epsilon^2)^{\frac{d}{2}},$$

and let $a^\epsilon = e^{\gamma^\epsilon}$. Using this definition, it is easy to see that the smoothed random field γ^ϵ has a covariance function defined as

$$\text{cov}_\nu^\epsilon(\mathbf{x}, \mathbf{x}') = \mathbb{E}[\gamma^\epsilon(\mathbf{x}, \cdot)\gamma^\epsilon(\mathbf{x}', \cdot)] = \phi_\epsilon(\mathbf{x}) * \text{cov}_\nu(\mathbf{x}, \mathbf{x}') * \phi_\epsilon(\mathbf{x}'). \tag{20}$$

Clearly, the smaller the parameter ϵ is, the more correlated the two random fields γ and γ^ϵ are, as it can be seen in Fig. 1; consistently, $u^\epsilon \to u$ when $\epsilon \to 0$.

Next, let us assume for the moment that we know exactly the mean of the control variate $\mathbb{E}[u^\epsilon]$ and define

$$\tilde{u}^{CV} = u - u^\epsilon + \mathbb{E}[u^\epsilon].$$

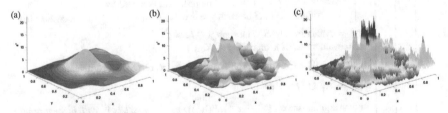

Fig. 1 Three different regularizations of the same realization of a. $\nu = 0.5$, $L_c = 0.5$, $\sigma = 1$. (a) $\epsilon = 1/2^4$. (b) $\epsilon = 1/2^6$. (c) Original field

This new variable is such that $\mathbb{E}[\tilde{u}^{CV}] = \mathbb{E}[u]$ and

$$\mathbb{V}\mathrm{ar}(\tilde{u}^{CV}) = \mathbb{V}\mathrm{ar}(u) + \mathbb{V}\mathrm{ar}(u^\epsilon) - 2\mathrm{cov}(u, u^\epsilon), \tag{21}$$

showing that the more positively correlated the quantities of interest are, the larger the variance reduction achievable. Although we do not have the exact mean of $u^\epsilon(\mathbf{y})$ at our disposal, we can successfully compute it with a sparse grid method, since $a^\epsilon(\mathbf{x}, \mathbf{y})$ has smooth realizations, and hence the coefficients of the KL expansion are rapidly decreasing, as long as the smoothing parameter ϵ remains sufficiently large. The final variable on which we will actually apply our MC algorithm is therefore

$$u^{CV} = u - u^\epsilon + Q^m_{\mathbf{I}(w)}[u^\epsilon], \tag{22}$$

and the associated MC control variate estimator (MCCV) is defined as

$$\hat{u}^{MCCV}_M = \frac{1}{M} \sum_{i=1}^M u^{CV}(\omega_i) = \frac{1}{M} \sum_{i=1}^M (u(\omega_i) - u^\epsilon(\omega_i)) + Q^m_{\mathbf{I}(w)}[u^\epsilon], \tag{23}$$

where $u^{CV}(\omega_i)$ are i.i.d. realizations of the control variate and M is the sample size. Note that

$$\mathbb{V}\mathrm{ar}(u^{CV}) = \mathbb{V}\mathrm{ar}(\tilde{u}^{CV}). \tag{24}$$

Observe that care must be taken from a computational point of view when generating the samples $u_i(\omega) - u^\epsilon_i(\omega)$. We propose to generate realizations of $u - u^\epsilon$ starting from the Fourier expansions of γ and γ^ϵ: indeed, the Fourier expansion is very convenient when expansions over several random variables are needed, as the basis functions are known analytically; moreover, the Fourier expansions of γ and γ^ϵ share the same basis functions and differ only by the coefficients. On the other hand, to compute $Q^m_{\mathbf{I}(w)}[u^\epsilon]$, it is more convenient to start from a Karhunen–Loève expansion of γ^ϵ, that needs less variables than a Fourier expansion but whose basis functions need to be determined solving an eigenvalue problem (which is however doable for γ^ϵ given that it is a smooth field). In other words, two expansions of γ^ϵ and one expansion of γ will be used simultaneously. In particular, we have considered the following truncated Fourier expansion over an hypercube of size $(2L)^d$, containing the domain D, with $L = \max(6L_c, diam(D))$:

$$\gamma(\mathbf{x}, \mathbf{y}) = \sum_{\mathbf{k} \in \mathbf{K}} \sqrt{c_{\mathbf{k}}} \sum_{\mathbf{n} \in \{0,1\}^d} y^{\mathbf{n}}_{\mathbf{k}}(\omega) \prod_{l=1}^d \cos\left(\frac{\pi k_l}{L} x_l\right)^{n_l} \sin\left(\frac{\pi k_l}{L} x_l\right)^{1-n_l},$$

where $\mathbf{K} \subset \mathbb{N}^d$ is a suitable multi-index set having cardinality K, the resulting vector of i.i.d. standard normal random variables is $\mathbf{y} = \{y^{\mathbf{n}}_{\mathbf{k}}, \ \mathbf{k} \in \mathbf{K}, \ \mathbf{n} \in \{0, 1\}^d\}$, and the coefficients $c_{\mathbf{k}}$ are the positive coefficients of the cosine expansion of the covariance function $\text{cov}_\nu(\mathbf{x}, \mathbf{x}')$ on $[-L, L]^d$, namely

$$\text{cov}_\nu(\mathbf{x}, \mathbf{x}') = \sum_{\mathbf{k} \in \mathbb{N}^d} c_{\mathbf{k}} \prod_{l=1}^{d} \cos\left(\frac{\pi k_l}{L}(x_l - x_l')\right).$$

Consistently, the realization $u_i^\epsilon(\omega)$ will be computed starting from a truncated Fourier expansion of γ^ϵ over the same index-set $\mathbf{K}$ and using the same realization $\mathbf{y}(\omega_i)$ used to generate $u_i(\omega)$.

Concerning the mean square error associated to the estimator (23), namely $e(\hat{u}_M^{MCCV})^2 = \mathbb{E}\left[(\hat{u}_M^{MCCV} - \mathbb{E}[u])^2\right]$, the following result holds:

Proposition 1 *The mean square error of the estimator (23) can be split as*

$$e(\hat{u}_M^{MCCV})^2 = \frac{\mathbb{V}ar(u^{CV})}{M} + \left(\mathbb{E}[u^\epsilon] - Q_{\mathbf{I}(w)}^m[u^\epsilon]\right)^2. \tag{25}$$

Proof We have

$$e(\hat{u}_M^{MCCV})^2 = \mathbb{E}[(\hat{u}_M^{MCCV} - \mathbb{E}[u])^2]$$

$$= \mathbb{E}\left[\left(\sum_{i=1}^{M} \frac{u_i - u_i^\epsilon}{M} + Q_{\mathbf{I}(w)}^m[u^\epsilon] \pm \mathbb{E}[u^\epsilon] - \mathbb{E}[u]\right)^2\right]$$

$$= \mathbb{E}\left[\left(\frac{1}{M}\sum_{i=1}^{M}(u_i - u_i^\epsilon - \mathbb{E}[u] + \mathbb{E}[u^\epsilon])\right)^2\right] + \mathbb{E}[(Q_{\mathbf{I}(w)}^m[u^\epsilon] - \mathbb{E}[u^\epsilon])^2]$$

$$= \frac{\mathbb{V}ar(u^{CV})}{M} + \left(\mathbb{E}[u^\epsilon] - Q_{\mathbf{I}(w)}^m[u^\epsilon]\right)^2. \qquad \square$$

The first term on the right hand side of (25) represents the variance of the estimator $\hat{u}_M^{MCCV}$, i.e. the error coming from the MCCV sampling, and it is expected to be significantly smaller than the variance of the standard MC estimator thanks to the presence of the control variate, cf. Eqs. (24) and (21); the second term represents instead the error due to the approximation of the mean of the smoothed quantity of interest u^ϵ with a sparse grid scheme. As already hinted, when ϵ goes to 0 the term $\mathbb{V}ar(u^{CV})/M$ vanishes, and more precisely, the following result (which is a simplified version of Theorem 5.1 in [27]) holds:

Proposition 2 *Let γ and γ^ϵ be two Gaussian random fields having covariance functions respectively defined as in (17) and (20); assume $\partial D_D = \partial D$, $f \in L^2(D)$ and $L \in H^{-1}(D)$. Then, $\mathbb{P}$-a.s. in Ω it holds*

$$|u - u^\epsilon|(\omega) \leq C(\nu, \omega)\epsilon^{\min(2,\alpha)}, \quad \forall \alpha < \nu,$$

where the constant $C(v, \omega)$ is $L^q_{\mathbb{P}}$-integrable for any $q > 0$ so the bound can also be expressed as

$$\|u - u^\epsilon\|_{L^q_{\mathbb{P}}(\Omega)} \leq C(v, q)\epsilon^{\min(2,\alpha)}, \quad \forall \alpha < v.$$

In particular, $\mathbb{V}ar(u^{CV}) \leq C^2(v, 2)\epsilon^{2\min(2,\alpha)}, \quad \forall \alpha < v.$

On the other hand an accurate approximation of $\mathbb{E}[u^\epsilon]$ by a sparse grid scheme might becomes non-advantageous if $\epsilon \to 0$. The parameter ϵ should therefore be chosen so as to have a good variance reduction while still keeping a manageable sparse grid approximation problem.

Remark 1 In this work we do not address the issue of the spatial approximation of Problem 2. In general all the results previously presented still hold if a finite dimensional subspace $V_h \subset V$, e.g. a finite element space, is considered in order to approximate functions in V.

5 Numerical Results

In this section we present the convergence results obtained for the Darcy problem on the unit square $D = (0, 1)^2$ with $f = 0$, Dirichlet boundary conditions $g(\mathbf{x}) = 1 - x_1$ on $\partial D_D = \{\mathbf{x} \in \partial D : x_1 = 0 \text{ or } x_1 = 1\}$, and homogeneous Neumann conditions on the remaining part of ∂D; the spatial approximation of the Darcy problem is done by piecewise linear finite elements defined on a structured mesh.

We will consider two cases: first we will solve Problem 2 with a smooth random field a, corresponding to the choice $v = 2.5$ in (17), and then we will move to the rough random field corresponding to $v = 0.5$, in which case we will consider the MCCV approach. In both cases we set $\sigma = 1$ and $L_c = 0.5$, while the mesh over $\overline{D}$ consists of 33×33 vertices in the case $v = 2.5$ and 65×65 vertices in the case $v = 0.5$; both meshes have been verified to be sufficiently refined for our purposes. In particular we will be interested in approximating the expected value of the functional

$$u(\omega) = \int_0^1 a(1, x_2, \omega) \frac{\partial p}{\partial x_1}(1, x_2, \omega)dx_2, \tag{26}$$

which represents the mass flow on the outlet. The aims of this section are:

1. establish whether using non-nested points in an adaptive sparse grid framework might be convenient or not;
2. verify the performance of adaptive sparse grids built with different profit indicators;
3. compare the performance of the adaptive sparse grids with that of the quasi-optimal sparse grids (note that our previous numerical experiences suggest that

indeed these two sparse grid constructions behave similarly when used to solve UQ problems depending on uniform random variables if nested univariate points are used, see [4, 24]);

4. test the effectiveness of using an adaptive (or quasi-optimal) sparse grid construction as control variate in a MC framework in order to tackle also problems depending on rough coefficients.

All results have been obtained using the Matlab package [30], available for download.

5.1   Smooth Case: $v = 2.5$

In this case we deal with an input random field with twice differentiable realizations; therefore the eigenvalues of the Karhunen–Loève expansion decay quickly enough to justify the use of the N-adaptive sparse grid algorithm to approximate the QoI.

For this test, we consider as a reference solution the approximation of the QoI obtained with a quasi-optimal sparse grid with approximately 8300 quadrature points base of Gauss–Hermite abscissas, for which 45 out of the first 50 random variables of the KL expansion are active: observe that this is sufficient to take into account 99.99 % of the total variability of the permeability field, i.e. there is essentially no KL truncation error. We monitor the convergence of the error measured as

$$err(w) \approx |Q^m_{\mathbf{I}(w)}[u] - Q^m_{\mathbf{I}(w_{ref})}[u]|,$$

i.e. the absolute value of the sparse grid quadrature error, where $\mathbf{I}(w), w = 0, 1, 2, \ldots,$ are the sequences of multi-index sets generated either by the adaptive or the quasi-optimal sparse grid scheme and $\mathbf{I}(w_{ref})$ is the multi-index set corresponding to the above-mentioned reference solution. More specifically, the sets $\mathbf{I}(w)$ for the adaptive strategies are obtained by stopping the algorithm as soon as at least $W_{max}(w)$ points have been added to the sparse grid (including the points needed for the exploration of the reduced margin), with $W_{max}(w) = \{1, 20, 50, 100, 250, 500, 1000, 2000, 4000\}$, for $w = 0, \ldots, 8$. As for the quasi-optimal sparse grids, the sets $\mathbf{I}(w)$ are defined as

$$\mathbf{I}(w) = \left\{ \mathbf{i} \in \mathbb{N}^N_+ : P(\mathbf{i}) \geq e^{-w} \right\} \tag{27}$$

with $w = 0, 1, \ldots, 5$, the reference solution being obtained with $w = 6$. We recall that the profits $P(\mathbf{i})$ are defined as the ratios between the error and work contributions, $P(\mathbf{i}) = \Delta E(\mathbf{i})/\Delta W(\mathbf{i})$, where $\Delta E(\mathbf{i})$ are estimated combining Eqs. (19) and (9), and $\Delta W(\mathbf{i})$ are defined either as (10) or (13).

The computational cost associated to each sparse grid is expressed in terms of number of linear system solves. For the adaptive sparse grids, this count also includes the cost of the exploration of the reduced margin. Moreover, when using non-nested points we also take into account the system solves related to the points that have been included and then excluded from the sparse grid, cf. Eq. (12). As for the quasi-optimal sparse grids, their construction requires some additional solves to estimate the parameters $g_1, \dots, g_N$ in (19), cf. [4, 24]. More precisely, the n-th rate is estimated by fixing all variables but y_n to their expected value, computing the value of the QoI increasing the number of collocation points along y_n and then fitting the resulting interpolation error: in practice, this amounts to solving 25 linear systems per random variable, which are included in the work count.

We start our discussion from Fig. 2, where we show the convergence results obtained with the dimension-adaptive Algorithm 2 varying the choice of profit indicators (cf. Algorithm 1) and the choice of interpolation points, i.e. Genz–Keister versus Gauss–Hermite points, the latter denoted by a suffix NN in the plot, as per "non nested" (cf. Table 1); in this test, we have set the buffer size to $N_b = 10$. More specifically, we used the "deltaint-based" profit indicators in Fig. 2a (D and D/NP in the plots, where NP stands for "divided by number of points") and "weighted L^∞-based" profit indicators in Fig. 2b (WLinf and WLinf/NP in the plots). In both cases we observe that there is not much difference between the profit indicators that take into account the number of points and the ones which do not; also the choice of nested or non nested nodes does not seems to affect the convergence.

The numbers next to each point give information about the shape of the multi-index sets $\mathbf{I}(w)$ generated by the adaptive algorithm, and consequently on the distribution of the sparse grid points on the $\widetilde{N}$-dimensional parameter space. The first number (out of the brackets) indicates the number of active directions, while the second number (in the brackets) denotes the maximum number of directions that have been activated at the same time, i.e. the highest dimensionality among all the tensor grids composing the sparse grid, cf. Eq. (4). Here and in the following, green labels refer to grids with nested points, while red labels to grids with non-nested points: we show only two series of labels per plot, due to the fact that

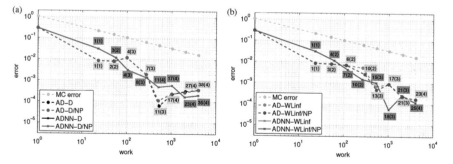

Fig. 2 Case $v = 2.5$, adaptive sparse grids error. (**a**) Profit indicators: deltaint (AD-D and ADNN-D), deltaint/new points (AD-D/NP and ADNN-D/NP). (**b**) Profit indicator: Weighted Linf (AD-WLinf and ADNN-WLinf), Weighted Linf/new points (AD-WLinf/NP and ADNN-WLinf/NP)

accounting for the work contributions in the profit definition does not seem to play a role in this test and, consequently, the sequences of sets $\mathbf{I}(w)$ generated by AD-D and AD-D/NP are essentially identical (and the same for ADNN-D and ADNN-D/NP). Observe that after ≈ 20 problem solves the algorithm has activated "only" 1 variable due to the fact the at the beginning of the algorithm N_b variables must be explored, requiring $1 + 2N_b = 21$ solver calls, in order to decide which variable should be activated as second; moreover the number of "active" variables is always smaller than $N = 45$, which is the number of "active" variables for the reference solution. In Fig. 2a, b we have also added the convergence curve for a plain MC approximation. This has been generated as $\sqrt{\mathrm{Var}(u)/M(w)}$, with $\mathrm{Var}(u)$ estimated as $\mathrm{Var}(u) \approx Q^m_{\mathbf{I}(w_{ref})}[u^2] - (Q^m_{\mathbf{I}(w_{ref})}[u])^2$.

In Fig. 3a we show instead the errors obtained by using quasi-optimal sparse grid approximations of the QoI built on Genz–Keister and Gauss–Hermite knots (labeled OPT and OPT NN respectively). Observe that since we build the sets $\mathbf{I}(w)$ in (27) again with a "buffered" procedure analogous to the one described in Sect. 4.1, the rate g_n is computed only at the level w for which y_n enters the buffer of random variables, and such work is thus accounted for at level w; this explains the initial plateau that can be seen in the convergence. Observe also that the Lebesgue constant $\mathbb{M}^{m(i_n)}_n$, introduced in Sect. 3.1 and needed for computations, has been proven in [24] to be identically equal to one in the case of Gauss-Hermite abscissas, while can be numerically estimated in the case of Genz–Keister points. Again, the labels next to each point represent the number of active variables (outside the brackets) and the number of variables activated at the same time (in the brackets). These numbers suggest that, for the same work, the adaptive sparse grids seem to activate a slightly smaller number of variables than the quasi-optimal ones, while the tensor grids dimensionality seems to be comparable. Also for the quasi-optimal sparse grids the number of "active" variables is always smaller than $N = 45$.

Figure 3b shows a comparison between the quasi-optimal and the adaptive schemes; among the adaptive schemes presented we take into account for this comparison the profit indicators deltaint and Weighted Linf/new points. We can

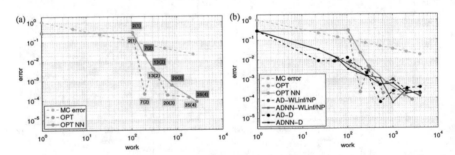

Fig. 3 Case $v = 2.5$, quasi-optimal sparse grids error (*left*) and a comparison between adaptive and quasi-optimal schemes (*right*). (**a**) Quasi-optimal sparse grids error. (**b**) Comparison adaptive/quasi-optimal

observe that, except for small values of work for which the cost needed to compute the parameters g_i in Eq. (19) largely dominates the cost needed to actually compute the quasi-optimal sparse grid approximation, the quasi-optimal and the adaptive schemes behave similarly.

Finally, we test numerically the pointwise approximation properties of the sparse grids methodology. In an unbounded domain context, the main difficulty encountered in performing pointwise approximation is that the Lagrangian polynomials on which the sparse grid construction is based are not uniformly bounded (regardless of the choice of collocation points). On the other hand, the solution of the Darcy problem is unbounded as well, and indeed it can only be shown that $p \in C_\pi^0(\Gamma, V)$ with an exponentially decaying weight, as we have discussed earlier: it would be therefore only reasonable to measure the convergence with respect to such proper weighting function. Thus, we will verify numerically the behaviour of the quantity $\left\| u - S_{\mathbf{I}}^m[u] \right\|_{C_\pi^0(\Gamma)}$ both for the exponential weight π and the gaussian weight ϱ, by sampling the difference between the exact value of the Quantity of Interest u and its sparse grid approximation over a set of 10,000 points randomly sampled from a multivariate standard gaussian distribution:

$$\left\| u - S_{\mathbf{I}}^m[u] \right\|_{C_\pi^0(\Gamma)} \approx \max_{\mathbf{y} \in \mathcal{R}} \left| \left(u(\mathbf{y}) - S_{\mathbf{I}}^m[u](\mathbf{y}) \right) \pi(\mathbf{y}) \right|.$$

Observe that the number of random variables considered by $S_{\mathbf{I}}^m[u]$ is increasing as the sparse grids algorithm keeps running; thus, we choose $\mathcal{R} \subset \mathbb{R}^{N^*}$, with N^* sufficiently larger than the number of random variables activated by the most refined sparse grids. The results are shown in Fig. 4, and indicate that the sparse grid pointwise weighted approximation error is indeed decreasing. The results suggest again that the various sparse grid construction techniques considered in this work behave similarly, and in particular "weighted L^∞" based sparse grids do not

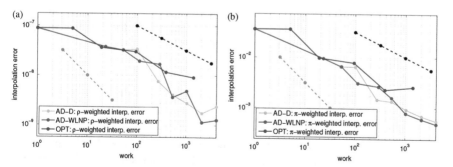

Fig. 4 Case $\nu = 2.5$. Interpolation error with different weight functions for the Quasi Optimal (OPT) and Adaptive (profit indicators Deltaint (AD-D) and Weighted Linf/new points (AD-WLNP)) cases. work $= W_{\mathbf{I},m}$, $N^* = 60$, card($\mathcal{R}$) $= 10,000$. *Dashed lines* represent the slopes -0.5 *(black)* and -1 *(orange)*. (a) $\varrho(\mathbf{y}) = \prod_{n=1}^{N^*} exp(-\frac{y_n^2}{2})$. (b) $\pi(\mathbf{y}) = \prod_{n=1}^{N^*} exp(-|y_n|\sqrt{\lambda_n} \|\psi_n\|_{L^\infty(D)})$

show particular gains with respect to the "deltaint"-based ones. The sparse grids considered in this test are the same ones used to obtain the results shown in Figs. 2 and 3; again we remark that the number of active variables remains significantly smaller than the number of variables $N^* = 60$ used to compute our approximated sample space $\mathcal{R}$. We actually observe convergence in the norms $\|\cdot\|_{C^0_\varrho(\Gamma)}$ and $\|\cdot\|_{C^0_\pi(\Gamma)}$; moreover the "weighted L^∞" norms of the differences between the exact sample u and its sparse grid reconstruction $S^m_{\mathbf{I}}[u]$ converge with a similar rate when using the exponential weight π and the gaussian one ϱ.

5.2 Rough Case: $\nu = 0.5$

In this case we deal with a rough input random field a that has realizations which are not even differentiable: thus, the slow decay of the eigenvalues of the Karhunen–Loève expansion may render unfavorable a sparse grid approach, even considering advanced techniques like the adaptive or the quasi-optimal schemes. Therefore we now solve the problem by the MCCV approach introduced in Sect. 4.2, using $\epsilon = 2^{-5}$ as smoothing parameter.

At each sparse grid approximation level w we use $M(w) = W_{\mathbf{I}(w),m}$ samples in the MCCV estimator, i.e. we balance the work of the sparse grid and that of the MC sampling so that the total work is $2W_{\mathbf{I}(w),m}$; other work splitting, e.g. balancing the two error contributions of the method detailed in Proposition 1, could be considered as well.

To obtain an error convergence curve, we will approximate the sparse grid component of the error by considering a reference solution obtained with a quasi-optimal sparse grid built with approximately 86,500 nodes based on Gauss–Hermite abscissas with $N = 163$ active random variables, that takes into account 99.99 % of the total variability of the smoothed field, and the sampling component by estimating $\mathbb{V}\text{ar}(\tilde{u}^{CV}) \approx \widehat{u^{2MCCV}_{M(w)}} - (\hat{u}^{MCCV}_{M(w)})^2$; as mentioned in Sect. 4.2, the sampling component is based on a Fourier expansion of the non-smoothed field γ, that has been truncated after $129 \times 129 = 16{,}641$ random variables. To summarize, we have

$$
err(w) \approx \sqrt{\frac{\widehat{u^{2MCCV}_{M(w)}} - (\hat{u}^{MCCV}_{M(w)})^2}{M(w)}} + \left| Q^m_{\mathbf{I}(w)}[u^\epsilon] - Q^m_{\mathbf{I}(w_{ref})}[u^\epsilon] \right|.
$$

In Fig. 5 we show the performance of the MCCV algorithm with adaptive sparse grids. Since we are running a sampling method, we also add to the plot error bars indicating the interval spanning ± 3 standard deviations of the error from its average value, assessed over 4 runs of the method. The considerations that can be made by looking at these plots are similar to the ones we did in the case $\nu = 2.5$, i.e. there is basically no difference between the profit indicators that take into account the number of points and those which do not; also changing the family of nodes does

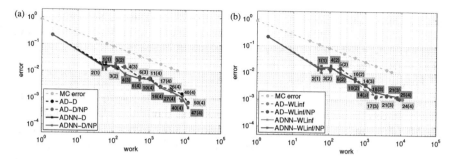

Fig. 5 Case $v = 0.5$, MCCV-adaptive sparse grid mean error. Bars represent 3 standard deviation of the sampling error. (**a**) Profit indicators: deltaint (AD-D and ADNN-D), deltaint/new points (AD-D/NP and ADNN-D/NP). (**b**) Profit indicator: Weighted Linf (AD-WLinf and ADNN-WLinf), Weighted Linf/new points (AD-WLinf/NP and ADNN-WLinf/NP)

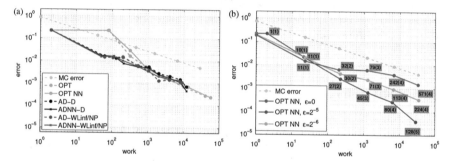

Fig. 6 Case $v = 0.5$. *Left*: comparison between MCCV-adaptive and MCCV-quasi-optimal schemes *Right*: Quasi-optimal sparse grid error for different values of ϵ. (**a**) Comparison MCCV-adaptive/MCCV-quasi-optimal sparse grid mean error. (**b**) Quasi-optimal sparse grid error for different values of ϵ

not seem to have a substantial impact on the quality of the approximation. Observe that since we are balancing the works of the Monte Carlo sampling and of the sparse grid, the observed convergence rate is larger than the MC rate $1/2$ for little values of work, where the sparse grid error dominates the sampling error and converges with a faster rate than $1/2$ (remember that the sparse grid is applied to a smoothed problem). For large w, the sampling error dominates the sparse grid error and one essentially recovers the MC rate $1/2$, however with a much smaller constant than MC due to the presence of the control variate.

Figure 6a shows the convergence of the MCCV method combined with quasi-optimal sparse grids and compares the results obtained with the adaptive sparse grids procedures. Again, among the adaptive schemes presented we consider the profit indicators deltaint and Weighted Linf/new points. For both quasi-optimal and adaptive schemes we only plot the average error of the Quantity of Interest over four runs. As in the smooth case, for sufficiently large values of work, all the schemes perform similarly. Note that in the quasi-optimal case, the quantities g_n are actually computed for the first 50 random variables only, after which we instead set

$g_n = \sqrt{\lambda_n}$, cf. Eq. (18), i.e. we approximate g_n with the value of the corresponding coefficient of the KL expansion. Indeed, these random variables have a moderate impact on the solution and numerical cancellations effects may significantly affect the results of the fitting procedure.

Finally, Fig. 6b shows the convergence of the quasi-optimal sparse grid error $|Q^m_{\mathbf{I}(w)}[u^\epsilon] - Q^m_{\mathbf{I}(w_{ref})}[u^\epsilon]|$ for different values of ϵ. It is clearly visible that the convergence rate deteriorates as ϵ decreases, thus motivating the introduction of the MCCV approach. Observe that for the sake of comparison, in this plot the work needed to determine the rates (that would cause an initial plateau in the convergence plot) has been neglected. We also observe that in this case there is a significant difference between the number of random variables activated by the quasi-optimal and adaptive sparse grid schemes. In fact, the latter tends to activate less variables than the former, adding conversely more points on the activated ones.

As a last remark, we mention here that the Control Variate technique is inherently a quadrature procedure, and extending it to the interpolation context goes beyond the scope of the current work.

6 Conclusions

In this work we have proposed an improved version of the classical adaptive sparse grid algorithm, that can handle non-nested collocation points and unbounded domains, and can be used for an arbitrary large number of random variables, assuming that a "rough ordering" of the variables according to their importance is available. We have also implemented several indicators to drive the adaptation process.

We have then used this algorithm to solve a Darcy equation with random log-normal permeability, and compared the results obtained by changing collocation points and adaptivity indicators against those obtained by the quasi-optimal sparse grids algorithm. The computational analysis has been performed first on a case with smooth permeability realizations, and then in the case of rough realizations: in the latter case, we have actually considered the sparse grid in a Monte Carlo Control Variate approach, in which the sparse grids are applied to a smoothed problem and the results serve as control variate for a Monte Carlo sampling for the rough problem. In the case of the smooth permeability realizations, we have also tested numerically the convergence of the sparse grid approximation in weighted maximum norm. The numerical results seem to suggest that

1. using non-nested points in an adaptive sparse grid framework yields results that are comparable to those obtained by nested points, at least in the log-normal context;
2. changing the indicator driving the adaptivity process does not have a dramatic impact on the quality of the solution; this however may be due to the specific choice of the QoI considered here, and more testing should be performed;

3. the sparse grid approximation will also converge with respect to suitably weighted maximum norms;
4. the adaptive and the quasi-optimal sparse grids perform similarly on lognormal problems, in agreement with our previous findings on uniform random variables;
5. in the case of smooth log-permeability fields the adaptive and the quasi-optimal sparse grids give quite satisfactory results;
6. in the case of rough fields the adaptive/quasi-optimal sparse grids *alone* have a performance asymptotically similar to a standard MC (with just a slight improvement on the constant) and we do not advocate their use in such a case; on the other hand the results are satisfactory if the sparse grids are used as control variate in a MC sampling.

Acknowledgements F. Nobile, F. Tesei and L. Tamellini have received support from the Center for ADvanced MOdeling Science (CADMOS) and partial support by the Swiss National Science Foundation under the Project No. 140574 "Efficient numerical methods for flow and transport phenomena in heterogeneous random porous media". R. Tempone is a member of the KAUST SRI Center for Uncertainty Quantification in Computational Science and Engineering.

References

1. I. Babuška, F. Nobile, R. Tempone, A stochastic collocation method for elliptic partial differential equations with random input data. SIAM Rev. **52**(2), 317–355 (2010)
2. J. Bäck, F. Nobile, L. Tamellini, R. Tempone, Stochastic spectral Galerkin and collocation methods for PDEs with random coefficients: a numerical comparison, in *Spectral and High Order Methods for Partial Differential Equations*, ed. by J. Hesthaven, E. Ronquist. Volume 76 of Lecture Notes in Computational Science and Engineering (Springer, Berlin/Heidelberg, 2011), pp. 43–62. Selected papers from the ICOSAHOM'09 conference, 22–26 June 2009, Trondheim
3. V. Barthelmann, E. Novak, K. Ritter, High dimensional polynomial interpolation on sparse grids. Adv. Comput. Math. **12**(4), 273–288 (2000)
4. J. Beck, F. Nobile, L. Tamellini, R. Tempone, On the optimal polynomial approximation of stochastic PDEs by Galerkin and collocation methods. Math. Models Methods Appl. Sci. **22**(09), 1250023 (2012)
5. J. Beck, F. Nobile, L. Tamellini, R. Tempone, A Quasi-optimal sparse grids procedure for groundwater flows, in *Spectral and High Order Methods for Partial Differential Equations – ICOSAHOM'12*, ed. by M. Azaïez, H. El Fekih, J. S. Hesthaven. Volume 95 of Lecture Notes in Computational Science and Engineering (Springer International Publishing, Switzerland, 2014), pp. 1–16. Selected papers from the ICOSAHOM'12 conference
6. H. Bungartz, M. Griebel, Sparse grids. Acta Numer. **13**, 147–269 (2004)
7. J. Charrier, Strong and weak error estimates for elliptic partial differential equations with random coefficients. SIAM J. Numer. Anal. **50**(1), 216–246 (2012)
8. A. Chkifa, A. Cohen, C. Schwab, High-dimensional adaptive sparse polynomial interpolation and applications to parametric PDEs. *Found. Comput. Math.* **14**, 601–633 (2014)
9. K. Cliffe, M. Giles, R. Scheichl, A. Teckentrup, Multilevel monte carlo methods and applications to elliptic PDEs with random coefficients. Comput. Vis. Sci. **14**(1), 3–15 (2011)
10. B.A. Davey, H.A. Priestley, *Introduction to Lattices and Order*, 2nd edn. (Cambridge University Press, New York, 2002)
11. P. Diggle, P.J. Ribeiro, *Model-Based Geostatistics* (Springer, New York, 2007)

12. M.S. Eldred, J. Burkardt, Comparison of non-intrusive polynomial chaos and stochastic collocation methods for uncertainty quantification. American Institute of Aeronautics and Astronautics Paper 2009–0976 (2009)
13. H.C. Elman, C.W. Miller, E.T. Phipps, R.S. Tuminaro, Assessment of collocation and Galerkin approaches to linear diffusion equations with random data. Int. J. Uncertain. Quantif. 1(1), 19–33 (2011)
14. O. Ernst, B. Sprungk, Stochastic collocation for elliptic PDEs with random data: the lognormal case, in Sparse Grids and Applications – Munich 2012, ed. by J. Garcke, D. Pflüger. Volume 97 of Lecture Notes in Computational Science and Engineering (Springer International Publishing, Switzerland, 2014), pp. 29–53
15. J. Foo, X. Wan, G. Karniadakis, The multi-element probabilistic collocation method (ME-PCM): error analysis and applications. J. Comput. Phys. 227(22), 9572–9595 (2008)
16. A. Genz, B.D. Keister, Fully symmetric interpolatory rules for multiple integrals over infinite regions with Gaussian weight. J. Comput. Appl. Math. 71(2), 299–309 (1996)
17. T. Gerstner, M. Griebel, Dimension-adaptive tensor-product quadrature. Computing 71(1), 65–87 (2003)
18. C.J. Gittelson, Stochastic Galerkin discretization of the log-normal isotropic diffusion problem. Math. Models Methods Appl. Sci. 20(2), 237–263 (2010)
19. H. Harbrecht, M. Peters, M. Siebenmorgen, Multilevel accelerated quadrature for PDEs with log-normal distributed random coefficient. Preprint 2013–18 (Universität Basel, 2013)
20. J.D. Jakeman, R. Archibald, D. Xiu, Characterization of discontinuities in high-dimensional stochastic problems on adaptive sparse grids. J. Comput. Phys. 230(10), 3977–3997 (2011)
21. A. Klimke, Uncertainty modeling using fuzzy arithmetic and sparse grids, PhD thesis, Universität Stuttgart, Shaker Verlag, Aachen, 2006
22. S. Martello, P. Toth, Knapsack Problems: Algorithms and Computer Implementations. Wiley-Interscience Series in Discrete Mathematics and Optimization (Wiley, New York, 1990)
23. A. Narayan, J.D. Jakeman, Adaptive leja sparse grid constructions for stochastic collocation and high-dimensional approximation. SIAM J. Sci. Comput. 36(6), A2952–A2983 (2014)
24. F. Nobile, L. Tamellini, R. Tempone, Convergence of quasi-optimal sparse-grid approximation of Hilbert-space-valued functions: application to random elliptic PDEs. Numerische Mathematik, doi:10.1007/s00211-015-0773-y
25. F. Nobile, L. Tamellini, R. Tempone, Comparison of Clenshaw–Curtis and Leja quasi-optimal sparse grids for the approximation of random PDEs, in Spectral and High Order Methods for Partial Differential Equations – ICOSAHOM'14, ed. by R.M. Kirby, M. Berzins, J.S. Hesthaven. Volume 106 of Lecture Notes in Computational Science and Engineering (Springer International Publishing, Switzerland, 2015)
26. F. Nobile, R. Tempone, C. Webster, An anisotropic sparse grid stochastic collocation method for partial differential equations with random input data. SIAM J. Numer. Anal. 46(5), 2411–2442 (2008)
27. F. Nobile, F. Tesei, A Multi Level Monte Carlo method with control variate for elliptic PDEs with log-normal coefficients. Stoch PDE: Anal Comp (2015) 3:398–444
28. C. Schillings, C. Schwab, Sparse, adaptive Smolyak quadratures for Bayesian inverse problems. Inverse Probl. 29(6), 065011 (2013)
29. S. Smolyak, Quadrature and interpolation formulas for tensor products of certain classes of functions. Dokl. Akad. Nauk SSSR 4, 240–243 (1963)
30. L. Tamellini, F. Nobile, Sparse Grids Matlab kit v.15-8. http://csqi.epfl.ch, 2011–2015
31. G. Wasilkowski, H. Wozniakowski, Explicit cost bounds of algorithms for multivariate tensor product problems. J. Complex. 11(1), 1–56 (1995)
32. G. Zhang, D. Lu, M. Ye, M. Gunzburger, C. Webster, An adaptive sparse-grid high-order stochastic collocation method for bayesian inference in groundwater reactive transport modeling. Water Resour. Res. 49(10), 6871–6892 (2013)

A New Subspace-Based Algorithm for Efficient Spatially Adaptive Sparse Grid Regression, Classification and Multi-evaluation

David Pfander, Alexander Heinecke, and Dirk Pflüger

Abstract As data has become easier to collect and precise sensors have become ubiquitous, data mining with large data sets has become an important problem. Because sparse grid data mining scales only linearly in the number of data points, large data mining problems have been successfully addressed with this method. Still, highly efficient algorithms are required to process very large problems within a reasonable amount of time.

In this paper, we introduce a new algorithm that can be used to solve regression and classification problems on spatially adaptive sparse grids. Additionally, our approach can be used to efficiently evaluate a spatially adaptive sparse grid function at multiple points in the domain. In contrast to other algorithms for these applications, our algorithm fits well to modern hardware and performs only few unnecessary basis function evaluations.

We evaluated our algorithm by comparing it to a highly efficient implementation of a streaming algorithm for sparse grid regression. In our experiments, we observed speedups of up to $7\times$, being faster in all experiments that we performed.

1 Introduction

A well-known problem in the area of data mining is regression. Regression deals with the representation of an unknown function based on given data points that are interpreted as evaluations of that unknown function. The data points that are given to learn the function form the training data set.

For regression problems many approaches exist [2]. Most of the algorithms have the property that their complexity is at least $O(m^2)$ where m is the number of training instances. In contrast, the sparse grid approach is linear in the number of training

D. Pfander (✉) • D. Pflüger
University of Stuttgart, IPVS, Stuttgart, Germany
e-mail: David.Pfander@ipvs.uni-stuttgart.de; Dirk.Pflueger@ipvs.uni-stuttgart.de

A. Heinecke
Intel Cooperation, Santa Clara, CA, USA
e-mail: alexander.heinecke@intel.com

© Springer International Publishing Switzerland 2016
J. Garcke, D. Pflüger (eds.), *Sparse Grids and Applications – Stuttgart 2014*,
Lecture Notes in Computational Science and Engineering 109,
DOI 10.1007/978-3-319-28262-6_9

221

instances. Because the sparse grid approach is build upon spatial discretization, it can only be used for problems of moderate dimension.

To address higher-dimensional problems successfully and to increase efficiency for lower-dimensional problems, spatial adaptivity can be used. With spatial adaptivity enabled, sparse grids were successfully applied to problems with up to 166 dimensions [4, 14]. In this paper, we differentiate between adaptive and non-adaptive sparse grids. Non-adaptive sparse grids are called regular sparse grids.

Sparse grid data mining has been used in high performance computing in the past. Highly efficient algorithms have been developed and implemented for large shared-memory machines, clusters, and for accelerator cards [9–11]. An algorithm that was shown to be well-suited for high-performance data mining is the streaming algorithm for sparse grid data mining. This algorithm can be implemented highly efficiently. But to achieve perfect streaming properties, it is doing a lot of unnecessary work. Because of its overall high performance, we used a highly efficient implementation of the streaming algorithm [10] as a reference in our experiments. The streaming algorithm is targeted at spatially adaptive sparse grids. For grids that make further assumptions about the distribution of the grid points, even faster algorithms exist [5, 13].

In this paper, we present a new algorithm to perform data mining on spatially adaptive sparse grids. Our algorithm is able to avoid most unnecessary computations that are performed by the streaming algorithm. To map our basic algorithm to modern processors, new algorithmic ideas were required. We introduce a subspace-skipping scheme to avoid unnecessary computations. Furthermore, we propose a data point blocking scheme that was developed to improve the data locality of our algorithm. As memory is an important issue for this approach, we present data structures for spatially adaptive sparse grids that reduce the required storage space.

Our algorithm assumes that the hierarchical predecessors of each grid point are also part of the grid. To this end, we use regular grids, for which this property is fulfilled, and an adaptivity criterion that adds the hierarchical predecessors of each new grid point. This assumption could be dropped in exchange for a lower performance.

The paper is structured as follows. In Sect. 2, we give a brief introduction to data mining on sparse grids. We continue in Sect. 3 with a presentation of the streaming algorithm and a comparison of different approaches. Then we describe our new algorithm in Sect. 4. The optimizations that are required to map our algorithm to modern processors are introduced in Sect. 5. In Sect. 6, we explain our experimental setup and present the results of our experiments. There, we will also discuss the memory requirements of the algorithm. Section 7 finally concludes the paper.

2 Sparse Grids and Regression

In this section, we present a very condensed introduction to sparse grids and regression on sparse grids. For sparse grids in general, we recommend the paper by Bungartz and Griebel [3]. More thorough presentations of sparse grid data mining are given by Garcke [7] and Pflüger [14].

2.1 Sparse Grids

A sparse grid is a grid in the hypercube $[0, 1]^d$ that is hierarchically constructed based on subgrids Ω_l that are anisotropic grids with a discretization level l. The grid points on a subgrid are enumerated by an index set I_l,

$$I_l := \{(i_1, \ldots, i_d) : 0 < i_k < 2^{l_k}, i_k \text{ odd}\}. \tag{1}$$

We use this definition to further define the subgrids Ω_l as

$$\Omega_l := \{\mathbf{x}_{l,i} = (i_1 2^{-l_1}, \ldots, i_d 2^{-l_d}) : \mathbf{i} \in I_l\}. \tag{2}$$

Now, we can define the hierarchical subspaces used to construct the sparse grid function space. For a level l and a basis function $\phi_{l,i}$ for each grid point, a subspace W_l is defined as

$$W_l := \text{span}\{\phi_{l,i} : \mathbf{i} \in I_l\}. \tag{3}$$

We note that for basis functions with pairwise disjoint interior of the support, the correct evaluation of a function $g \in W_l$ involves only the evaluation of the basis function that belongs to a single grid points $\mathbf{x}_{l,i} \in \Omega_l$.

Based on these definitions, we can now formally define a sparse grid $V_n^{(1)}$ with an overall discretization level n,

$$V_n^{(1)} := \bigoplus_{|l|_1 \leq n+d-1} W_l. \tag{4}$$

For our basis functions, we employ a tensor-product approach with d-dimensional hat functions. With this well-known approach, the basis functions $\phi_{l,i}$ are given by

$$\phi_{l,i}(\mathbf{x}) := \prod_{j=1}^{d} \phi_{l_j,i_j}(x_j), \tag{5}$$

with the scaled and translated 1-dimensional hat functions

$$\phi_{l,i}(x) := \max(0, 1 - |2^l x - i|). \tag{6}$$

A sparse grid function $f \in V_n^{(1)}$ can now be written as

$$f(\mathbf{x}) = \sum_{|\mathbf{l}|_1 \leq n+d-1} \sum_{\mathbf{i} \in h} \alpha_{\mathbf{l},\mathbf{i}} \phi_{\mathbf{l},\mathbf{i}}(\mathbf{x}) = \underbrace{\sum_{i=1}^{N} \alpha_i \phi_i(\mathbf{x})}_{\text{shorthand}}. \tag{7}$$

The coefficients $\boldsymbol{\alpha}$ of the sparse grid functions are called surpluses. Additionally, we denote the total number of grid points by N.

2.2 Regression on Sparse Grids

We assume that a normalized training data set

$$T := \{(\mathbf{x}_i, y_i) : \mathbf{x}_i \in [0,1]^d, y_i \in \mathbb{R}\}_{i=1}^m \tag{8}$$

is given. It consists of m data points that can be interpreted as function evaluations of an unknown function $f : [0,1]^d \to \mathbb{R}$. The goal of the regression problem is to construct an approximation $f^* \approx f$ so that for every tuple in the data set is holds that $f^*(\mathbf{x}_i) \approx y_i$ [2].

Such an approximation $f^* \in V_n^{(1)}$ of the function f can be constructed with a penalized least-squares approach [7, 8, 14],

$$f^* := \arg\min_{g \in V_n^{(1)}} \left(\frac{1}{m} \sum_{i=1}^m (y_i - g(\mathbf{x}_i))^2 + \lambda \sum_{j=1}^N \alpha_j^2 \right). \tag{9}$$

We use a weight decay regularizer [2] to enforce a certain degree of smoothness in the obtained sparse grid function [7].

The minimum f^* can be calculated by setting the gradient of the term to minimize to zero. This leads to a system of linear equations for the surpluses $\boldsymbol{\alpha}$ of the sparse grid function [2, 14],

$$\left(\frac{1}{m} BB^T + \lambda I \right) \boldsymbol{\alpha} = \frac{1}{m} B\mathbf{y}, \tag{10}$$

$$B_{ij} = \phi_i(\mathbf{x}_j). \tag{11}$$

Here, I denotes the identity matrix.

To solve this system of linear equations, we have used the conjugate gradient algorithm [16]. We note that the computational effort is concentrated in calculating the two matrix-vector-products $\mathbf{v} := B^T\boldsymbol{\alpha}$ and $\mathbf{v}' := B\mathbf{v}$ in each iteration. Therefore, to improve the overall performance of this sparse grid regression approach, efficient algorithms for these matrix-vector-products are required.

2.3 Closely Related Problems

The new algorithm for sparse grid regression proposed in this paper also addresses two further problems. In a binary classification problem, we assume that a normalized training data set

$$T = \{(\mathbf{x}_i, c_i) : \mathbf{x}_i \in [0, 1]^d, c_i \in \mathrm{K}\}_{i=1}^m \qquad (12)$$

is given. Here, the data points $\mathbf{x}$ are associated with a class label from a set of labels $\mathrm{K} = \{k_0, k_1\}$. The goal is to construct a classification function f^* that can classify unseen data points correctly [7, 14].

This binary classification problem can be reduced to a regression problem by replacing the two class labels k_0 and k_1 with the values -1 and 1. This modified data set is now treated as a regression problem. By solving this regression problem, we obtain a sparse grid function f^*. To classify a data point $\mathbf{x}$, we first evaluate the function f^* at the data point $\mathbf{x}$. This value is then used to select the class that fits best:

$$c(\mathbf{x}) = \begin{cases} k_0 & \text{if } f^*(\mathbf{x}) < 0, \\ k_1 & \text{if } f^*(\mathbf{x}) \geq 0. \end{cases} \qquad (13)$$

This approach can be easily extended to work with non-binary classification problems [14].

A second closely related problem is the evaluation of a sparse grid function at many points in the domain. This operation is called multi-evaluation. When the system of linear equations of the regression algorithm is solved, the matrix-vector-product $B^T\boldsymbol{\alpha}$ is calculated. With the definition of B this leads to the sums

$$(B^T\boldsymbol{\alpha})_i = \sum_{j=1}^N \phi_j(\mathbf{x}_i)\alpha_j, \quad i \in \{1, \ldots, m\}. \qquad (14)$$

This is just an evaluation for m given data points. In this paper, we therefore address multi-evaluation by providing efficient algorithms for the transposed operator $B^T\boldsymbol{\alpha}$.

3 Relationship to Previous Approaches

In this section, we describe the streaming algorithm for regression and the recursive algorithm for multi-evaluation. Both algorithms are well-suited for spatially adaptive sparse grids. Because the streaming algorithm treats both matrix-vector-products $\mathbf{v} := B^T \boldsymbol{\alpha}$ and $\mathbf{v}' := B\mathbf{v}$ similarly, we only describe the multi-evaluation operation $\mathbf{v} := B^T \boldsymbol{\alpha}$.

The streaming algorithm for regression on sparse grids performs the most direct approach to sparse grid function evaluation. For each data point, it evaluates every basis function and accumulates the result [10]. For this algorithm, the grid can be represented as a list of tuples with one tuple for each grid point. Each tuple consists of the level $\mathbf{l}$, the index $\mathbf{i}$, and the surplus value $\alpha_{\mathbf{l},\mathbf{i}}$. This algorithm is shown in pseudo code in Algorithm 1.

This brute-force approach has several properties that can be exploited on modern processors. First, each evaluation of a data point is completely independent from every other evaluation. Additionally, as the algorithm simply iterates through the grid points for each evaluation, the operations are exactly the same for each data point. This enables a straightforward and highly efficient parallelization and vectorization of the algorithm. Thus, the streaming algorithm maps to modern processors very well.

On the other hand, the streaming algorithm performs many unnecessary basis function evaluations. The support of the basis functions on each subspace partitions the domain. Therefore, only one basis function per subspace has to be evaluated for each data point. However, the streaming approach evaluates all basis functions on a subspace.

A different approach to calculate the multi-evaluation operation $v := B^T \boldsymbol{\alpha}$ is the recursive algorithm. For this algorithm, the grid is stored as a tree that represents the hierarchical relationship between the grid points so that every grid point can be reached by exactly one parent grid point, if a parent grid point exists [15]. An example for such a tree in 2 dimensions is shown in Fig. 1. The algorithm traverses the tree from the root to the leaf nodes. It evaluates the basis function associated with the current node and then performs a recursive call for each child grid point

---

Algorithm 1: The streaming algorithm for the operator $v := B^T \boldsymbol{\alpha}$ performs a separate evaluation for each data point. To evaluate a data point, it evaluates every basis function. Due to the support of the basis functions, many zero-evaluations of basis functions happen

```
for xⱼ ∈ T do
    vⱼ ← 0;
    for (l, i, αₗ,ᵢ) ∈ gridPoints do
        vⱼ ← vⱼ + φₗ,ᵢ(xⱼ) αₗ,ᵢ;
```

Fig. 1 A tree that represents
a sparse grid for the recursive
algorithm. The grid point
marked in *red* is the root node
of the tree

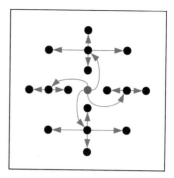

Algorithm 2: The recursive algorithm for the operation $B^T \alpha$

Function eval (**x**, node)
    $\mathbf{l}, \mathbf{i}, \alpha_{\mathbf{l},\mathbf{i}}$ ←extractFromNode (node);
    result ←$\phi_{\mathbf{l},\mathbf{i}}(\mathbf{x}) \alpha_{\mathbf{l},\mathbf{i}}$;
    successors ←successorsWithSupport (node);
    for successor ∈ successors **do**
        result ← result+eval (**x**, successor);
    return result;
for $\mathbf{x}_j \in T$ **do**
    v_j ←eval ($\mathbf{x}_j$, root);

---

the data point has support on. Therefore, it only evaluates one basis function per
subspace. The algorithm is shown in pseudo code in Algorithm 2.

For an implementation of this algorithm that uses a pointer-based data structure,
the memory accesses are expensive. Furthermore, the recursive structure makes it
more difficult to vectorize and parallelize the algorithm which has become important
on modern processors with wide vector registers. For these reasons, the recursive
algorithm was evaluated to be significantly slower than the streaming algorithm for
large data sets [10].

The algorithm that we present in this paper is related to the recursive approach
as it also evaluates at most one basis function per subspace. At the same time, our
approach maps well to modern processors. Thus, we realize the optimal complexity
of the recursive approach while retaining the good hardware utilization of the
streaming algorithm.

4 The Subspace-Based Approach

In this section, we introduce a new subspace-based algorithm for spatially adaptive
sparse grids. However, before turning to spatially adaptive sparse grids, we will first
introduce a simpler subspace-based algorithm for regular sparse grids (Fig. 2).

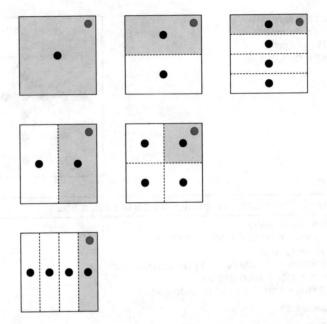

Fig. 2 A regular sparse grid is evaluated at a data point (*red dot*) with the subspace-based algorithm for regular sparse grids. We display the hierarchical components of the grid by showing the subgrids with the support of the corresponding basis functions. The corresponding subspaces can be processed in arbitrary order

4.1 A Subspace-Based Algorithm for Regular Sparse Grids

Like the recursive algorithm, the subspace-based algorithm for regular sparse grids processes the data points individually. For every data point, it iterates the list of subspaces. The subspaces are modeled as a tuple consisting of the level l and an array that holds the surpluses for the basis functions. On each subspace, the algorithm calculates the index of the basis function on which the data point has support. Then it fetches the corresponding surplus value $\alpha_{l,i}$ from the surplus array. Finally, the basis function is evaluated and the algorithm proceeds with the next subspace. This approach is outlined in Algorithm 3. An explanatory evaluation of a single data point is shown in Fig. 2.

In comparison to the streaming algorithm, the evaluation of a single grid point is more expensive. Both the index calculation and the surplus fetch operation have to be efficiently performed to be competitive.

First, we explain how to efficiently calculate the index of the relevant basis function on a single subspace for a given data point. To this end, we look at the formula

$$x_j = i_j 2^{-l_j} \tag{15}$$

Algorithm 3: The subspace-based algorithm for the operation $\mathbf{v} := B^T \alpha$ for regular sparse grids

---

for $\mathbf{x}_j \in T$ **do**
  $v_j \leftarrow 0$;
  for $(\mathbf{l}, \text{surplusArray}) \in$ subspaces **do**
    $\mathbf{i} \leftarrow$ calculateIndex $(\mathbf{l}, \mathbf{x}_j)$;
    $\alpha_{\mathbf{l},\mathbf{i}} \leftarrow$ fetchSurplus $(\text{surplusArray}, \mathbf{l}, \mathbf{i})$;
    $v_j \leftarrow v_j + \phi_{\mathbf{l},\mathbf{i}}(\mathbf{x}_j)\,\alpha_{\mathbf{l},\mathbf{i}}$;

---

that describes the relationship between level $\mathbf{l}$, index $\mathbf{i}$, and location $\mathbf{x}_{\mathbf{l},\mathbf{i}}$ of a grid point in the individual dimensions. By solving this equation by the index i_j, we get the formula

$$i_j = x_j 2^{l_j}. \tag{16}$$

We can now infer the relevant index $\mathbf{i}'$ for a data point $\mathbf{x}'$ by replacing the component x_j of the grid point with the components x'_j of the data point and then rounding to the nearest odd number in each dimension,

$$i'_j = \text{nearestOdd}(x'_j 2^{l_j}). \tag{17}$$

Thereby, we get the index of the closest grid point. Due to the support of the 1-dimensional hat functions, the closest grid point always belongs to a basis function the data point has support on. Additionally, data points that are exactly in the middle between two grid points in at least one dimension will evaluate to zero for any basis function and we therefore obtain the correct value for these data points, too.

For this algorithm, the surpluses can be stored efficiently in a d-dimensional array for each subgrid as all subgrids are d-dimensional anisotropic grids. Usually, a d-dimensional array is implemented using a 1-dimensional array and an address calculation scheme. This approach was used for our algorithms as well. The index in a linear array for a d-dimensional level-index-vector can be calculated with the formula

$$\text{linearIndex}(\mathbf{l}, \mathbf{i}) := \left(\left(\left(\left\lfloor \frac{i_1}{2} \right\rfloor 2^{l_2 - 1} + \left\lfloor \frac{i_2}{2} \right\rfloor\right) 2^{l_3 - 1} + \left\lfloor \frac{i_3}{2} \right\rfloor\right) \cdots + \left\lfloor \frac{i_d}{2} \right\rfloor\right). \tag{18}$$

The divisions and rounding operations are required to skip the even indices correctly.

With this formula, the calculation of the linear index that is used to access the array surpluses can be done in $O(d)$ operations. We note that this index calculation scheme is efficient as only multiplications, additions, and shift operations are required. Because this algorithm deals with regular grids, each array component represents a grid point and all array components are used. Therefore, there is no memory overhead for the storage of the grid.

Algorithm 4: Basic spatially adaptive algorithm for the operator $\mathbf{v} := B^T \alpha$

---

for $\mathbf{x}_j \in T$ **do**
  $\quad v_j \leftarrow 0;$
  for $(\mathbf{l}, \text{surplusArray}) \in \text{subspaces}$ **do**
    $\quad\quad \mathbf{i} \leftarrow \text{calculateIndex}(\mathbf{l}, \mathbf{x}_j);$
    $\quad\quad \alpha_{\mathbf{l},\mathbf{i}} \leftarrow \text{fetchSurplus}(\mathbf{l}, \mathbf{i}, \text{surplusArray});$
    $\quad\quad$ **if** $\neg \text{isNaN}(\alpha_{\mathbf{l},\mathbf{i}})$ **then**
      $\quad\quad\quad v_j \leftarrow v_j + \phi_{\mathbf{l},\mathbf{i}}(\mathbf{x}_j)\,\alpha_{\mathbf{l},\mathbf{i}};$

4.2 Towards Spatially Adaptive Sparse Grids

The algorithm for regular grids introduced in Sect. 4.1 can be extended easily to work with spatially adaptive sparse grids. We can use the same data structure and store the surpluses as if the grid were a regular sparse grid. Because the algorithm for regular sparse grids does not make any assumptions about the specific subspaces involved, a modification is only required to treat some of the grid points correctly. For a spatially adaptive sparse grid, a subgrid does not necessarily contain all grid points that it could contain. We mark these missing grid points with a Not-A-Number (NaN) value from the IEEE 754 floating point standard in the surplus array of the subgrid.

The basic algorithm for spatially adaptive sparse grids is shown in Algorithm 4. With the described way of storing the surpluses, the only difference to Algorithm 3 is the special treatment of the missing grid points. Here, the NaN values are used to skip a specific evaluation.

As a potential optimization, non-existing grid points could be encoded with the value zero. As a result, the conditional statement in the algorithm could be omitted. But if zero would be used instead of NaN, it would be impossible to differentiate between an existing grid point with a surplus value of zero and a non-existing one. However, this information will be important when the algorithm is further improved in Sect. 5.1. Therefore, the value NaN was chosen.

Filling up the non-empty subspaces wastes a lot of memory. This problem is addressed in Sect. 5.5 as we first focus on improving the performance of the basic algorithm and then discuss the memory usage.

Similar to the streaming algorithm, the parallelization of this algorithm is straightforward. Because all evaluations are independent, the algorithm can be parallelized by assigning a data point to a thread. This is an efficient kind of parallelization when dealing with large data sets as the degree of parallelism easily exceeds what is required by modern shared-memory systems.

4.3 The Operator $\mathbf{v}' := B\mathbf{v}$

To iteratively solve the system of linear equations, an efficient algorithm for the second operator $\mathbf{v}' := B\mathbf{v}$ is required as well. Unfortunately, the operation $\mathbf{v}' := B\mathbf{v}$ is not a function evaluation. Without an evaluation-like structure, we can no longer avoid the unnecessary basis function evaluations easily.

To make use of the subspace structure again, we first observe that an evaluation-like structure is still present in the columns of B. We therefore run through the columns instead of the rows. But as the values in the columns all belong to different components in the result vector $\mathbf{v}'$, the results cannot be simply accumulated. Instead, after each basis function evaluation, the corresponding component of $\mathbf{v}'$ is updated.

As we still want to evaluate only a single basis function per subspace, we have to find the entry in the result vector that is associated with the currently processed basis function. Fortunately, a data structure that efficiently fetches a value associated with a basis function was introduced to store the surpluses for the operation $\mathbf{v} := B^T\alpha$. We use this data structure again, but this time to store the result vector $\mathbf{v}'$. To this end, a preprocessing and a postprocessing step are required.

In the preprocessing step, the surplus data structure is initialized by setting the values to zero for existing grid points and to NaN for missing grid points. Now, the surplus data structure can be used to accumulate the results. Furthermore, a differentiation between existing and missing grid points is possible as well. After the computation, the grid has to be traversed again to write the results from the surplus data structure back into a vector. Because pre- and postprocessing require only a single iteration through the grid, these operations are cheap compared to the actual matrix-vector-product calculation. The algorithm for this operator is shown in pseudo code in Algorithm 5.

Algorithm 5: Basic spatially adaptive algorithm for the operator $\mathbf{v}' := B\mathbf{v}$ with pre- and postprocessing only indicated. The data structure for the surpluses now holds the partial results instead of actual surplus values. We use the same concepts that we used for the operator $\mathbf{v} := B^T\alpha$ to make it explicit that the data structure and the operations are identical

```
initSurpluses(subspaces);
for j ← 1; j ≤ |T|; j ← j + 1 do
    x ← T[j];
    for (l, surplusArray) ∈ subspaces do
        i ← calculateIndex(l, x);
        partial ← fetchSurplus(l, i, surplusArray);
        if ¬isNaN(partial) then
            partial ← partial + φ_{l,i}(x) v_j;
            setSurplus(partial, l, i, surplusArray);
v' ← toVector(subspaces)
```

Similar to the operation $\mathbf{v} := B^T\boldsymbol{\alpha}$, this algorithm can be easily parallelized by assigning a data point to a thread and the thread then performs the evaluation-like operation for the data point. While the calculation of $\mathbf{v}' := B\mathbf{v}$ is still more expensive than $\mathbf{v} := B^T\boldsymbol{\alpha}$, its runtime is only about 10 % larger. Additionally, because we can treat both operators in a very similar way, all optimizations that were developed were integrated in both operators. For these reasons, we do not specifically address the operator $\mathbf{v}' := B\mathbf{v}$ in the presentation of our improved algorithm.

5 An Improved Subspace-Based Algorithm

In this section, we present improvements for the basic subspace-based algorithm to obtain an algorithm that is competitive with a highly efficient implementation of the streaming algorithm. We will introduce two major algorithmic improvements: subspace-skipping to further reduce the number of grid point evaluations and a data point blocking scheme to improve the data locality of the algorithm. At the end of the section, we discuss some implementation details and present additional smaller improvements.

5.1 Subspace-Skipping

A subspace-based approach already reduces the number of required evaluations significantly by evaluating only one basis function per subspace. But there are still unnecessary evaluations performed. This is best illustrated with an example.

Figure 3 shows the evaluation of a 1-dimensional grid at a point x. The grid point on level 1 exists and has already been evaluated. The next subspace evaluated is the subspace with level 2. Here, a matching grid point does not exist. The original algorithm would proceed with the evaluation of the subspace with level 3. Due to the hierarchical structure of the grid and by employing a refinement criterion that ensures that all predecessors of a grid point are part of the grid, we know that there

Fig. 3 A data point x is evaluated with a sparse grid function. Only the *black grid points* are part of the grid. The *gray grid points* are the grid points that are considered for evaluation. As the grid point on level 2 is missing, the subspace with level 3 can be skipped

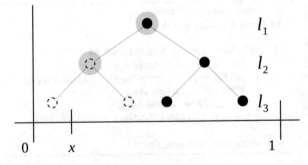

cannot be a non-zero evaluation for x on the subspace with level 3. We can therefore skip this subspace.

In one dimension the skipping process is straightforward. We iterate the subspaces with ascending level and stop the iteration as soon as we encounter the first subspace in which we cannot find a matching grid point. This is more complicated in higher dimensions.

To deal with higher dimensions, we first define a reflexive parent-child relation between subspaces. A subspace with level $\mathbf{l}$ is a child of another subspace with level $\mathbf{l'}$ if

$$(\mathbf{l}, \mathbf{l'}) \in \text{IsChild} \Leftrightarrow l_j \geq l'_j, \forall j \in \{1, \ldots, d\}. \tag{19}$$

As all predecessors of a grid point are part of the grid, it holds for higher dimensional cases that all child subspaces can be skipped. However, as each subspace has up to d direct successors, the algorithm cannot stop at the first encounter of a non-existent grid point as in the 1-dimensional case.

To get to a subspace-skipping algorithm for higher dimensions, we order the subspaces lexicographically according to their level. This preprocessing step is cheap as the number of subspaces is usually much smaller than the number of grid points and data points. In principle, if a grid point does not exist on a specific subspace, all child subspaces of that subspace can be skipped. However, it would require additional effort to track which child subspaces should be skipped. To avoid this, we skip only the child subspaces that immediately follow the current subspace in the iteration. An example for the algorithm with subspace-skipping enabled is shown in Fig. 4.

5.2 Data Point Blocking to Improve Cache-Efficiency

Modern processors depend on efficient memory access patterns as the gap between the computational resources and the bandwidth as well as the latency of the memory has widened. Caches are implemented to mitigate this problem to some extent. But in the case of sparse grid regression, the data set and the grid can be large and typically do not fit completely into the last level cache of the processor.

As the surpluses are stored in one array per subspace, the basic algorithm utilizes the cache of the processor very inefficiently. During the evaluation of a data point, there is only one access per array as only one basis function per subspace is evaluated. The next access to the same array happens only when the next data point is evaluated and even then it is not guaranteed that a surplus in the same cache-line is accessed. We therefore conclude that this algorithm has bad spatial and temporal data locality.

The algorithm can be improved significantly by subscribing multiple data points to a single thread. Thereby, a single thread evaluates a batch of data points with one pass through the subspaces, instead of evaluating only a single data point. An

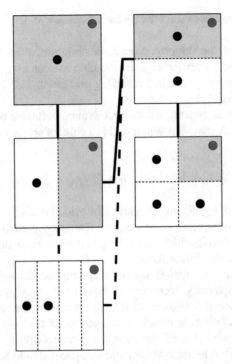

Fig. 4 A 2-dimensional sparse grid function is evaluated at a point (*red*) in the domain. The connections between the subspaces show the order in which the subspaces are iterated (*dashed*), the iteration starts with the subspace with level (1, 1). The *black line* shows the actually processed subspaces. As no matching grid point exists on the second subspace, the third subspace is skipped. One additional zero-evaluation is still performed as only immediately following child subspaces are skipped

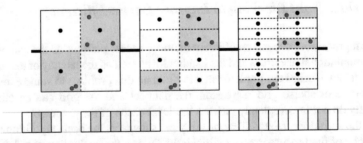

Fig. 5 A set of data points (*red*) is processed on some subspaces of a grid. On some subspaces they reside on the support of the same grid point and share the surplus value

evaluation of multiple data points which partially share the same surpluses is shown in Fig. 5. This approach is a data blocking scheme with a corresponding block size.

The benefit of this scheme is that if the subgrid of a subspace is small compared to the block size, there is a good chance that memory accesses to the same cache line occur. This enables a reusage of some data that was already loaded into the cache.

Therefore, the spatial and temporal data locality of the algorithm is improved for smaller subgrids.

5.3 Caching Intermediate Values

As a further optimization, some intermediate results throughout the evaluation process are cached. If the evaluation of a subspace is followed by the evaluation of a subspace whose level **l** changed only in one component, only one component in the index **i** will change. For example, the basis function with the index $(1, 1, 1)$ on the subspace with level $(2, 2, 2)$ is evaluated and the next subspace has the level $(2, 2, 3)$. Then, there are only two possible values for the index vector. Either the next grid point has the index $(1, 1, 1)$ or the index $(1, 1, 3)$. In general, only the index components have to be recalculated for which the level vector of the subspace changed.

The same idea can be applied to the basis function evaluation and the calculation of the linear index. Due to the tensor-product approach, only those 1-dimensional basis functions will yield different results for which the index changed. For the linear index calculation, caching can be enabled slightly differently. As the linear index calculation is done with a formula similar to a Horner scheme, the linear index has to be recomputed starting with the first component that changed. For an efficient implementation, all caching was implemented so that the values were recalculated starting with the first component that changed.

5.4 Vectorization

The implementation of our algorithm uses the Advanced Vector Instructions (AVX) and the Streaming SIMD Extensions (SSE) for a better utilization of our hardware resources [12]. These are vector instruction sets that are available on most recent processors. Vectorization was relatively straightforward as it was combined with the data point blocking scheme discussed in Sect. 5.2. The set of data points introduced to improve the data locality can be used to enable parallel evaluations by combining groups of 4 data points and perform the calculations with vector instruction.

Both instruction sets had to be combined as AVX does not contain integer operations and SSE offers instructions for packed groups of four 32-bit integers. The integer instructions are used to speed up the calculation of the index **i** and the linear index used to access the surpluses.

While there was a benefit of about 20 % in performance, the gain through vectorization was relatively small compared to the maximum factor of 4 as some non-continuous memory accesses are required and AVX and SSE have no gather and no scatter support.

5.5 Representation in Memory

The basic subspace-based algorithm stores all subgrids as if a regular grid was processed. It therefore allocates memory for all grid points that could exist on the subgrids. However, through spatially adaptive refinement, very large subgrids can be reached after some refinement steps while the grid actually contains only few grid points. This results in a large memory overhead.

This issue can be mitigated by representing different subgrids in different ways. We differentiate between two types of subgrids:

$$\text{A subgrid } \Omega_l \text{ is a list subgrid} \Leftrightarrow \frac{\text{existingGridPoints}(\Omega_l)}{|\Omega_l|} \leq t_{\text{repr}}. \quad (20)$$

$$\text{A subgrid } \Omega_l \text{ is an array subgrid} \Leftrightarrow \frac{\text{existingGridPoints}(\Omega_l)}{|\Omega_l|} > t_{\text{repr}}. \quad (21)$$

Here, the value $t_{\text{repr}} \in [0, 1]$ is the threshold used to determine how a subgrid is represented. In our implementation t_{repr} was set to 0.2.

Surpluses of array subgrids are stored using d-dimensional arrays as if the subgrid would contain all possible grid points, i.e., the same way the basic algorithm stores the surpluses. For the array subgrids, the algorithm does not have to be modified.

Surpluses of list subgrids are stored as a list of pairs. We use the linear index of the corresponding grid point as the first component of the pair. The surplus value of the corresponding grid point is used as the second component of the pair. When a list subgrid is processed, the tuples are used to temporarily construct an array representation. With this representation, list subgrids with few existing grid points can be stored efficiently.

A limitation of our algorithm is introduced by the temporary array that is required to unpack the list representation. To accommodate all possible subgrids, the temporary array has to be of the size of the largest list subgrid. Furthermore, for an efficient parallel implementation, we use a temporary array for each thread.

The maximum number of existing grid points of a subgrid Ω_l is given by

$$\prod_{i=1}^{d} 2^{l_i - 1} = 2^{(\sum_{i=1}^{d} l_i) - d}. \quad (22)$$

We define the value

$$s_l := \left(\sum_{i=1}^{d} l_i \right) - d \quad (23)$$

to discuss the memory usage independent from the dimensionality of the problem.

If it holds for the largest subgrid that $s_l = 30$, then $2^{30} \cdot 8\,\text{Bytes} = 8\,\text{GB}$ of memory per thread are required. Therefore, the practical limit for this kind of approach will be around $s_l = 30$ for current hardware. There are ways to avoid this limitation, e.g., by falling back to a streaming algorithm for subgrids with s_l greater than a threshold value.

A further potential limitation is the cost associated with the construction of the temporary array representation. However, the cost required to set up a temporary array is usually low compared the cost of the evaluations on the subspace. Additionally, after the temporary arrays are set up, the surpluses are now in the cache and can be cheaply accessed in the following evaluations. In our experiments, we did not measure any significant overhead for the construction of the temporary arrays.

6 Results

Experiments were performed to evaluate the presented subspace-based algorithms. In this section, we introduce the data sets used for our comparisons and the computational environment. We then compare the basic subspace-based algorithm and the improved subspace-based algorithm to the streaming algorithm. In the end, we evaluate some improvements in detail and discuss the measured memory requirements.

6.1 Data Sets and Experimental Setup

We used four data sets in our experiments with 90,000 to almost 400,000 data points. These are the same data sets that were used to evaluate the streaming algorithm [10]. The most important properties of these data sets are summarized in Table 1.

The DR5 data set is based on the fifth data release of the Sloan Digital Sky Survey and enables the prediction of the redshift estimation of galaxies based on photometric data [1]. It is a real-world 5-dimensional data set that results in a grid with a very irregular structure. This is the most interesting data set for our

Table 1 The data sets used in the experiments and a description of the grids at the end of the regression experiments

Name	Dim	Data points	Grid	Grid points
DR5	5	371,907	Very irregular	57,159
Friedman 4d	4	90,000	Somewhat irregular	25,915
Friedman 10d	10	90,000	More regular	37,838
Chess 5d	5	262,143	Very regular	42,896

experiments as we especially want to achieve a high performance for spatially adaptive grids and this data set produces a very irregular grid,

The two Friedman data sets [6] are synthetic data sets that model intermediate cases. The Friedman 4d data set is a 4-dimensional data set that is based on the function

$$\text{fried}_4(x) = \left(x_1^2 + (x_2 x_3 - (x_2 x_4)^{-1})^2\right)^{1/2} + \epsilon. \tag{24}$$

The additional noise ϵ is normally distributed, $\mathcal{N}(0, 125)$.

The Friedman 10d data set results in a more regular grid and was included as a higher dimensional data set. It is based on the function

$$\text{fried}_{10}(\mathbf{x}) = 10\sin(\pi x_1 x_2) + 20(x_3 - 0.5)^2 + 10x_4 + 5x_5 + \epsilon. \tag{25}$$

There are five additional dimensions that contain normally distributed noise, $\mathcal{N}(0, 1)$.

The Chess 5d data set is a 5-dimensional data set and results in a very regular grid. This data sets is used to model a near best-case scenario for our algorithm. It was constructed using the function

$$\text{chess}(\mathbf{x}) = \prod_{k=1}^{5} \begin{cases} -1 & 1/3 < x \le 2/3, \\ 1 & \text{otherwise,} \end{cases} \tag{26}$$

with data points drawn uniformly from $[0, 1]^5$. This data set does not contain any additional noise. However, it has 3^5 different regions to detect [10].

The Friedman 4d, Friedman 10d, and Chess 5d data sets were obtained by sampling the underlying function and then normalizing the results to $[0, 1]^d$ in case of the Friedman data sets. For the DR5 data set, a subset of the data included in the fifth data release of the Sloan Digital Sky Survey was used. This data set had to be normalized as well. Details on the construction of the data sets are given in the literature [14].

All experiments were conducted on a dual-socket platform. This machine was equipped with two Intel® Xeon® E5-2650v2 processors with 8 cores each that are clocked at 2.6 GHz. As the processors support Hyper-Threading and as this feature was enabled, there are 32 threads overall. The Turbo-Boost feature of the processor was enabled as well. While 128 GB of RAM were installed, we required only a small fraction of the available memory in all of our experiments.

The data mining process was always started with a regular grid of level 2 and then 20 refinement steps were performed. After creating the initial grid and after each refinement step, the regression problem was solved with 120 iterations of a conjugate gradient solver. The parameter λ for the regularization operator was set to 10^{-5} in all experiments. It was shown in prior work that sparse grids are well suited to learn these data sets [14] and this work focuses on the performance of the data mining algorithms rather than accuracy. We therefore designed our experiments to be similar to real world data mining scenarios while keeping the runtime low. Furthermore, we started with a grid of a low level and performed many refinement steps to study the performance for spatially adaptive sparse grids. To ensure the correctness of our algorithm, we compared the data mining results to a well-tested implementation of the streaming algorithm.

In the refinement steps, a surplus-based refinement criterion was used. The algorithm first orders the grid points according to the absolute value of their surpluses. Already refined grid points are excluded from this list. Additionally, an upper limit for the number of grid points in the list has to be specified. The grid points in the list are then refined by adding all hierarchical successors of these grid points to the grid. Furthermore, the hierarchical predecessors of the newly added grid points are added as well. Thereby, the predecessors of all grid points are always part of the grid. This is a commonly used refinement strategy that has been successfully used in data mining in the past [14]. In our experiments, the criterion was configured so that up to 80 grid points were refined in each refinement step.

The size of the sparse grids at the end of the experiments is shown in Table 1. The experiments were designed to ensure that the number of grid points did not exceed the number of data points in the data set.

We compared our algorithms against an implementation of the streaming algorithm described in [10]. This highly efficient implementation is parallelized with OpenMP and vectorized for AVX. Furthermore, it makes use of manual loop-unrolling to improve the pipeline utilization.

6.2 The Performance of the Basic Subspace-Based Approach

We performed our data mining experiments with the basic subspace-based algorithm and compared the runtime to the streaming algorithm. The results of these experiments are listed in Table 2. While the performance is acceptable for the Chess 5d data set, the basic algorithm shows a much lower performance than the streaming algorithm for the other data sets. Several reasons contribute to these results.

The streaming algorithm accesses the memory in a very efficient way. It reads sequentially through lists utilizing all data it reads. In contrast, the basic subspace-

Table 2 Overall runtime for the experiments with the basic subspace-based algorithm and the streaming algorithm as comparison. Based on the runtime, the speedup of the basic subspace-based algorithm is calculated. Additionally, the speedup for learning after the last refinement step is provided

Data set	Duration (s)	Duration streaming (s)	Speedup overall	Speedup last step
Chess 5d	2379.46	3530.26	1.48	1.81
Friedman 4d	1966.09	339.19	0.17	0.13
Friedman 10d	5746.21	1679.68	0.29	0.33
DR5	13,515.8	3002.22	0.22	0.24

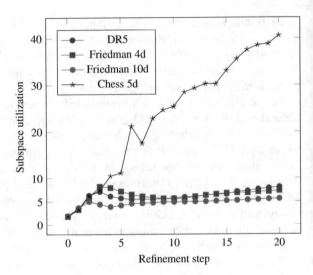

Fig. 6 The development of the average number of grid points per subgrid throughout the refinement process

based approach accesses a single surplus value while processing each subspace. This results in bad data locality, as was explained in Sect. 5.2. Additionally, the evaluation of a single grid point is more expensive due to the calculation of the index **i** and the linear index used to access the surplus array.

The possible speedup of the basic subspace-based algorithm compared to the streaming algorithm depends on the structure of the grid. As we deal with spatially adaptive sparse grids, the number of grid points on a subgrid can be low. Therefore, the additional work that the streaming algorithm performs can be low as well. We conducted experiments to estimate the possible speedup.

The graph in Fig. 6 shows the average number of grid points per subgrid throughout the refinement process. To calculate this measure, the initial grid and the grids after each refinement step were used. For a sparse grid g, the average number of grid points per subgrid is given by the subgrid utilization of a sparse grid,

$$\frac{\text{countGridPoints}(g)}{\text{countSubgrids}(g)}. \tag{27}$$

The basic subspace-based approach always evaluates one grid point per subgrid while the streaming algorithm evaluates all grid points on a subgrid. Therefore, the subgrid utilization describes the possible speedup of the basic subspace-based approach compared to the streaming algorithm.

In our experiments, the subgrid utilization varied widely for different data sets and different refinement steps. For the Chess 5d data set a subgrid utilization of 40.6 was calculated after the last refinement step. That means that in the last refinement step a subspace-based approach could be 40.6 times faster than the streaming algorithm. However, the Friedman 10d data set has only a subgrid utilization of 5.57 after the last refinement step which limits the possible speedup to 5.57×.

Table 3 Overall runtime for the experiments with the improved subspace-based algorithm and the streaming algorithm as comparison. Based on the runtime, the speedup of the improved subspace-based algorithm is calculated. Additionally, the speedup for learning after the last refinement step is provided

Data set	Duration (s)	Duration streaming (s)	Speedup overall	Speedup last step
Chess 5d	465.99	3309.05	7.1	10.25
Friedman 4d	112.05	340.43	3.04	3.16
Friedman 10d	806.15	1683.09	2.09	2.47
DR5	1657.79	3002.56	1.81	2.31

6.3 The Performance of the Improved Subspace-Based Approach

To evaluate the improved subspace-based approach, we compared the runtime of the improved subspace-based algorithm to the streaming algorithm. The speedups calculated from these experiments are shown in Table 3.

These results show that the improved subspace-based algorithm is significantly faster than the streaming algorithm for all data sets. However, the magnitude of the speedup significantly depends on the data set and the resulting structure of the grid. A speedup of 7.1× was observed for the Chess 5d data set with its very regular grid. For the data sets with less regular grids, the Friedman 4d and Friedman 10d data sets, speedups of 3.04× and 2.09× were observed. The data set with the most irregular grid is the DR5 data set. Still, the improved subspace-based approach is faster than the streaming algorithm for the DR5 data set with a speedup of 1.81–2.31×.

Throughout our experiments, we observed a tendency for the performance of our algorithm to improve with larger grids. Figure 7 shows the development of the speedup throughout the refinement process. This graph suggests that the speedup is not yet saturated and even higher speedups would be observed after further refinement steps. This was tested for the DR5 data set in an additional experiment where the number of refinement steps was increased to 30. In this experiment, the overall speedup increased to 2×. But as the subspace-based algorithm depends on the structure of the grid, the performance can also decrease with further refinement steps. This was observed in the experiments with the Friedman 4d data set. Here, the performance started to decrease after 16 refinement steps.

6.4 Important Improvements in Detail

To quantify the benefit of the individual improvements of the improved subspace-based approach, we implemented the individual improvements so that they can be turned off by setting a compiler option. That way we were able to use our fastest implementation and to compare it to the same implementation with a single feature

Fig. 7 The performance of the improved subspace-based algorithm after each refinement step for three of the four data sets

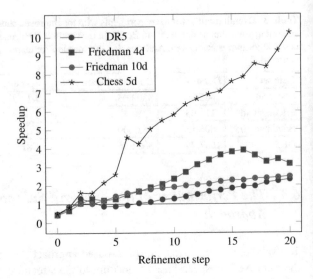

Table 4 Contribution of subspace-skipping to the final algorithm. The speedups were calculated by using the same algorithm with subspace-skipping turned off as a baseline

Experiment	Speedup
DR5	1.79
Friedman 4d	1.50
Friedman 10d	0.98
Chess 5d	1.17

turned off. However, as some code fragments of the improvements could not be temporarily removed, a small bias in the baseline performance is possible.

6.4.1 Subspace-Skipping

Subspace-skipping is designed to improve the performance of the evaluation process for more irregular grids. The results in Table 4 reflect this intention. Because the DR5 data set results in the most irregular grid, the highest benefit was measured for this data set. The other data sets also showed benefits, except for the Friedman 10d data set which showed basically no change in runtime.

6.4.2 Data Point Blocking

In Sect. 5.2, a blocked evaluation scheme was introduced. Because a good choice for the block size has to be made, we varied the block size to calculate the benefit of this improvement. As our algorithm makes use of vectorization with a vector width of 4, we set the minimum block size to 4 to disable the blocking scheme, but still have

Fig. 8 Speedup of the blocking scheme depending on the size of the set of data points that are evaluated by a single thread. The baseline for this comparison is the same algorithm with a set size of four

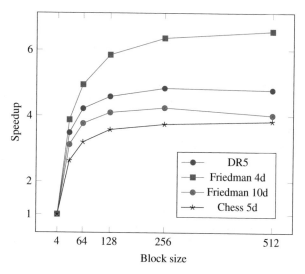

Table 5 Memory usage after the last refinement step. The memory estimate includes the grid points and temporary array of each thread. It does not include the data set itself

Data set	Grid points	Array only	With list	Memory estimate (MB)
DR5	52,294	114,949,887	60,255	256.46
Friedman 4d	25,915	117,181,057	28,342	256.22
Friedman 10d	37,838	442,449	66,731	0.57
Chess 5d	42,896	194,271	63,545	0.73

vectorization enabled. The algorithm with a block size of 4 was used as a baseline for our comparisons. The results of these experiments are shown in Fig. 8.

A significantly improved performance was observed for all data sets, with speedups ranging from 3.8× for the Chess 5d data set to 6.6× for the Friedman 4d data set. Based on these results, the size of the data point set was set to 256 elements.

The curve displays saturation for values larger than 256 and the speedup even decreases for the Friedman 10d data set. In general, we want to hold as many data points and surpluses as possible in the cache. But there is a trade-off. If the size of the set for the blocking scheme is too high, fewer surpluses can be held in the cache. But if the value is too low, the data locality for the surplus access is not improved. Therefore, an intermediate value for the set size maximizes the performance.

6.5 Evaluating the Memory Usage

The introduction of the list representation led to a significant reduction in the number of grid points that are stored. Table 5 shows the stored grid points and memory usage for the different data sets. The table lists the number of grid points that are part of the grid, the number of grid points that were stored with only the array representation, and the number of grid points with the list representation added.

The Friedman 4d data set is of special interest here. After the last refinement step, $\approx 10^8$ grid points had to be stored without the list representation, even though the grid contained only 25,915 grid points. With the list representation enabled, 28,342 grid points had to be stored. Even if we consider that we have to store the linear index in addition to the surplus value for the list representation, this is a reduction in stored grid points by four orders of magnitude. A similar reduction in stored grid points could be observed for the DR5 data set.

To estimate the impact of the temporary arrays, we calculated the memory usage at the end of the refinement process for the DR5 data set. We required 256 MB to store the temporary arrays for our 32 thread system, which amounts to 8 MB per thread. While we think that better data structures should be constructed that deal with large subgrids more efficiently, we expect that the storage requirements of our algorithm can be met even for large problems with irregular grids.

7 Conclusions

We presented algorithms for two matrix-vector-product operations to improve the performance of data mining on spatially adaptive sparse grids. Furthermore, we presented an efficient multi-evaluation algorithm, as this operation is one of the matrix-vector-products in our data mining algorithm.

Our subspace-based algorithm avoids many unnecessary computations compared to the frequently used streaming algorithm. Because mapping a subspace-based approach to modern processors architectures is more difficult, we had to develop additional algorithmic improvements. Subspace-skipping further reduced the amount of basis functions that are evaluated. The data point blocking improved the temporal and spatial data locality of our algorithm. Overall, we obtained a highly efficient algorithm that showed a higher performance than the streaming algorithm in all experiments.

Achieving this result was more difficult than we assumed based on the number of basis function evaluations. There are two main reasons for this. First, we used a highly efficient implementation of the streaming approach in our experiments as a challenging baseline. Second, in three of four of our experiments, we observed grids with a subgrid utilization that was lower than we expected. It is clear that a streaming approach is highly efficient if a subgrid contains only very few grid points. Moreover, if a subgrid contains all possible grid points, a subspace-based approach is most efficient except for very small subgrids.

Optimization Notice: Software and workloads used in performance tests may have been optimized for performance only on Intel microprocessors. Performance tests, such as SYSmark and MobileMark, are measured using specific computer systems, components, software, operations and functions. Any change to any of those factors may cause the results to vary. You should consult other information and performance tests to assist you in fully evaluating your contemplated purchases, including the performance of that product when combined with other products. For more information go to http://www.intel.com/performance. Intel, Xeon, and Intel Xeon Phi are trademarks of Intel Corporation in the U.S. and/or other countries.

Acknowledgements This work was financially supported by the Juniorprofessurenprogramm of the Landesstiftung Baden-Württemberg.

References

1. J.K. Adelman-McCarthy et al., The fifth data release of the Sloan digital sky survey. Astrophys. J. Suppl. Ser. **172**(2), 634 (2007)
2. C.M. Bishop, *Pattern Recognition and Machine Learning (Information Science and Statistics)* (Springer, New York/Secaucus, 2006)
3. H.-J. Bungartz, M. Griebel, Sparse grids. Acta Numer. **13**, 1–123 (2004)
4. H.-J. Bungartz, D. Pflüger, S. Zimmer, Adaptive sparse grid techniques for data mining, in *Modelling, Simulation and Optimization of Complex Processes 2006, Proceedings of the International Conference HPSC*, Hanoi, ed. by H. Bock, E. Kostina, X. Hoang, R. Rannacher (Springer, 2008), pp. 121–130
5. G. Buse, D. Pflüger, R. Jacob, Efficient pseudorecursive evaluation schemes for non-adaptive sparse grids, in *Sparse Grids and Applications – Munich 2012*. Lecture Notes in Computational Science and Engineering (Springer, Heidelberg, 2013), http://link.springer.com/chapter/10.1007%2F978-3-319-04537-5_1
6. J.H. Friedman, Multivariate adaptive regression splines. Ann. Stat. **19**(1), 1–67 (1991)
7. J. Garcke, Maschinelles Lernen durch Funktionsrekonstruktion mit verallgemeinerten dünnen Gittern. PhD thesis, Universität Bonn, Institut für Numerische Simulation (2004)
8. J. Garcke, M. Griebel, M. Thess, Data mining with sparse grids. Computing, **67**(3), 225–253 (2001)
9. A. Heinecke, D. Pflüger, Multi- and many-core data mining with adaptive sparse grids, in *Proceedings of the 8th ACM International Conference on Computing Frontiers CF '11*, New York (ACM, 2011), pp. 29:1–29:10
10. A. Heinecke, D. Pflüger, Emerging architectures enable to boost massively parallel data mining using adaptive sparse grids. Int. J. Parallel Programm. **41**(3), 357–399 (2012)
11. A. Heinecke, R. Karlstetter, D. Pflüger, H.-J. Bungartz, Data mining on vast datasets as a cluster system benchmark. Concurr. Comput. Pract. Exp. (2015). ISSN: 1532-0634, doi:10.1002/cpe.3514
12. Intel Cooperation, *Intel® 64 and IA-32 Architectures Optimization Reference Manual* (2014), http://www.intel.com/content/dam/www/public/us/en/documents/manuals/64-ia-32-architectures-optimization-manual.pdf
13. A. Murarasu, G. Buse, J. Weidendorfer, D. Pflüger, A. Bode, fastsg: a fast routines library for sparse grids, in *Proceedings of the International Conference on Computational Science (ICCS 2012)*, Omaha (Procedia Computer Science, 2012)
14. D. Pflüger, *Spatially Adaptive Sparse Grids for High-Dimensional Problems* (Verlag Dr.Hut, Munich, 2010)

15. D. Pflüger, Spatially adaptive refinement, in *Sparse Grids and Applications*, ed. by J. Garcke, M. Griebel. Lecture Notes in Computational Science and Engineering (Springer, Berlin/Heidelberg, 2012), pp. 243–262
16. J.R. Shewchuk, An introduction to the conjugate gradient method without the agonizing pain. Technical report, School of Computer Science, Carnegie Mellon University (1994)

High-Dimensional Stochastic Design Optimization by Adaptive-Sparse Polynomial Dimensional Decomposition

Sharif Rahman, Xuchun Ren, and Vaibhav Yadav

Abstract This paper presents a novel adaptive-sparse polynomial dimensional decomposition (PDD) method for stochastic design optimization of complex systems. The method entails an adaptive-sparse PDD approximation of a high-dimensional stochastic response for statistical moment and reliability analyses; a novel integration of the adaptive-sparse PDD approximation and score functions for estimating the first-order design sensitivities of the statistical moments and failure probability; and standard gradient-based optimization algorithms. New analytical formulae are presented for the design sensitivities that are simultaneously determined along with the moments or the failure probability. Numerical results stemming from mathematical functions indicate that the new method provides more computationally efficient design solutions than the existing methods. Finally, stochastic shape optimization of a jet engine bracket with 79 variables was performed, demonstrating the power of the new method to tackle practical engineering problems.

1 Introduction

Uncertainty quantification of complex systems, whether natural or man-made, is an important ingredient in numerous fields of science and engineering. For practical applications, encountering hundreds of input variables or more is not uncommon, where an output function of interest, often defined algorithmically

S. Rahman (✉)
College of Engineering and Program of Applied Mathematical & Computational Sciences, The University of Iowa, Iowa City, IA 52242, USA
e-mail: sharif-rahman@uiowa.edu

X. Ren
Department of Mechanical Engineering, Georgia Southern University, Statesboro, GA 30458, USA
e-mail: xren@georgiasouthern.edu

V. Yadav
Department of Aerospace Engineering, San Diego State University, San Diego, CA 92182, USA
e-mail: vyadav@mail.sdsu.edu

© Springer International Publishing Switzerland 2016
J. Garcke, D. Pflüger (eds.), *Sparse Grids and Applications – Stuttgart 2014*,
Lecture Notes in Computational Science and Engineering 109,
DOI 10.1007/978-3-319-28262-6_10

via finite-element analysis (FEA), is all too often expensive to evaluate. Modern surrogate methods, comprising stochastic collocation [1], polynomial chaos expansion [13], and sparse-grid quadrature [3], are known to offer significant computational advantages over crude Monte Carlo simulation (MCS). However, for truly high-dimensional systems, they require astronomically large numbers of terms or coefficients, succumbing to the curse of dimensionality. Therefore, alternative computational methods capable of exploiting low effective dimensions of multivariate functions, such as the polynomial dimensional decomposition (PDD) methods [7, 9], including a recently developed adaptive-sparse PDD method, are desirable [15]. Although PDD and PCE contain the same measure-consistent orthogonal polynomials, a recent work reveals that the error committed by the PDD approximation cannot be worse than that perpetrated by the PCE approximation for identical expansion orders [9].

An important application of uncertainty quantification is stochastic design optimization, which can be grouped in two principal classes: (1) design optimization for robustness [10], which minimizes the propagation of input uncertainty to output responses of interest, leading to an insensitive design; and (2) design optimization for reliability [11], which concentrates on attaining an optimal design by ensuring sufficiently low risk of failure. Depending on the objective set forth by a designer, uncertainty can be effectively mitigated by either class of design optimization. Indeed, with new formulations and methods appearing almost every year, stochastic design optimization in conjunction with FEA are becoming increasingly relevant and perhaps necessary for realistic design of complex structures and systems.

This paper presents an adaptive-sparse PDD method for stochastic design optimization of complex systems. The method is based on (1) an adaptive-sparse PDD approximation of a high-dimensional stochastic response for statistical moment and reliability analyses; (2) a novel integration of the adaptive-sparse PDD approximation and score functions for calculating the first-order sensitivities of the statistical moments and failure probability with respect to the design variables; and (3) standard gradient-based optimization algorithms. Section 2 formally defines two general variants of stochastic design optimization, including their concomitant mathematical statements. Section 3 starts with a brief exposition of the adaptive-sparse PDD approximation, leading to statistical moment and reliability analyses. Exploiting score functions, the section explains how the effort required to perform stochastic analyses also delivers the design sensitivities, sustaining no additional cost. The section also describes a coupling between stochastic analyses and design sensitivity analysis, resulting in an efficient optimization algorithm for solving both variants of the design optimization problem. Section 4 presents two numerical examples, including solving a large-scale shape design optimization problem. Finally, the conclusions are drawn in Sect. 5.

2 Stochastic Design Optimization

Consider a measurable space $(\Omega_{\mathbf{d}}, \mathcal{F}_{\mathbf{d}})$, where $\Omega_{\mathbf{d}}$ is a sample space and $\mathcal{F}_{\mathbf{d}}$ is a σ-field on $\Omega_{\mathbf{d}}$. Defined over $(\Omega_{\mathbf{d}}, \mathcal{F}_{\mathbf{d}})$, let $\{P_{\mathbf{d}} : \mathcal{F}_{\mathbf{d}} \to [0, 1]\}$ be a family of probability measures, where for $M \in \mathbb{N} := \{1, 2, \cdots\}$ and $N \in \mathbb{N}$, $\mathbf{d} = (d_1, \cdots, d_M) \in \mathcal{D}$ is an $\mathbb{R}^M$-valued design vector with non-empty closed set $\mathcal{D} \subseteq \mathbb{R}^M$ and let $\mathbf{X} := (X_1, \cdots, X_N) : (\Omega_{\mathbf{d}}, \mathcal{F}_{\mathbf{d}}) \to (\mathbb{R}^N, \mathcal{B}^N)$ be an $\mathbb{R}^N$-valued input random vector with $\mathcal{B}^N$ representing the Borel σ-field on $\mathbb{R}^N$, describing the statistical uncertainties in input variables of a complex system. The probability law of $\mathbf{X}$ is completely defined by a family of the joint probability density functions $\{f_{\mathbf{X}}(\mathbf{x}; \mathbf{d}), \mathbf{x} \in \mathbb{R}^N, \mathbf{d} \in \mathcal{D}\}$ that are associated with probability measures $\{P_{\mathbf{d}}, \mathbf{d} \in \mathcal{D}\}$, so that the probability triple $(\Omega_{\mathbf{d}}, \mathcal{F}_{\mathbf{d}}, P_{\mathbf{d}})$ of $\mathbf{X}$ depends on $\mathbf{d}$. A design variable d_k can be any distribution parameter or a statistic—for instance, the mean or standard deviation—of X_i.

Let $y_l(\mathbf{X})$, $l = 0, 1, \cdots, K$, be a collection of $K+1$ real-valued, square-integrable, measurable transformations on $(\Omega_{\mathbf{d}}, \mathcal{F}_{\mathbf{d}})$, describing performance functions of a complex system. It is assumed that $y_l : (\mathbb{R}^N, \mathcal{B}^N) \to (\mathbb{R}, \mathcal{B})$ is not an explicit function of $\mathbf{d}$, although y_l implicitly depends on $\mathbf{d}$ via the probability law of $\mathbf{X}$. This is not a major limitation, as most design optimization problems involve means and/or standard deviations of random variables as design variables. There exist two prominent variants of design optimization under uncertainty: (1) design optimization for robustness and (2) design optimization for reliability. Their mathematical formulations, comprising an objective function $c_0 : \mathbb{R}^M \to \mathbb{R}$ and constraint functions $c_l : \mathbb{R}^M \to \mathbb{R}$, $l = 1, \cdots, K$, $1 \le K < \infty$, entail finding an optimal design solution $\mathbf{d}^*$ as follows.

- **Design for Robustness** [10]

$$\mathbf{d}^* = \underset{\mathbf{d} \in \mathcal{D} \subseteq \mathbb{R}^M}{\arg\min} \, c_0(\mathbf{d}) := w_1 \frac{\mathbb{E}_{\mathbf{d}}[y_0(\mathbf{X})]}{\mu_0^*} + w_2 \frac{\sqrt{\text{var}_{\mathbf{d}}[y_0(\mathbf{X})]}}{\sigma_0^*},$$

$$\text{subject to } c_l(\mathbf{d}) := \alpha_l \sqrt{\text{var}_{\mathbf{d}}[y_l(\mathbf{X})]} - \mathbb{E}_{\mathbf{d}}[y_l(\mathbf{X})] \le 0, \, l = 1, \cdots, K, \tag{1}$$

where $\mathbb{E}_{\mathbf{d}}[y_l(\mathbf{X})] := \int_{\mathbb{R}^N} y_l(\mathbf{x}) f_{\mathbf{X}}(\mathbf{x}; \mathbf{d}) d\mathbf{x}$ is the mean of $y_l(\mathbf{X})$ with $\mathbb{E}_{\mathbf{d}}$ denoting the expectation operator with respect to the probability measure $P_{\mathbf{d}}, \mathbf{d} \in \mathcal{D}$, $\text{var}_{\mathbf{d}}[y_l(\mathbf{X})] := \mathbb{E}_{\mathbf{d}}[\{y_l(\mathbf{X}) - \mathbb{E}_{\mathbf{d}}[y_l(\mathbf{X})]\}^2]$ is the variance of $y_l(\mathbf{X})$, $w_1 \in \mathbb{R}_0^+ := [0, \infty)$ and $w_2 \in \mathbb{R}_0^+$ are two non-negative, real-valued weights, satisfying $w_1 + w_2 = 1$, $\mu_0^* \in \mathbb{R} \setminus \{0\}$ and $\sigma_0^* \in \mathbb{R}_0^+ \setminus \{0\}$ are two non-zero, real-valued scaling factors, and $\alpha_l \in \mathbb{R}_0^+$, $l = 0, 1, \cdots, K$, are non-negative, real-valued constants associated with the probabilities of constraint satisfaction. For most applications, equal weights are chosen, but they can be distinct and biased, depending on the objective set forth by a designer. By contrast, the scaling factors are relatively arbitrary and chosen to better condition, such as normalize, the objective function. In (1), $c_0(\mathbf{d})$ describes the objective robustness, whereas $c_l(\mathbf{d})$, $l = 1, \cdots, K$, describe the feasibility robustness of a given design. The evaluations of both robustness measures involve the first two moments of various stochastic responses, consequently demanding statistical moment analysis.

- **Design for Reliability** [11]

$$\mathbf{d}^* = \arg \min_{\mathbf{d} \in \mathcal{D} \subseteq \mathbb{R}^M} c_0(\mathbf{d}),$$

$$\text{subject to } c_l(\mathbf{d}) := P_{\mathbf{d}} \left[\mathbf{X} \in \Omega_{F,l}(\mathbf{d}) \right] - p_l \le 0, \ l = 1, \cdots, K, \tag{2}$$

where $\Omega_{F,l}$ is the lth failure domain, $0 \le p_l \le 1$ is the lth target failure probability. In (2), the objective function c_0 is commonly prescribed as a deterministic function of $\mathbf{d}$, describing relevant system geometry, such as area, volume, and mass. In contrast, the constraint functions c_l, $l = 1, \cdots, K$, depending on the failure domain $\Omega_{F,l}$, require component or system reliability analyses. For a component reliability analysis, the failure domain is often adequately described by a single performance function $y_l(\mathbf{X})$, for instance, $\Omega_{F,l} := \{\mathbf{x} : y_l(\mathbf{x}) < 0\}$, whereas multiple, interdependent performance functions $y_{l,i}(\mathbf{x})$, $i = 1, 2, \cdots$, are required for a system reliability analysis, leading, for example, to $\Omega_{F,l} := \{\mathbf{x} : \cup_i y_{l,i}(\mathbf{x}) < 0\}$ and $\Omega_{F,l} := \{\mathbf{x} : \cap_i y_{l,i}(\mathbf{x}) < 0\}$ for series and parallel systems, respectively.

The solution of a stochastic design optimization problem, whether in conjunction with robustness or reliability, mandates not only statistical moment and reliability analyses, but also the evaluations of gradients of moments and failure probability with respect to the design variables. The focus of this work is to solve a general high-dimensional design optimization problem described by (1) or (2) for arbitrary square-integrable functions $y_l(\mathbf{X})$, $l = 1, 2, \cdots, K$, and for an arbitrary probability density $f_{\mathbf{X}}(\mathbf{x}; \mathbf{d})$ of $\mathbf{X}$, provided that a few regularity conditions are met.

3 Adaptive-Sparse Polynomial Dimensional Decomposition Method

Let $y(\mathbf{X}) := y(X_1, \cdots, X_N)$ represent any one of the random functions y_l, $l = 0, 1, \cdots, K$, introduced in Sect. 2 and let $\mathcal{L}_2(\Omega_{\mathbf{d}}, \mathcal{F}_{\mathbf{d}}, P_{\mathbf{d}})$ represent a Hilbert space of square-integrable functions y with respect to the probability measure $f_{\mathbf{X}}(\mathbf{x}; \mathbf{d})d\mathbf{x}$ supported on $\mathbb{R}^N$. Assuming independent coordinates, the joint probability density function of $\mathbf{X}$ is expressed by the product, $f_{\mathbf{X}}(\mathbf{x}; \mathbf{d}) = \prod_{i=1}^{i=N} f_{X_i}(x_i; \mathbf{d})$, of marginal probability density functions $f_{X_i} : \mathbb{R} \to \mathbb{R}_0^+$ of X_i, each defined on its probability triple $(\Omega_{i,\mathbf{d}}, \mathcal{F}_{i,\mathbf{d}}, P_{i,\mathbf{d}})$ with a bounded or an unbounded support on $\mathbb{R}$, $i = 1, \cdots, N$. Then, for a given subset $u \subseteq \{1, \cdots, N\}$, $f_{\mathbf{X}_u}(\mathbf{x}_u; \mathbf{d}) := \prod_{p=1}^{|u|} f_{X_{i_p}}(x_{i_p}; \mathbf{d})$ defines the marginal density function of the subvector $\mathbf{X}_u = \{X_{i_1}, \cdots, X_{i_{|u|}}\}^T$ of $\mathbf{X}$.

Let $\{\psi_{ij}(X_i; \mathbf{d}); \ j = 0, 1, \cdots\}$ be a set of univariate orthonormal polynomial basis functions in the Hilbert space $\mathcal{L}_2(\Omega_{i,\mathbf{d}}, \mathcal{F}_{i,\mathbf{d}}, P_{i,\mathbf{d}})$ that is consistent with the probability measure $P_{i,\mathbf{d}}$ of X_i for a given design $\mathbf{d}$, where $i = 1, \cdots, N$. For a given

$u = \{i_1, \cdots, i_{|u|}\} \subseteq \{1, \cdots, N\}$, $1 \leq |u| \leq N$, $1 \leq i_1 < \cdots < i_{|u|} \leq N$, denote by $(\times_{p=1}^{p=|u|} \Omega_{i_p,\mathbf{d}}, \times_{p=1}^{p=|u|} \mathcal{F}_{i_p,\mathbf{d}}, \times_{p=1}^{p=|u|} P_{i_p,\mathbf{d}})$ the product probability triple of the subvector $\mathbf{X}_u$. Since the probability density function of $\mathbf{X}_u$ is separable (independent), the product polynomial $\psi_{u\mathbf{j}_{|u|}}(\mathbf{X}_u; \mathbf{d}) := \prod_{p=1}^{|u|} \psi_{i_p j_p}(X_{i_p}; \mathbf{d})$, where $\mathbf{j}_{|u|} = (j_1, \cdots, j_{|u|}) \in \mathbb{N}_0^{|u|}$, $\mathbb{N}_0 := \mathbb{N} \cup \{0\}$, is a $|u|$-dimensional multi-index, constitutes an orthonormal basis in $\mathcal{L}_2(\times_{p=1}^{p=|u|} \Omega_{i_p,\mathbf{d}}, \times_{p=1}^{p=|u|} \mathcal{F}_{i_p,\mathbf{d}}, \times_{p=1}^{p=|u|} P_{i_p,\mathbf{d}})$.

The PDD of a square-integrable function y represents a hierarchical expansion [7, 9]

$$y(\mathbf{X}) = y_\emptyset(\mathbf{d}) + \sum_{\emptyset \neq u \subseteq \{1, \cdots, N\}} \sum_{\substack{\mathbf{j}_{|u|} \in \mathbb{N}_0^{|u|} \\ j_1, \cdots, j_{|u|} \neq 0}} C_{u\mathbf{j}_{|u|}}(\mathbf{d}) \psi_{u\mathbf{j}_{|u|}}(\mathbf{X}_u; \mathbf{d}) \tag{3}$$

in terms of random multivariate orthonormal polynomials, where

$$y_\emptyset(\mathbf{d}) = \int_{\mathbb{R}^N} y(\mathbf{x}) f_{\mathbf{X}}(\mathbf{x}; \mathbf{d}) d\mathbf{x} \tag{4}$$

and

$$C_{u\mathbf{j}_{|u|}}(\mathbf{d}) := \int_{\mathbb{R}^N} y(\mathbf{x}) \psi_{u\mathbf{j}_{|u|}}(\mathbf{x}_u; \mathbf{d}) f_{\mathbf{X}}(\mathbf{x}; \mathbf{d}) d\mathbf{x}, \quad \emptyset \neq u \subseteq \{1, \cdots, N\}, \quad \mathbf{j}_{|u|} \in \mathbb{N}_0^{|u|} \tag{5}$$

are various expansion coefficients. The condition $j_1, \cdots, j_{|u|} \neq 0$ used in (3) and equations throughout the remainder of this paper implies that $j_k \neq 0$ for all $k = 1, \cdots, |u|$. Derived from the ANOVA dimensional decomposition [4], (3) provides an exact representation because it includes all main and interactive effects of input variables. For instance, $|u| = 0$ corresponds to the constant component function $y_\emptyset$, representing the mean effect of y; $|u| = 1$ leads to the univariate component functions, describing the main effects of input variables, and $|u| = S$, $1 < S \leq N$, results in the S-variate component functions, facilitating the interaction among at most S input variables $X_{i_1}, \cdots, X_{i_S}$, $1 \leq i_1 < \cdots < i_S \leq N$. Further details of PDD are available elsewhere [7, 9].

Equation (3) contains an infinite number of coefficients, emanating from infinite numbers of orthonormal polynomials. In practice, the number of coefficients must be finite, say, by retaining finite-order polynomials and reduced-degree interaction among input variables. For instance, an S-variate, mth-order PDD approximation [7, 9]

$$\tilde{y}_{S,m}(\mathbf{X}) = y_\emptyset(\mathbf{d}) + \sum_{\substack{\emptyset \neq u \subseteq \{1, \cdots, N\} \\ 1 \leq |u| \leq S}} \sum_{\substack{\mathbf{j}_{|u|} \in \mathbb{N}_0^{|u|}, \|\mathbf{j}_{|u|}\|_\infty \leq m \\ j_1, \cdots, j_{|u|} \neq 0}} C_{u\mathbf{j}_{|u|}}(\mathbf{d}) \psi_{u\mathbf{j}_{|u|}}(\mathbf{X}_u; \mathbf{d}) \tag{6}$$

is generated, where $\|\mathbf{j}_{|u|}\|_\infty := \max(j_1, \cdots, j_{|u|})$ defines the ∞-norm and the integers $0 \leq S \leq N$ and $1 \leq m < \infty$ represent the largest degree of interactions among input variables and the largest order of orthogonal polynomials retained in a concomitant truncation of the sum in (3). It is important to clarify that the right side of (6) contains sums of at most S-dimensional PDD component functions of y. Therefore, the term "S-variate" used for the PDD approximation should be interpreted in the context of including at most S-degree interaction of input variables, even though $\tilde{y}_{S,m}$ is strictly an N-variate function. When $S \to N$ and $m \to \infty$, $\tilde{y}_{S,m}$ converges to y in the mean-square sense, generating a hierarchical and convergent sequence of approximations [7, 9].

3.1 Adaptive-Sparse PDD Approximation

For practical applications, the dimensional hierarchy or nonlinearity of a stochastic response, in general, is not known a priori. Therefore, indiscriminately assigning values of the truncation parameters S and m is not desirable. Nor is it possible to do so when a stochastic solution is obtained via complex numerical algorithms. In which case, one must perform these truncations automatically by progressively drawing in higher-variate or higher-order contributions as appropriate. Based on the authors' past experience, an S-variate PDD approximation, where $S \ll N$, is adequate, when solving real-world engineering problems, with the computational cost varying polynomially (S-order) with respect to the number of variables [7, 9]. As an example, consider the selection of $S = 2$ for solving a stochastic problem in 100 dimensions by a bivariate PDD approximation, comprising $100 \times 99/2 = 4950$ bivariate component functions. If all such component functions are included, then the computational effort for even a full bivariate PDD approximation may exceed the computational budget allocated to solving this problem. But many of these component functions contribute little to the probabilistic characteristics sought and can be safely ignored. Similar conditions may prevail for higher-variate component functions. Henceforth, define an S-variate, partially adaptive-sparse PDD approximation [15]

$$\bar{y}_S(\mathbf{X}) := y_\emptyset(\mathbf{d}) + \sum_{\substack{\emptyset \neq u \subseteq \{1,\cdots,N\} \\ 1 \leq |u| \leq S}} \sum_{m_u=1}^{\infty} \sum_{\substack{\|\mathbf{j}_{|u|}\|_\infty = m_u, j_1,\cdots j_{|u|} \neq 0 \\ \tilde{G}_{u,m_u} > \epsilon_1, \Delta \tilde{G}_{u,m_u} > \epsilon_2}} C_{u\mathbf{j}_{|u|}}(\mathbf{d}) \psi_{u\mathbf{j}_{|u|}}(\mathbf{X}_u; \mathbf{d}) \qquad (7)$$

of $y(\mathbf{X})$, where

$$\tilde{G}_{u,m_u} := \frac{1}{\sigma^2(\mathbf{d})} \sum_{\substack{\mathbf{j}_{|u|} \in \mathbb{N}_0^{|u|}, \|\mathbf{j}_{|u|}\|_\infty \leq m_u \\ j_1,\cdots j_{|u|} \neq 0}} C_{u\mathbf{j}_{|u|}}^2(\mathbf{d}), \ m_u \in \mathbb{N}, \ 0 < \sigma^2(\mathbf{d}) < \infty,$$

defines the approximate m_uth-order approximation of the global sensitivity index of $y(\mathbf{X})$ for a subvector $\mathbf{X}_u$, $\emptyset \neq u \subseteq \{1, \cdots, N\}$, of input variables $\mathbf{X}$ and

$$
\Delta \tilde{G}_{u,m_u} := \begin{cases} \infty & \text{if } m_u = 1 \text{ or } (m_u \geq 2, \tilde{G}_{u,m_u-1} = 0, \tilde{G}_{u,m_u} \neq 0), \\ 0 & \text{if } m_u \geq 2, \tilde{G}_{u,m_u-1} = 0, \tilde{G}_{u,m_u} = 0, \\ \dfrac{\tilde{G}_{u,m_u} - \tilde{G}_{u,m_u-1}}{\tilde{G}_{u,m_u-1}} & \text{if } m_u \geq 2, \tilde{G}_{u,m_u-1} \neq 0 \end{cases}
$$

defines the relative change in the approximate global sensitivity index when the largest polynomial order increases from $m_u - 1$ to m_u. The non-trivial definition applies when $m_u \geq 2$ and $\tilde{G}_{u,m_u-1} \neq 0$. When $m_u = 1$ or $m_u = 1$ or $(m_u \geq 2, \tilde{G}_{u,m_u-1} = 0, \tilde{G}_{u,m_u} \neq 0)$, the infinite value of $\Delta \tilde{G}_{u,m_u}$ guarantees that the m_uth-order contribution of y_u to y is preserved in the adaptive-sparse approximation. When $m_u \geq 2$, $\tilde{G}_{u,m_u-1} = 0$, and $\tilde{G}_{u,m_u} = 0$, the *zero* value of $\Delta \tilde{G}_{u,m_u}$ implies that there is no contribution of the m_uth-order contribution of y_u to y. Here,

$$
\sigma^2(\mathbf{d}) = \sum_{\emptyset \neq u \subseteq \{1, \cdots, N\}} \sum_{\substack{\mathbf{j}_{|u|} \in \mathbb{N}_0^{|u|} \\ j_1, \cdots j_{|u|} \neq 0}} C_{u\mathbf{j}_{|u|}}^2(\mathbf{d}) \tag{8}
$$

is the variance of $y(\mathbf{X})$. Then the sensitivity indices $\tilde{G}_{u,m_u}$ and $\Delta \tilde{G}_{u,m_u}$ provide an effective means to truncate the PDD in (3) both adaptively and sparsely. Equation (7) is attained by subsuming at most S-variate component functions, but fulfilling two inclusion criteria: (1) $\tilde{G}_{u,m_u} > \epsilon_1$ for $1 \leq |u| \leq S \leq N$, and (2) $\Delta \tilde{G}_{u,m_u} > \epsilon_2$ for $1 \leq |u| \leq S \leq N$, where $\epsilon_1 \geq 0$ and $\epsilon_2 \geq 0$ are two user-defined tolerances. The resulting approximation is partially adaptive because the truncations are restricted to at most S-variate component functions of y. When $S = N$, (7) becomes the fully adaptive-sparse PDD approximation [15]. The algorithmic details of numerical implementation associated with either fully or partially adaptive-sparse PDD approximation are available elsewhere [15].

The determination of PDD expansion coefficients $y_\emptyset(\mathbf{d})$ and $C_{u\mathbf{j}_{|u|}}(\mathbf{d})$ is vitally important for statistical moment and reliability analyses, including design sensitivities. As defined in (4) and (5), the coefficients involve N-dimensional integrals over $\mathbb{R}^N$. For large N, a multivariate numerical integration employing an N-dimensional tensor product of a univariate quadrature formula is computationally prohibitive and is, therefore, ruled out. An attractive alternative approach entails dimension-reduction integration [14], where the N-variate function y in (4) and (5) is replaced by an R-variate ($1 \leq R \leq N$) referential dimension decomposition at a chosen reference point. For instance, given a reference point $\mathbf{c} = (c_1, \cdots, c_N) \in \mathbb{R}^N$, the expansion coefficients $C_{u\mathbf{j}_{|u|}}$ are approximated by [14]

$$
C_{u\mathbf{j}_{|u|}}(\mathbf{d}) \cong \sum_{i=0}^{R} (-1)^i \binom{N-R+i-1}{i} \sum_{\substack{v \subseteq \{1, \cdots, N\} \\ |v| = R-i, u \subseteq v}} \int_{\mathbb{R}^{|v|}} y(\mathbf{x}_v, \mathbf{c}_{-v}) \psi_{u\mathbf{j}_{|u|}}(\mathbf{x}_u; \mathbf{d}) f_{\mathbf{X}_v}(\mathbf{x}_v; \mathbf{d}) d\mathbf{x}_v,
$$

$$
\tag{9}
$$

requiring evaluations of at most R-dimensional integrals. The estimation of $y_\emptyset(\mathbf{d})$ is similar. The reduced integration facilitates calculation of the coefficients approaching their exact values as $R \to N$, and is significantly more efficient than performing one N-dimensional integration, particularly when $R \ll N$. Hence, the computational effort is significantly lowered using the dimension-reduction integration. For instance, when $R = 1$ or 2, (9) involves one-, or at most, two-dimensional integrations, respectively. Nonetheless, numerical integrations are still required for performing various $|v|$-dimensional integrals over $\mathbb{R}^{|v|}$, where $0 \leq |v| \leq R$. When $R > 1$, the multivariate integrals involved can be subsequently approximated by a sparse-grid quadrature, such as the fully symmetric interpolatory rule [5], as implemented by Yadav and Rahman [15].

3.2 Stochastic Analysis

3.2.1 Statistical Moments

Applying the expectation operator on $\bar{y}_S(\mathbf{X})$ and recognizing the *zero*-mean and orthogonal properties of PDD component functions, the mean

$$\mathbb{E}_\mathbf{d}\left[\bar{y}_S(\mathbf{X})\right] = y_\emptyset(\mathbf{d}) \tag{10}$$

of the partially adaptive-sparse PDD approximation agrees with the exact mean $\mathbb{E}_\mathbf{d}[y(\mathbf{X})] = y_\emptyset$ for any ϵ_1, ϵ_2, and S [15]. However, the variance, obtained by applying the expectation operator on $(\bar{y}_S(\mathbf{X}) - y_\emptyset)^2$, varies according to [15]

$$\bar{\sigma}_S^2(\mathbf{d}) := \mathbb{E}_\mathbf{d}\left[(\bar{y}_S(\mathbf{X}) - \mathbb{E}[\bar{y}_S(\mathbf{X})])^2\right] = \sum_{\substack{\emptyset \neq u \subseteq \{1,\cdots,N\} \\ 1 \leq |u| \leq S}} \sum_{\substack{m_u = 1 \\ \tilde{G}_{u,m_u} > \epsilon_1, \Delta\tilde{G}_{u,m_u} > \epsilon_2}}^{\infty} \sum_{\substack{\|\mathbf{j}_{|u|}\|_\infty = m_u, j_1, \cdots j_{|u|} \neq 0}} C_{u\mathbf{j}_{|u|}}^2(\mathbf{d}),$$

$$\tag{11}$$

where the squares of the expansion coefficients are summed following the same two pruning criteria discussed in the preceding subsection. Equation (11) provides a closed-form expression of the approximate second-moment properties of any square-integrable function y in terms of the PDD expansion coefficients. When $S = N$ and $\epsilon_1 = \epsilon_2 = 0$, the right side of (11) coincides with that of (8). In consequence, the variance from the partially adaptive-sparse PDD approximation $\bar{y}_S(\mathbf{X})$ converges to the exact variance of $y(\mathbf{X})$ as $S \to N$, $\epsilon_1 \to 0$, and $\epsilon_2 \to 0$.

3.2.2 Failure Probability

A fundamental problem in reliability analysis entails calculation of the failure probability

$$P_F(\mathbf{d}) := P_{\mathbf{d}}\left[\mathbf{X} \in \Omega_F\right] = \int_{\mathbb{R}^N} I_{\Omega_F}(\mathbf{x}) f_{\mathbf{X}}(\mathbf{x}; \mathbf{d}) d\mathbf{x} =: \mathbb{E}_{\mathbf{d}}\left[I_{\Omega_F}(\mathbf{X})\right],$$

where $I_{\Omega_F}(\mathbf{x})$ is the indicator function associated with the failure domain Ω_F, which is equal to *one* when $\mathbf{x} \in \Omega_F$ and *zero* otherwise. Depending on component or system reliability analysis, let $\bar{\Omega}_{F,S} := \{\mathbf{x} : \bar{y}_S(\mathbf{x}) < 0\}$ or $\bar{\Omega}_{F,S} := \{\mathbf{x} : \cup_i \bar{y}_{i,S}(\mathbf{x}) < 0\}$ or $\bar{\Omega}_{F,S} := \{\mathbf{x} : \cap_i \bar{y}_{i,S}(\mathbf{x}) < 0\}$ be an approximate failure set as a result of S-variate, adaptive-sparse PDD approximations $\bar{y}_S(\mathbf{X})$ of $y(\mathbf{X})$ or $\bar{y}_{i,S}(\mathbf{X})$ of $y_i(\mathbf{X})$. Then the adaptive-sparse PDD estimate of the failure probability $P_F(\mathbf{d})$ is

$$\bar{P}_{F,S}(\mathbf{d}) = \mathbb{E}_{\mathbf{d}}\left[I_{\bar{\Omega}_{F,S}}(\mathbf{X})\right] = \lim_{L \to \infty} \frac{1}{L} \sum_{l=1}^{L} I_{\bar{\Omega}_{F,S}}(\mathbf{x}^{(l)}), \tag{12}$$

where L is the sample size, $\mathbf{x}^{(l)}$ is the lth realization of $\mathbf{X}$, and $I_{\bar{\Omega}_{F,S}}(\mathbf{x})$ is another indicator function, which is equal to *one* when $\mathbf{x} \in \bar{\Omega}_{F,S}$ and *zero* otherwise.

Note that the simulation of the PDD approximation in (12) should not be confused with crude MCS commonly used for producing benchmark results. The crude MCS, which requires numerical calculations of $y(\mathbf{x}^{(l)})$ or $y_i(\mathbf{x}^{(l)})$ for input samples $\mathbf{x}^{(l)}$, $l = 1, \cdots, L$, can be expensive or even prohibitive, particularly when the sample size L needs to be very large for estimating small failure probabilities. In contrast, the MCS embedded in the adaptive-sparse PDD approximation requires evaluations of simple polynomial functions that describe $\bar{y}_S(\mathbf{x}^{(l)})$ or $\bar{y}_{i,S}(\mathbf{x}^{(l)})$. Therefore, a relatively large sample size can be accommodated in the adaptive-sparse PDD method even when y or y_i is expensive to evaluate.

3.3 Design Sensitivity Analysis

When solving design optimization problems employing gradient-based optimization algorithms, at least the first-order derivatives of the first two moments and failure probability with respect to each design variable are required. In this subsection, the adaptive-sparse PDD method for statistical moment and reliability analyses is expanded for design sensitivity analysis. For such sensitivity analysis, the following regularity conditions are assumed: (1) The domains of design variables $d_k \in \mathcal{D}_k \subset \mathbb{R}$, $k = 1, \cdots, M$, are open intervals of $\mathbb{R}$; (2) the probability density function $f_{\mathbf{X}}(\mathbf{x}; \mathbf{d})$ of $\mathbf{X}$ is continuous. In addition, the partial derivative $\partial f_{\mathbf{X}}(\mathbf{x}; \mathbf{d})/\partial d_k$, $k = 1, \cdots, M$, exists and is finite for all $\mathbf{x} \in \mathbb{R}^N$ and $d_k \in \mathcal{D}_k$. Furthermore, the statistical moments of y and failure probability are differentiable functions of

$\mathbf{d} \in \mathcal{D} \subseteq \mathbb{R}^M$; and (3) there exists a Lebesgue integrable dominating function $z(\mathbf{x})$ such that $|y^r(\mathbf{x})\partial f_{\mathbf{X}}(\mathbf{x};\mathbf{d})/\partial d_k| \leq z(\mathbf{x})$ and $|I_{\Omega_F}(\mathbf{x})\partial f_{\mathbf{X}}(\mathbf{x};\mathbf{d})/\partial d_k| \leq z(\mathbf{x})$, where $r = 1, 2$, and $k = 1, \cdots, M$.

3.3.1 Score Function

Let

$$h(\mathbf{d}) := \mathbb{E}_{\mathbf{d}}[g(\mathbf{X})] := \int_{\mathbb{R}^N} g(\mathbf{x})f_{\mathbf{X}}(\mathbf{x};\mathbf{d})d\mathbf{x} \tag{13}$$

be a generic probabilistic response, where $h(\mathbf{d})$ and $g(\mathbf{x})$ are either the rth-order raw moment $m^{(r)}(\mathbf{d}) := \mathbb{E}_{\mathbf{d}}[y_S^r(\mathbf{X})]$ ($r = 1, 2$) and $y^r(\mathbf{x})$ for statistical moment analysis or $P_F(\mathbf{d})$ and $I_{\Omega_F}(\mathbf{x})$ for reliability analysis. Suppose that the first-order derivative of $h(\mathbf{d})$ with respect to a design variable d_k, $1 \leq k \leq M$, is sought. Taking the partial derivative of $h(\mathbf{d})$ with respect to d_k and then applying the Lebesgue dominated convergence theorem [2], which permits the differential and integral operators to be interchanged, yields the first-order sensitivity

$$\begin{aligned}
\frac{\partial h(\mathbf{d})}{\partial d_k} &:= \frac{\partial \mathbb{E}_{\mathbf{d}}[g(\mathbf{X})]}{\partial d_k} \\
&= \frac{\partial}{\partial d_k} \int_{\mathbb{R}^N} g(\mathbf{x})f_{\mathbf{X}}(\mathbf{x};\mathbf{d})d\mathbf{x} \\
&= \int_{\mathbb{R}^N} g(\mathbf{x}) \frac{\partial \ln f_{\mathbf{X}}(\mathbf{x};\mathbf{d})}{\partial d_k} f_{\mathbf{X}}(\mathbf{x};\mathbf{d})d\mathbf{x} \\
&=: \mathbb{E}_{\mathbf{d}}\left[g(\mathbf{X})s_{d_k}^{(1)}(\mathbf{X};\mathbf{d}) \right],
\end{aligned} \tag{14}$$

provided that $f_{\mathbf{X}}(\mathbf{x};\mathbf{d}) > 0$ and the derivative $\partial \ln f_{\mathbf{X}}(\mathbf{x};\mathbf{d})/\partial d_k$ exists. In the last line of (14), $s_{d_k}^{(1)}(\mathbf{X};\mathbf{d}) := \partial \ln f_{\mathbf{X}}(\mathbf{X};\mathbf{d})/\partial d_k$ is known as the first-order score function for the design variable d_k [8, 12]. According to (13) and (14), the generic probabilistic response and its sensitivities have both been formulated as expectations of stochastic quantities with respect to the same probability measure, facilitating their concurrent evaluations in a single stochastic simulation or analysis.

3.3.2 Sensitivity of Statistical Moments

Selecting $h(\mathbf{d})$ and $g(\mathbf{x})$ to be $m^{(r)}(\mathbf{d})$ and $y^r(\mathbf{x})$, respectively, and then replacing $y(\mathbf{x})$ with its S-variate adaptive-sparse PDD approximation $\bar{y}_S(\mathbf{x})$ in the last line of (14), the resultant approximation of the sensitivities of the rth-order moment is obtained as

$$\mathbb{E}_{\mathbf{d}}\left[\bar{y}_S^r(\mathbf{X})s_{d_k}^{(1)}(\mathbf{X};\mathbf{d}) \right] = \int_{\mathbb{R}^N} \bar{y}_S^r(\mathbf{x})s_{d_k}^{(1)}(\mathbf{x};\mathbf{d})f_{\mathbf{X}}(\mathbf{x};\mathbf{d})d\mathbf{x}. \tag{15}$$

The N-dimensional integral in (15) can be estimated by the same or similar dimension-reduction integration as employed for estimating the PDD expansion coefficients. Furthermore, if $s_{d_k}^{(1)}$ is square-integrable, then it can be expanded with respect to the same orthogonal polynomial basis functions, resulting in a closed-form expression of the design sensitivity [8]. Finally, setting $r = 1$ or 2 in (15) delivers the approximate sensitivity of the first or second moment.

3.3.3 Sensitivity of Failure Probability

Selecting $h(\mathbf{d})$ and $g(\mathbf{x})$ to be $P_F(\mathbf{d})$ and $I_{\Omega_F}(\mathbf{x})$, respectively, and then replacing $y(\mathbf{x})$ with its S-variate adaptive-sparse PDD approximation $\bar{y}_S(\mathbf{x})$ in the last line of (14), the resultant approximation of the sensitivities of the failure probability is obtained as

$$
\mathbb{E}_{\mathbf{d}}\left[I_{\bar{\Omega}_{F,S}}(\mathbf{X})s_{d_k}^{(1)}(\mathbf{X};\mathbf{d})\right] = \lim_{L\to\infty}\frac{1}{L}\sum_{l=1}^{L}\left[I_{\bar{\Omega}_{F,S}}(\mathbf{x}^{(l)})s_{d_k}^{(1)}(\mathbf{x}^{(l)};\mathbf{d})\right], \tag{16}
$$

where L is the sample size, $\mathbf{x}^{(l)}$ is the lth realization of $\mathbf{X}$, and $I_{\bar{\Omega}_{F,S}}(\mathbf{x})$ is the adaptive-sparse PDD-generated indicator function. Again, the sensitivity in (16) is easily and inexpensively determined by sampling elementary polynomial functions that describe $\bar{y}_S$ and $s_{d_k}^{(1)}$.

Remark 1 The PDD expansion coefficients depend on the design vector $\mathbf{d}$. Naturally, a PDD approximation, whether obtained by truncating arbitrarily or adaptively, is also dependent on $\mathbf{d}$, unless the approximation exactly reproduces the function $y(\mathbf{X})$. It is important to clarify that the approximate sensitivities in (15) and (16) are obtained not by taking partial derivatives of the approximate moments in (10) and (11) and approximate failure probability in (12) with respect to d_k. Instead, they result from replacing $y(\mathbf{x})$ with $\bar{y}_S(\mathbf{x})$ in the expectation describing the last line of (14).

Remark 2 The score function method has the nice property that it requires differentiating only the underlying probability density function $f_{\mathbf{X}}(\mathbf{x};\mathbf{d})$. The resulting score functions can be easily and, in most cases, analytically determined. If the performance function is not differentiable or discontinuous—for example, the indicator function that comes from reliability analysis—the proposed method still allows evaluation of the sensitivity if the density function is differentiable. In reality, the density function is often smoother than the performance function, and therefore the proposed sensitivity methods are able to calculate sensitivities for a wide variety of complex mechanical systems.

Algorithm 1: Proposed adaptive-sparse PDD for stochastic design optimization

---

Input: an initial design $\mathbf{d}_0$, S, $\epsilon > 0$, $\epsilon_1 > 0$, $\epsilon_2 > 0$, $q = 0$
Output: an approximation $\mathbf{d}_S^*$ of optimal design $\mathbf{d}^*$
$\mathbf{d}^{(q)} \leftarrow \mathbf{d}_0$;
repeat

    $\mathbf{d} \leftarrow \mathbf{d}^{(q)}$;
    Generate adaptive-sparse PDD approximations $\bar{y}_{l,S}(\mathbf{X})$ at current design $\mathbf{d}$ of all $y_l(\mathbf{X})$
    in (1) or (2), $l = 0, 1, \cdots, K$;
    if *design for robustness* **then**

        compute moments $\mathbb{E}_\mathbf{d}[\bar{y}_{l,S}(\mathbf{X})]$ and $\bar{\sigma}_{l,S}^2(\mathbf{d})$ of $\bar{y}_{l,S}(\mathbf{X})$;
        ;   /* *from* (10) *and* (11) */
        estimate design sensitivity of moments ;
        ;   /* *from* (15) */

    else if *design for reliability* **then**

        compute failure probability $\bar{P}_{F,S}(\mathbf{d})$ for $\bar{y}_{l,S}(\mathbf{X})$;
        ;   /* *from* (12) */
        ;
        estimate design sensitivity of failure probability ;
        ;   /* *from* (16) */

    endif;
    Evaluate objective and constraint functions in (1) or (2) and their sensitivities at $\mathbf{d}$;
    Using a gradient-based algorithm, obtain the next design $\mathbf{d}^{(q+1)}$;
    Set $q = q + 1$;
until $||\mathbf{d}^{(q)} - \mathbf{d}^{(q-1)}||_2 < \epsilon$;
$\mathbf{d}_S^* \leftarrow \mathbf{d}^{(q)}$

3.4 Optimization Algorithm

The adaptive-sparse PDD approximations described in the preceding subsections provide a means to evaluate the objective and constraint functions, including their design sensitivities, from a single stochastic analysis. An integration of statistical moment analysis, reliability analysis, design sensitivity analysis, and a suitable optimization algorithm should render a convergent solution of the design optimization problems in (1) or (2). Algorithm 1 describes the computational flow of the adaptive-sparse PDD method for stochastic design optimization.

4 Numerical Examples

Two examples are presented to illustrate the adaptive-sparse PDD method for design optimization under uncertainty, where the objective and constraint functions are either elementary mathematical constructs or relate to complex engineering problems. Orthonormal polynomials consistent with the probability distributions of input random variables were used as bases. The PDD expansion coefficients were estimated using dimension-reduction integration and sparse-grid quadrature entailing an extended fully symmetric interpolatory rule [5, 15]. The sensitivities of moments and failure probability were evaluated using dimension-reduction integration and embedded MCS of the adaptive-sparse PDD approximation, respectively. The optimization algorithm selected is sequential quadratic programming in both examples.

4.1 Example 1: Mathematical Functions

The first example entails design optimization for robustness, which calls for finding

$$
\mathbf{d}^* = \arg\min_{\mathbf{d}\in\mathcal{D}\subseteq\mathbb{R}^M} c_0(\mathbf{d}) := 0.5\frac{\mathbb{E}_{\mathbf{d}}\left[y_0(\mathbf{X})\right]}{10} + 0.5\frac{\sqrt{\mathrm{var}_{\mathbf{d}}\left[y_0(\mathbf{X})\right]}}{2},
$$
$$
\text{subject to } c_1(\mathbf{d}) := 3\sqrt{\mathrm{var}_{\mathbf{d}}\left[y_1(\mathbf{X})\right]} - \mathbb{E}_{\mathbf{d}}\left[y_1(\mathbf{X})\right] \le 0,
$$
$$
c_2(\mathbf{d}) := 3\sqrt{\mathrm{var}_{\mathbf{d}}\left[y_2(\mathbf{X})\right]} - \mathbb{E}_{\mathbf{d}}\left[y_2(\mathbf{X})\right] \le 0,
$$

where $\mathbf{d} \in \mathcal{D} = [0.00002, 0.002] \times [0.1, 1.6]$ and

$$
y_0(\mathbf{X}) = X_3 X_1 \sqrt{1 + X_2^2}
$$

and

$$
y_l(\mathbf{X}) = 1 - \frac{5X_4\sqrt{1 + X_2^2}}{\sqrt{65}X_5}\left(\frac{8}{X_1} + (-1)^{l+1}\frac{1}{X_1 X_2}\right), \; l = 1, 2,
$$

are three random response functions of five independent random variables. The first two variables X_1 and X_2 follow Gaussian distributions with respective means $d_1 = \mathbb{E}_{\mathbf{d}}[X_1]$ and $d_2 = \mathbb{E}_{\mathbf{d}}[X_2]$ and coefficients of variations both equal to 0.02. The remaining variables, X_3, X_4, and X_4, follow Beta, Gumbel, and Lognormal distributions with respective means of 10,000, 0.8, and 1050 and respective coefficients of variations of 0.2, 0.25 and 0.238. The initial design vector is $\mathbf{d}_0 = (0.001, 1)$. In this example, $N = 5$, $M = 2$, and $K = 2$.

Table 1 presents detailed optimization results from two distinct adaptive-sparse PDD approximations, entailing univariate ($S = 1$) and bivariate ($S = 2$) truncations, employed to solve this optimization problem. The optimal solutions by these two approximations are close to each other, both indicating that the first constraint is nearly active ($c_1 \cong 0$). The results of the bivariate approximations confirm that the univariate solution is adequate. However, the total numbers of function evaluations step up for the bivariate approximation, as expected.

Since this problem can also be solved by the non-adaptively truncated PDD [10] and tensor product quadrature (TPQ) [6] methods, comparing their solutions,

Table 1 Optimization results for Example 1

	Adaptive-sparse PDD		Truncated PDD [10]		
Results	Univariate	Bivariate	Univariate	Bivariate	TPQ [6]
d_1^* ($\times 10^{-4}$)	11.3902	11.5753	11.3921	11.5695	11.6476
d_2^*	0.3822	0.3780	0.3817	0.3791	0.3767
$c_0(\mathbf{d}^*)$	1.2226	1.2408	1.2227	1.2406	1.2480
$c_1(\mathbf{d}^*)$	0.0234	0.0084	0.0233	0.0084	0.0025
$c_2(\mathbf{d}^*)$	−0.4810	−0.4928	−0.4816	−0.4917	−0.4970
No. of iterations	12	13	12	14	10
Total no. of function evaluations	465	2374	696	6062	17,521

listed in the last three columns of Table 1, with the adaptive-sparse PDD solutions should be intriguing. It appears that the existing PDD truncated at the largest polynomial order of the adaptive-sparse PDD approximation, which is four, and TPQ methods are also capable of producing a similar optimal solution, but by incurring computational cost far more than the adaptive-sparse PDD methods. For instance, comparing the total numbers of function evaluations, the univariate adaptive-sparse PDD method is more economical than the existing univariate PDD and TPQ methods by factors of 1.5 and 37.7, respectively. The new bivariate adaptive-sparse PDD is more than twice as efficient as the existing non-adaptively truncated bivariate PDD.

4.2 Example 2: Shape Optimization of a Jet Engine Bracket

The final example demonstrates the usefulness of the adaptive-sparse PDD method in designing for reliability an industrial-scale mechanical component, known as jet engine bracket, as shown in Fig. 1a. Seventy-nine random shape parameters, X_i, $i = 1, \cdots, 79$, resulting from manufacturing variability, describe the shape of a jet engine bracket in three dimensions, including two quadrilateral holes introduced to reduce the mass of the jet engine bracket as much as possible. The design variables, $d_i = \mathbb{E}_{\mathbf{d}}[X_i]$, $i = 1, \cdots, 79$, are the means of these 79 independent random variables, with Fig. 1b–d depicting the initial design of the jet engine bracket geometry at mean values of the shape parameters. The centers of the four bottom circular holes are fixed; a deterministic force, $F = 43.091$ kN, was applied at the center of the top circular hole with a 48° angle from the horizontal line, as shown in Fig. 1c, and a deterministic torque, $T = 0.1152$ kN-m, was applied at the center of the top circular hole, as shown in Fig. 1d. The jet engine bracket is made of Titanium Alloy Ti-6Al-4V with deterministic material properties described elsewhere [11]. Due to their finite bounds, the random variables X_i, $i = 1, \cdots, 79$, were assumed to follow truncated Gaussian distributions [11].

The objective of this example is to minimize the mass of the engine bracket by changing the shape of the geometry such that the fatigue life $y(\mathbf{u}(\boldsymbol{\xi}; \mathbf{X}), \sigma(\boldsymbol{\xi}; \mathbf{X}))$

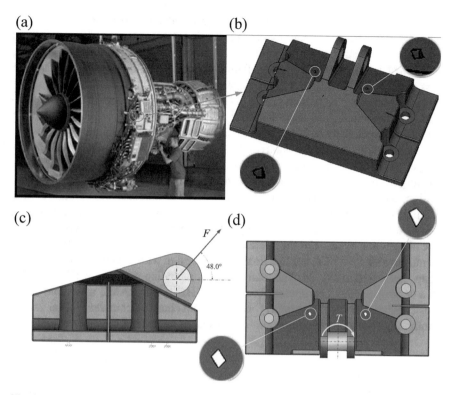

Fig. 1 A jet engine bracket; (**a**) a jet engine; (**b**) isometric view; (**c**) lateral view; (**d**) top view

exceeds a million loading cycles with 99.865 % probability. The underlying stochastic differential equations call for finding the displacement $\mathbf{u}(\boldsymbol{\xi};\mathbf{X})$ and stress $\boldsymbol{\sigma}(\boldsymbol{\xi};\mathbf{X})$ solutions at a spatial coordinate $\boldsymbol{\xi} = (\xi_1, \xi_2, \xi_3) \in \Omega \subset \mathbb{R}^3$, satisfying $P_{\mathbf{d}}$-almost surely

$$\begin{aligned}
\nabla \cdot \boldsymbol{\sigma}(\boldsymbol{\xi};\mathbf{X}) + \mathbf{b}(\boldsymbol{\xi};\mathbf{X}) &= \mathbf{0} \text{ in } \Omega \subset \mathbb{R}^3, \\
\boldsymbol{\sigma}(\boldsymbol{\xi};\mathbf{X}) \cdot \mathbf{n}(\boldsymbol{\xi};\mathbf{X}) &= \bar{\mathbf{t}}(\boldsymbol{\xi};\mathbf{X}) \text{ on } \partial\Omega_t, \\
\mathbf{u}(\boldsymbol{\xi};\mathbf{X}) &= \bar{\mathbf{u}}(\boldsymbol{\xi};\mathbf{X}) \text{ on } \partial\Omega_u,
\end{aligned} \tag{17}$$

such that $\partial\Omega_t \cup \partial\Omega_u = \partial\Omega$ and $\partial\Omega_t \cap \partial\Omega_u = \emptyset$ with $\nabla := (\partial/\partial\xi_1, \partial/\partial\xi_2, \partial/\partial\xi_3)$ and $\mathbf{b}(\boldsymbol{\xi};\mathbf{X})$, $\bar{\mathbf{t}}(\boldsymbol{\xi};\mathbf{X})$, $\bar{\mathbf{u}}(\boldsymbol{\xi};\mathbf{X})$, and $\mathbf{n}(\boldsymbol{\xi};\mathbf{X})$ representing the body force, prescribed traction on $\partial\Omega_t$, prescribed displacement on $\partial\Omega_u$, and unit outward normal vector, respectively. Mathematically, the problem entails finding an optimal design solution

$$\mathbf{d}^* = \arg \min_{\mathbf{d} \in \mathcal{D} \subseteq \mathbb{R}^{79}} c_0(\mathbf{d}) := \rho \int_{\Omega(\mathbf{d})} d\Omega$$

subject to $c(\mathbf{d}) := P_{\mathbf{d}} \left[y_{\min}(\mathbf{u}(\boldsymbol{\xi}_c;\mathbf{X}), \boldsymbol{\sigma}(\boldsymbol{\xi}_c;\mathbf{X})) \le 10^6 \right] \le 1 - 0.99865,$

where the objective function $c_0(\mathbf{d})$, with ρ representing the mass density of the material, describes the overall mass of the bracket; on the other hand, the constraint function $c(\mathbf{d})$ quantifies the probability of minimum fatigue crack-initiation life y_{min}, attained at a critical spatial point $\boldsymbol{\xi}_c$, failing to exceed a million loading cycles to be less than $(1 - 0.99865)$. Here, y_{min} depends on displacement and stress responses $\mathbf{u}(\boldsymbol{\xi}_c; \mathbf{X})$ and $\boldsymbol{\sigma}(\boldsymbol{\xi}_c; \mathbf{X})$, which satisfy (17). An FEA comprising 341,112 nodes and 212,716 ten-noded, quadratic, tetrahedral elements, was performed to solve the variational weak form of (17). Further details are available elsewhere [11].

The univariate ($S = 1$) adaptive-sparse PDD method was applied to solve this shape optimization problem. Figure 2a–d show the contour plots of the logarithm of fatigue life at mean shapes of several design iterations, including the initial design, throughout the optimization process. Due to a conservative initial design, with fatigue life contour depicted in Fig. 2a, the minimum fatigue crack-initiation life of 6.65×10^9 cycles is much larger than the required fatigue crack-initiation life of a million cycles. For the tolerance and subregion size parameters selected, 14 iterations and 2808 FEA led to a final optimal design with the corresponding mean shape presented in Fig. 2d. The total run time, including performing all 2808 FEA in a desktop personal computer (8 cores, 2.3 GHz, 16 GB RAM), was about 165 h. Most design variables have undergone significant changes from their initial values, prompting substantial modifications of the shapes or sizes of the outer boundaries,

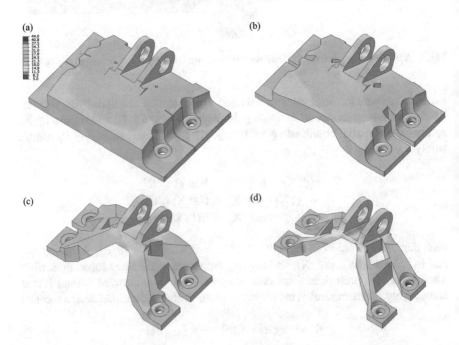

Fig. 2 Contours of logarithmic fatigue life at mean shapes of the jet engine bracket by the adaptive-sparse PDD method; (**a**) initial design; (**b**) iteration 3; (**c**) iteration 6; (**d**) iteration 14 (optimum)

quadrilateral holes, and bottom surfaces of the engine bracket. The mean optimal mass of the engine bracket is 0.48 kg—an almost 84 % reduction from the mean initial mass of 3.02 kg. At optimum, the constraint function $c(\mathbf{d})$ is practically zero and is, therefore, close to being active.

This example shows some promise of the adaptive-sparse PDD methods in solving industrial-scale engineering design problems with an affordable computational cost. However, an important drawback persists: given the computer resources available at the time of this work, only the univariate adaptive-sparse PDD approximation is feasible. The univariate result has yet to be verified with those obtained from bivariate or higher-variate adaptive-sparse PDD approximations. Therefore, the univariate "optimal" solution reported here should be guardedly interpreted.

Finally, it is natural to ask how much the bivariate adaptive-sparse PDD approximation will cost to solve this design problem. Due to quadratic computational complexity, the full bivariate PDD approximation using current computer resources of this study is prohibitive. However, a bivariate adaptive-sparse PDD approximation with a cost scaling markedly less than quadratic, if it can be developed, should be encouraging. In which case, a designer should exploit the univariate solution as the initial design to seek a better design using the bivariate adaptive-sparse PDD method, possibly, in fewer design iterations. The process can be repeated for higher-variate PDD methods if feasible. Clearly, additional research on stochastic design optimization, including more efficient implementation of the adaptive-sparse PDD methods, is required.

5 Conclusion

A new adaptive-sparse PDD method was developed for stochastic design optimization of high-dimensional complex systems commonly encountered in applied sciences and engineering. The method is based on an adaptive-sparse PDD approximation of a high-dimensional stochastic response for statistical moment and reliability analyses; a novel integration of the adaptive-sparse PDD approximation and score functions for estimating the first-order sensitivities of the statistical moments and failure probability with respect to the design variables; and standard gradient-based optimization algorithms, encompassing a computationally efficient design process. When blended with score functions, the adaptive-sparse PDD approximation leads to analytical formulae for calculating the design sensitivities. More importantly, the statistical moments, failure probability, and their respective design sensitivities are all determined concurrently from a single stochastic analysis or simulation. Numerical results stemming from a mathematical example indicate that the new method provides more computationally efficient design solutions than the existing methods. Finally, stochastic shape optimization of a jet engine bracket with 79 variables was performed, demonstrating the power of the new methods to tackle practical engineering problems.

Acknowledgements The authors acknowledge financial support from the U.S. National Science Foundation under Grant Nos. CMMI-0969044 and CMMI-1130147.

References

1. I. Babuska, F. Nobile, R. Tempone, A stochastic collocation method for elliptic partial differential equations with random input data. SIAM Rev. **52**, 317–355 (2010)
2. A. Browder, *Mathematical Analysis: An Introduction*. Undergraduate Texts in Mathematics (Springer, New York, 1996)
3. H.J. Bungartz, M. Griebel, Sparse grids. ACTA Numer. **13**, 147–269 (2004)
4. B. Efron, C. Stein, The jackknife estimate of variance. Ann. Stat. **9**(3), 586–596 (1981)
5. A. Genz, B.D. Keister, Fully symmetric interpolatory rules for multiple integrals over infinite regions with gaussian weight. J. Comput. Appl. Math. **71**(2), 299–309 (1996)
6. S. Lee, W. Chen, B. Kwak, Robust design with arbitrary distribution using gauss type quadrature formula. Struct. Multidiscip. Optim. **39**, 227–243 (2009)
7. S. Rahman, A polynomial dimensional decomposition for stochastic computing. Int. J. Numer. Methods Eng. **76**, 2091–2116 (2008)
8. S. Rahman, X. Ren, Novel computational method for high-dimensional stochastic sensitivity analysis. Int. J. Numer. Methods Eng. **98**, 881–916 (2014)
9. S. Rahman, V. Yadav, Orthogonal polynomial expansions for solving random eigenvalue problems. Int. J. Uncertain. Quantif. **1**, 163–187 (2011)
10. X. Ren, S. Rahman, Robust design optimization by polynomial dimensional decomposition. Struct. Multidiscip. Optim. **48**, 127–148 (2013)
11. X. Ren, V. Yadav, S. Rahman, Reliability-based design optimization by adaptive-sparse polynomial dimensional decomposition. Struct. Multidiscip. Optim. doi:10.1007/s00158-015-1337-6. (2015, Accepted)
12. R. Rubinstein, A. Shapiro, *Discrete Event Systems: Sensitivity Analysis and Stochastic Optimization by the Score Function Method*. Wiley Series in Probability and Mathematical Statistics (Wiley, New York, 1993)
13. D. Xiu, G. E. Karniadakis, The Wiener-Askey polynomial chaos for stochastic differential equations. SIAM J. Sci. Comput. **24**, 619–644 (2002)
14. H. Xu, S. Rahman, A generalized dimension-reduction method for multi-dimensional integration in stochastic mechanics. Int. J. Numer. Methods Eng. **61**, 1992–2019 (2004)
15. V. Yadav, S. Rahman, Adaptive-sparse polynomial dimensional decomposition for high-dimensional stochastic computing. Comput. Methods Appl. Mech. Eng. **274**, 56–83 (2014)

Efficient Spectral-Element Methods for the Electronic Schrödinger Equation

Jie Shen, Yingwei Wang, and Haijun Yu

Abstract Two efficient spectral-element methods, based on Legendre and Laguerre polynomials respectively, are derived for direct approximation of the electronic Schrödinger equation in one spatial dimension. Compared to existing literatures, a spectral-element approach is used to treat the singularity in nucleus-electron Coulomb potential, and with the help of Slater determinant, special basis functions are constructed to obey the antisymmetric property of the fermionic wavefunctions. Numerical tests are presented to show the efficiency and accuracy of the proposed methods.

1 Introduction

In this article we consider the *electronic Schrödinger equation* (ESE) in one spatial dimension

$$H\Psi(\mathbf{x}) = E\Psi(\mathbf{x}), \tag{1}$$

with the Hamiltonian operator

$$H = T + V_{ne} + V_{ee}, \tag{2}$$

J. Shen • Y. Wang (✉)
Department of Mathematics, Purdue University, West Lafayette, IN 47907, USA
e-mail: shen7@purdue.edu; wywshtj@gmail.com

H. Yu
LSEC, Institute of Computational Mathematics and Scientific/Engineering Computing, AMSS, Chinese Academy of Sciences, Beijing, 100190, China
e-mail: hyu@lsec.cc.ac.cn

© Springer International Publishing Switzerland 2016
J. Garcke, D. Pflüger (eds.), *Sparse Grids and Applications – Stuttgart 2014*,
Lecture Notes in Computational Science and Engineering 109,
DOI 10.1007/978-3-319-28262-6_11

265

where the kinetic energy T, nucleus-electron potential V_{ne} and electron-electron potential V_{ee} operators are

$$T = -\frac{1}{2}\sum_{i=1}^{N}\partial_{x_i}^2, \quad V_{ne} = N\sum_{i=1}^{N}|x_i|, \quad V_{ee} = -\sum_{i=1}^{N}\sum_{j>i}|x_i - x_j|. \tag{3}$$

Here N denotes the number of electrons in this system, $x_i \in \mathbb{R}$ the position of the i-th electron, and the solution $\Psi(\mathbf{x})$, with $\mathbf{x} = (x_1, \cdots, x_N)$, describes the wave function associated to the total energy E, and satisfies the boundary condition

$$\Psi(x_1, x_2, \cdots, x_N) \to 0, \quad \text{as } |x_j| \to \infty, \ j = 1, \cdots, N. \tag{4}$$

The electronic Schrödinger equation, in three spatial dimension, results from Born-Oppenheimer approximation to the general Schrödinger equation for a system of electrons and nuclei, which is one of the core problems in computational quantum chemistry [5, 13, 23]. However, except for very simple cases, there is no analytical solution available. Hence, it is essential to develop efficient and accurate numerical algorithms for this problem. While most applications of the ESE are in three spatial dimension, the one-dimensional formulation above does inherits some essential features, such as high-dimensionality and singular behavior, of the three dimensional case. Hence, developing a solver in one dimension is an important preliminary and calibrating step that serves as a prototype for solving the ESE in two or three spatial dimensions.

There are several major difficulties for solving the ESE (1). We summarize them below and describe our strategies.

(i) It is an N dimensional problem so it suffers from the so-called *curse of dimensionality* if classical numerical methods are employed. Therefore, various model approximations have been developed in quantum chemistry to reduced the computational complexity. We intend to discretize the ESE directly using sparse grids [3] which have proven to be useful for a class of high-dimensional problems, including in particular the ESE [25, 26]. For example, M. Griebel and J. Hamaekers proposed sparse grid methods for ESE based on Meyer wavelets [7], Fourier functions [8], adaptive Gaussian type orbitals basis sets [9]. On the other hand, we propose to use spectral sparse grid methods based on hyperbolic cross approximations [19–22].

(ii) The singularities of the Coulomb potentials shown in (3), called "*Coulomb singularity*" or "*Kato cusp condition*" [6, 12], deteriorate the convergence rates of global spectral methods. In order to treat the singularity in V_{ne} more effectively, we propose a spectral element framework to design basis functions which provide better approximations to the singularity.

(iii) The wave function $\Psi(\mathbf{x})$ has the additional constraint that it must be *antisymmetric* under exchange of variables, according to Pauli exclusion principle. We shall construct, using the antisymmetrizer and Slater determinant, basis functions which obey the antisymmetric property. We also propose an efficient implementation of inner products with respect to antisymmetric functions.

In our previous attempt for solving ESE [22], we used a global spectral method whose convergence rate is severely affected by the *Coulomb singularity*, and we did not enforce the antisymmetry so it resulted in a much larger number of unknowns than actually needed by the physical problem. The main purpose of this paper is to develop efficient procedures to address these two issues.

The rest of the paper is organized as follows. In Sects. 2 and 3, we propose two kinds of efficient spectral Galerkin methods based on Legendre and Laguerre polynomials respectively, including the basis functions for one or many electrons, full or sparse grids, and with or without the antisymmetric property. In Sect. 4, we present numerical results to illustrate the convergence of our methods for ESE calculations. Finally, conclusions and discussion of possible directions for future research are presented in Sect. 5.

2 A Spectral-Element Method for ESE

In this and next sections, we develop a spectral-element framework to discretize the ESE (1). First, we focus on the set of basis functions for one electron case. Then, we demonstrate the strategies for dealing with high dimensional problems and antisymmetric functions. In addition, we also briefly show how to generate the matrices required in Galerkin methods efficiently, involving mass, stiffness and various potential matrices.

2.1 One Electron Case

As a starting point, let us focus on the case with $N = 1$ in Eq. (1),

$$
\begin{cases}
-\dfrac{1}{2}\Psi''(x) + |x|\Psi(x) = E\Psi(x), & x \in \mathbb{R}, \\[2mm]
\lim_{x \to \pm\infty} \Psi(x) = 0.
\end{cases}
\tag{5}
$$

Let ξ be a truncation parameter. After a truncation from the unbounded interval $(-\infty, +\infty)$ to bounded one $[-2\xi, 2\xi]$, $\xi > 0$, and further a linear map from general interval $[-2\xi, 2\xi]$ to standard one $[-2, 2]$, we arrive at

$$
\begin{cases}
-\dfrac{1}{2\xi^2}\tilde{\Psi}''(x) + \xi|x|\tilde{\Psi}(x) = E\tilde{\Psi}(x), & x \in [-2, 2], \\[2mm]
\tilde{\Psi}(\pm 2) = 0.
\end{cases}
\tag{6}
$$

2.1.1 Galerkin Formulation

Let X_n be an approximation space and ω be the weight function. The spectral Galerkin method for the problems (5) or (6) can all be casted in the following form: Find $u_n \in X_n$ such that

$$c_1 \langle \partial_x u_n, \partial_x(\phi_n \omega) \rangle + c_2 \langle |x| u_n, \phi_n \rangle_\omega = \lambda \langle u_n, \phi_n \rangle_\omega, \quad \forall \, \phi_n \in X_n. \tag{7}$$

Note that $c_1 = \frac{1}{2}, c_2 = 1$ for problem (5), $c_1 = \frac{1}{2\xi^2}, c_2 = \xi$ for problem (6), and λ is the numerical estimate of E.

Let $\{\phi_k\}_{k=-n}^n$ be a set of basis functions for X_n. We denote

$$u_n(x) = \sum_{k=-n}^n \hat{u}_k \phi_k(x), \qquad \mathbf{u} = (\hat{u}_{-n}, \cdots, \hat{u}_n)^T, \tag{8}$$

$$s_{lk} = \langle \phi_k', (\phi_l \omega)' \rangle, \qquad \mathbf{S} = (s_{lk})_{-n \le l, k \le n}, \tag{9}$$

$$m_{lk} = \langle \phi_k, \phi_l \rangle_\omega, \qquad \mathbf{M} = (m_{lk})_{-n \le l, k \le n}, \tag{10}$$

$$p_{lk}^{ne} = \langle |x| \phi_l, \phi_k \rangle_\omega, \qquad \mathbf{P}^{ne} = (p_{lk}^{ne})_{-n \le l, k \le n}. \tag{11}$$

Thus, the Galerkin formulation (7) yields the following generalized eigenvalue problem

$$(c_1 \mathbf{S} + c_2 \mathbf{P}^{ne}) \mathbf{u} = \lambda \mathbf{M} \mathbf{u}, \tag{12}$$

where λ is the eigenvalue and $\mathbf{u}$ is the corresponding eigenvector.

2.1.2 Basis Functions

In classical spectral-Galerkin approach, Hermite functions are often served as the basis functions for the problem defined on the whole line [10, 18] while Legendre or Chebyshev polynomials are frequently used for the problem in bounded intervals [14, 15]. However, the nucleus-electron potential $V_{ne} = |x|$ in Eqs. (5) and (6) is not differentiable at the origin. Thus, the convergence rates are rather limited if classical spectral methods are employed here. Therefore, we split the interval at the origin into two subintervals, and use a spectral-element method [4] to deal with the singularity at the origin. The basis functions for the (two-elements) spectral-element methods are as follows:

(i) For the problem (6) in bounded domain $[-2, 2]$, the function space $X_n^b = \text{span}\{\phi_k^b : k = -n, \cdots, n\}$, the weight $\omega = 1$, and the basis functions $\{\phi_k^b\}_{k=-n}^n$

for one electron are chosen as

$$\phi_k^b(x) = \begin{cases} L_{k-1}(x-1) - L_{k+1}(x-1), & k > 0, \ x \in [0,2], \\ 2 - |x|, & k = 0, \ x \in [-2,2], \\ L_{|k|-1}(|x|-1) - L_{|k|+1}(|x|-1), & k < 0, \ x \in [-2,0], \end{cases} \tag{13}$$

where $L_k(x), x \in [-1,1]$, is the *Legendre polynomial* of degree k. For the functions defined in (13), $\phi_k^b(x) = 0$ on unspecified intervals, i.e. $k > 0, x \in [-2,0)$ and $k < 0, x \in (0,2]$. Those parts with zero values are not plotted in Fig. 1. By the property of Legendre polynomials, we know that

$$\phi_k^b(0) = \phi_k^b(2) = 0, \qquad\qquad k > 0,$$
$$\phi_0^b(0) = 1, \quad \phi_0^b(\pm 2) = 0, \qquad k = 0,$$
$$\phi_k^b(-2) = \phi_k^b(0) = 0, \qquad\qquad k < 0.$$

If the basis functions $\{\phi_k\}$ in Eq. (7) are chosen as $\phi_k(x) = \phi_k^b(x)$, then by using the properties of Legendre polynomials [17], the stiffness, mass and potential matrices defined in (9), (10) and (11) are diagonal, penta-diagonal and seven-diagonal matrices, respectively, and can be computed explicitly.

(ii) For the problem (5) in unbounded domain $(-\infty, +\infty)$, the function space $X_n^u = \text{span}\{\phi_k^u : k = -n, \cdots, n\}$, the weight $\omega = 1$, and the basis functions $\{\phi_k^u\}_{k=-n}^n$ for one electron are chosen as

$$\phi_k^u(x) = \begin{cases} \hat{L}_k(x) - \hat{L}_{k-1}(x), & k > 0, \ x \in [0, +\infty), \\ e^{-|x|/2}, & k = 0, x \in (-\infty, +\infty), \\ \hat{L}_{|k|}(|x|) - \hat{L}_{|k|-1}(|x|), & k < 0, \ x \in (-\infty, 0], \end{cases} \tag{14}$$

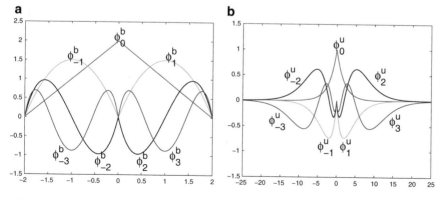

Fig. 1 First few basis functions for one electron case: Legendre and Laguerre basis sets. (**a**) Legendre basis in bounded domain. (**b**) Laguerre basis in unbounded domain

where $\hat{L}_k(x), x \in [0, +\infty)$, is the *Laguerre function* of degree k. In Eq. (14), $\phi_k^u(x) = 0$ on unspecified intervals, i.e. $k > 0, x \in (-\infty, 0)$ and $k < 0, x \in (0, +\infty)$. By the property of the Laguerre functions, we know that

$$
\begin{aligned}
\phi_k^u(0) &= 0, & k \neq 0, \\
\phi_0^u(0) &= 1, & k = 0, \\
\lim_{x \to \infty} \phi_k^u(x) &= \lim_{x \to -\infty} \phi_k^u(x) = 0, & \forall k.
\end{aligned}
$$

If the basis functions $\{\phi_k\}$ in Eq. (7) are chosen as $\phi_k(x) = \phi_k^u(x)$, then by using the properties of Laguerre polynomials [17], the stiffness, mass and potential matrices defined in (9), (10) and (11) are tri-diagonal, tri-diagonal and penta-diagonal matrices, respectively, and can be computed explicitly.

A few basis functions $\{\phi_k^b(x)\}$ and $\{\phi_k^u(x)\}$ for $k = -3, -2, -1, 0, 1, 2, 3$ defined above are illustrated in Fig. 1.

2.2 N-Electron Case

We first introduce some notations:

- For $N \in \mathbb{N}$, we use boldface lowercase letters to denote N-dimensional multi-indices and vectors, e.g., $\mathbf{k} = (k_1, \cdots, k_N) \in \mathbb{Z}^N$. Besides, we need following norms: $|\mathbf{k}|_1 = \sum_{j=1}^{N} |k_j|$, $|\mathbf{k}|_\infty = \max_{1 \leq j \leq N} |k_j|$, $|\mathbf{k}|_{mix} = \prod_{j=1}^{N} \max\{1, |k_j|\}$. Note that $|\mathbf{k}|_{mix} \geq 1$ for all $\mathbf{k} \in \mathbb{Z}^N$.
- $\Lambda \subset \mathbb{Z}^N$ is the set of indices and $|\Lambda|$ means its cardinality.

Now let us consider the ESE for the system with N electrons.

$$
\begin{cases}
-\dfrac{1}{2} \displaystyle\sum_{i=1}^{N} \partial_{x_i}^2 \Psi + N \sum_{i=1}^{N} |x_i| \Psi - \sum_{i=1}^{N} \sum_{j>i} |x_i - x_j| \Psi = E\Psi, & \mathbf{x} \in \mathbb{R}^N, \\
\lim_{x_j \to \pm\infty} \Psi(\mathbf{x}) = 0, & \forall j = 1, 2, \cdots N.
\end{cases}
\tag{15}
$$

Similarly as in the one electron case, after truncation and linear mapping, the problem in the unbounded domain is equivalent to the following in a bounded domain:

$$
\begin{cases}
-\dfrac{1}{2\xi^2} \displaystyle\sum_{i=1}^{N} \partial_{x_i}^2 \Psi + N\xi \sum_{i=1}^{N} |x_i| \Psi - \xi \sum_{i=1}^{N} \sum_{j>i} |x_i - x_j| \Psi = E\Psi, & \mathbf{x} \in \Omega, \\
\Psi(\mathbf{x})|_{\partial\Omega} = 0,
\end{cases}
\tag{16}
$$

where $\Omega = [-2, 2]^N$.

2.2.1 Galerkin Formulation

Similarly as in previous subsection, let X_n be the approximation space. The spectral Galerkin method for the problems (15) or (16) can all be casted in the following form: Find $u_\mathbf{n} \in X_n$ such that $\forall \Phi_\mathbf{n} \in X_n$,

$$c_1 \sum_{j=1}^{N} \langle \partial_{x_j} u_\mathbf{n}, \partial_{x_j} \Phi_\mathbf{n} \rangle + c_2 \left\langle \sum_{j=1}^{N} |x_j| u_\mathbf{n}, \Phi_\mathbf{n} \right\rangle + c_3 \left\langle \sum_{i=1}^{N} \sum_{j>i} |x_i - x_j| u_\mathbf{n}, \Phi_\mathbf{n} \right\rangle = \tag{17}$$

$$\lambda \langle u_\mathbf{n}, \Phi_\mathbf{n} \rangle .$$

Note that for problem (15), $c_1 = \frac{1}{2}, c_2 = N, c_3 = -1$ while for problem (16), $c_1 = \frac{1}{2\xi^2}, c_2 = N\xi, c_3 = -\xi$. λ, to be solved, is an approximation of E in X_n.

Let $\{\Phi_\mathbf{k}\}_{\mathbf{k} \in \Lambda}$ be a set of basis functions for X_n, where Λ is the set of indices to be determined. We denote

$$u_\mathbf{k}(x) = \sum_{\mathbf{k} \in \Lambda} \hat{u}_\mathbf{k} \Phi_\mathbf{k}(x), \qquad\qquad \mathbf{u} = \text{vec}\, (\hat{u}_\mathbf{k})_{\mathbf{k} \in \Lambda}\,, \tag{18}$$

$$s_{\hat{l},\hat{k}} = \sum_{j=1}^{N} \langle \partial_j \Phi_\mathbf{k}, \partial_j \Phi_\mathbf{l} \rangle, \qquad\qquad S = (s_{\hat{l},\hat{k}}), \tag{19}$$

$$m_{\hat{l},\hat{k}} = \langle \Phi_\mathbf{k}, \Phi_\mathbf{l} \rangle, \qquad\qquad M = (m_{\hat{l},\hat{k}}), \tag{20}$$

$$p_{\hat{l},\hat{k}}^{ne} = \langle \sum_{j=1}^{N} |x_j| \Phi_\mathbf{k}, \Phi_\mathbf{l} \rangle, \qquad\qquad P^{ne} = (p_{\hat{l},\hat{k}}^{ne}), \tag{21}$$

$$p_{\hat{l},\hat{k}}^{ee} = \langle \sum_{i=1}^{N} \sum_{j>i} |x_i - x_j| \Phi_\mathbf{k}, \Phi_\mathbf{l} \rangle, \qquad\qquad P^{ee} = (p_{\hat{l},\hat{k}}^{ee}). \tag{22}$$

where $\hat{k}$ is the corresponding order of $\mathbf{k} = (k_1, \cdots, k_N)$ in the set Λ and $\mathbf{u} = \text{vec}\,(\hat{u}_\mathbf{k})_{\mathbf{k} \in \Lambda}$ is a column vector with entries $\{\hat{u}_\mathbf{k}\}_{\mathbf{k} \in \Lambda}$. Suppose the cardinality of the set Λ be $|\Lambda|$, then $\mathbf{u}$ defined in (18) is a $|\Lambda|$-by-1 column vector and the matrices defined in (19), (20), (21) and (22) are $|\Lambda|$-by-$|\Lambda|$ square matrices.

Thus, the Galerkin formulation (17) gives the following generalized eigenvalue problem

$$(c_1 S + c_2 P^{ne} + c_3 P^{ee})\mathbf{u} = \lambda M \mathbf{u}, \tag{23}$$

where λ is the eigenvalue and $\mathbf{u}$ is the corresponding eigenvector.

2.2.2   Full Grid and Sparse Grid

The classical tensor-product basis function in N-dimensional space is

$$\Phi_{\mathbf{k}}(\mathbf{x}) = \prod_{j=1}^{N} \phi_{k_j}(x_j), \tag{24}$$

where $\mathbf{k} = (k_1, k_2, \cdots, k_N) \in \mathbb{Z}^N$, $\mathbf{x} = (x_1, x_2, \cdots, x_N) \in \mathbb{R}^N$, and $\phi_{k_j}(x_j)$ is the one-dimensional basis function considered in previous subsection, e.g. $\{\phi_k^b(x)\}$ defined in (13) or $\{\phi_k^u(x)\}$ defined in (14).

The approximation space in N-dimensional space is

$$X_n^N = \text{span } \{\Phi_{\mathbf{k}}(\mathbf{x}) : \mathbf{k} \in \Lambda_n\}, \quad n \in \mathbb{N}.$$

For different set of indices Λ_n, we have different space.

- The set of indices for *full grid* is

$$\Lambda_n^F = \{\mathbf{k} \in \mathbb{Z}^N : |\mathbf{k}|_\infty \leq n\}, \quad n \in \mathbb{N}. \tag{25}$$

- The set of indices for *sparse grid* of hyperbolic cross type is

$$\Lambda_n^S = \{\mathbf{k} \in \mathbb{Z}^N : |\mathbf{k}|_{mix} \leq n\}, \quad n \in \mathbb{N}. \tag{26}$$

Here are several remarks on the spectral-element basis function sets we use.

- The electronic eigenfunctions are proved to decay exponentially as the spatial variable goes to infinity in the sense that there exist positive constants A and B for which $|\phi(x)| \leq Ae^{-B\|x\|}$. (see Ref [1].) Hence, in the Legendre spectral method, the error caused by restriction from unbounded domain to bounded one would also be exponentially convergent as the parameter L goes to the infinity.
- The idea of spectral-element method to treat the nuclei-electron cusps could be easily generalized to the case with several nuclei. See Fig. 2 for two nuclei case.
- For ESE in three spatial dimension, it is known that the sparse grids based on hyperbolic cross fit the smoothness property of the eigenfunctions [3, 25, 26]. However, for the ESE in one spatial dimension considered in this paper, there are no theoretical results available in the literature. In Sect. 4, we will compare the numerical results obtained from full grids (25) and sparse grids (26).

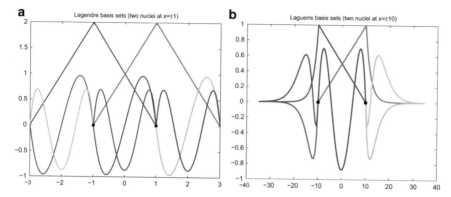

Fig. 2 Legendre and Laguerre basis sets for two nuclei case. (**a**) Legendre basis in bounded domain. (**b**) Laguerre basis in unbounded domain

3 Antisymmetry and Antisymmetric Inner Product

The electronic wavefunction $\Psi(\mathbf{x})$ for many body system must be *antisymmetric* with respect to electron positions $\mathbf{x} = (x_1, \cdots, x_N)$, i.e.

$$\Psi(x_1, \cdots, x_i, \cdots, x_j, \cdots, x_N) = -\Psi(x_1, \cdots, x_j, \cdots, x_i, \cdots, x_N). \tag{27}$$

It is obvious that $\Phi_{\mathbf{k}}(\mathbf{x})$ defined in (24) does not obey the antisymmetric property. A main difficulty is how to construct basis functions which satisfy the antisymmetry, and how to efficiently compute the inner products between them.

3.1 Antisymmetrizer and Slater Determinant

In order to enforce antisymmetry, we introduce a linear operator called *antisymmetrizer* [11], also called *skew symmetrization* or *alternation*, which is defined by

$$\mathcal{A} = \frac{1}{N!} \sum_{p \in S_N} (-1)^p \mathcal{P}, \tag{28}$$

where S_N is the permutation group on N elements. For the element $p \in S_N$, the operator $\mathcal{P}$ acts on a function by permuting its variables, as $\mathcal{P}\Psi(\gamma_1, \gamma_2, \cdots) = \Psi(\gamma_{p(1)}, \gamma_{p(2)}, \cdots)$. The sign $(-1)^p$ is -1 if p is an odd permutation and 1 if it is even. Applying $\mathcal{A}$ to the function $\Phi_{\mathbf{k}}(\mathbf{x})$ defined in (24) leads to the antisymmetric basis $\Phi_{\mathbf{k}}^A(\mathbf{x})$ expressed as a Slater determinant:

$$\Phi_{\mathbf{k}}^A(\mathbf{x}) := \mathcal{A}\Phi_{\mathbf{k}}(\mathbf{x})$$

$$= \frac{1}{N!} \begin{vmatrix} \phi_{k_1}(x_1) & \phi_{k_1}(x_2) & \cdots\cdots & \phi_{k_1}(x_N) \\ \phi_{k_2}(x_1) & \phi_{k_2}(x_2) & \cdots\cdots & \phi_{k_2}(x_N) \\ \cdots & \cdots & \cdots\cdots & \cdots \\ \phi_{k_N}(x_1) & \phi_{k_N}(x_2) & \cdots\cdots & \phi_{k_N}(x_N) \end{vmatrix}. \tag{29}$$

It is easy to check that the basis function $\Phi_{\mathbf{k}}^A(\mathbf{x})$ satisfies the antisymmetric property (27). Besides, if $k_i = k_j$, then the determinant in Eq. (29) would be zero. Thus, the set of indices for antisymmetric basis $\{\Phi_{\mathbf{k}}^A(\mathbf{x})\}_{\mathbf{k}\in\Lambda_n^A}$ should be

$$\Lambda_n^A = \{\mathbf{k} \in \mathbb{Z}^N : |\mathbf{k}|_\infty \le n, k_1 < k_2 < \cdots < k_N\}, \quad n \in \mathbb{N}. \tag{30}$$

It implies that the cardinality of antisymmetric basis set is about $\frac{1}{N!}$ times of the regular one.

Now we have four kinds of grids, namely *full grid* (**'F'**), *sparse grid* (**'S'**), *full grid with antisymmetric property* (**'FA'**) and *sparse grid with antisymmetric property* (**'SA'**). The cardinality of them are shown in the Table 1 and sketch for two dimensional case are shown Fig. 3.

3.2 Antisymmetric Inner Product and Löwdin's Rule

One of the main difficulties in implementation of spectral type methods based on antisymmetric grids (**'FA'** and **'SA'**) is the calculation of inner products between two Slater determinants. In this subsection, we briefly show how to compute the entries in the matrices S, M, P^{ne} and P^{ee} defined in (19), (20), (21), and (22) with respect to the antisymmetric basis functions $\{\Phi_{\mathbf{k}}^A(\mathbf{x})\}_{\mathbf{k}\in\Lambda_n^A}$.

Table 1 Four kinds of grids for N-dimensional problems

Grids	Set of indices	Cardinality[a]
'F'	Λ_n^F	$O\left((2n+1)^N\right)$
'FA'	$\Lambda_n^F \cap \Lambda_n^A$	$O\left(\frac{(2n+1)^N}{N!}\right)$
'S'	Λ_n^S	$O\left((2n+1)\log^{N-1}(2n+1)\right)$
'SA'	$\Lambda_n^S \cap \Lambda_n^A$	$O\left(\frac{(2n+1)\log^{N-1}(2n+1)}{N!}\right)$

[a] The cardinality of hyperbolic cross sparse grids can be found in [19]

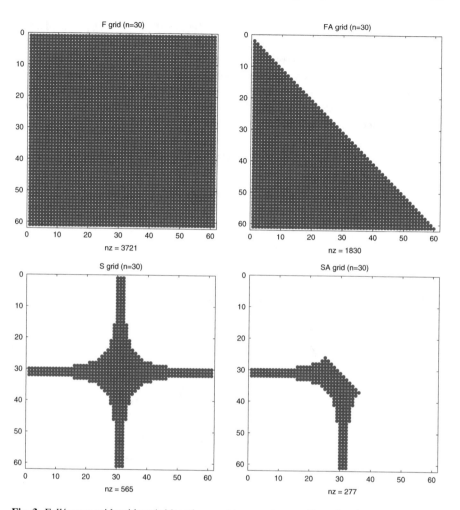

Fig. 3 Full/sparse grids without/with antisymmetric property: two dimensional case

• For the mass matrix M, we need to construct the following auxiliary matrix

$$\tilde{M}_{\mathbf{k},\mathbf{l}} = \begin{pmatrix} \langle \phi_{k_1}, \phi_{l_1} \rangle & \cdots & \langle \phi_{k_1}, \phi_{l_N} \rangle \\ \vdots & \ddots & \vdots \\ \langle \phi_{k_N}, \phi_{l_1} \rangle & \cdots & \langle \phi_{k_N}, \phi_{l_N} \rangle \end{pmatrix} \quad \forall \, \mathbf{k}, \mathbf{l} \in \Lambda_n^A. \tag{31}$$

Then each entry of M can be computed as

$$\langle \Phi_{\mathbf{k}}^A, \Phi_{\mathbf{l}}^A \rangle = \frac{1}{N!} \det(\tilde{M}_{\mathbf{k},\mathbf{l}}), \tag{32}$$

which is the so called *Löwdin's rule* [11]. Note that the matrix $\tilde{M}_{\mathbf{k},\mathbf{l}}$ defined above is a submatrix of the one-dimensional mass matrix M defined in (10), either tri-diagonal one in Laguerre case or penta-diagonal one in Legendre case, so its determinant can be computed efficiently. The denominator $N!$ need never be computed, since it will occur in every term in our equations and so cancels.

- For the stiffness matrix S and nucleus-electron potential matrix P^{ne}, we need to construct the following auxiliary matrices

$$\tilde{S}_{\mathbf{k},\mathbf{l},i} = \begin{pmatrix} \langle \phi_{k_1}, \phi_{l_1} \rangle & \cdots & \langle \phi'_{k_1}, \phi'_{l_i} \rangle & \cdots & \langle \phi_{k_1}, \phi_{l_N} \rangle \\ \vdots & \cdots & \vdots & \cdots & \vdots \\ \langle \phi_{k_N}, \phi_{l_1} \rangle & \cdots & \langle \phi'_{k_N}, \phi'_{l_i} \rangle & \cdots & \langle \phi_{k_N}, \phi_{l_N} \rangle \end{pmatrix},$$

$$\tilde{P}^{ne}_{\mathbf{k},\mathbf{l},i} = \begin{pmatrix} \langle \phi_{k_1}, \phi_{l_1} \rangle & \cdots & \langle |x|\phi_{k_1}, \phi_{l_i} \rangle & \cdots & \langle \phi_{k_1}, \phi_{l_N} \rangle \\ \vdots & \cdots & \vdots & \cdots & \vdots \\ \langle \phi_{k_N}, \phi_{l_1} \rangle & \cdots & \langle |x|\phi_{k_N}, \phi_{l_i} \rangle & \cdots & \langle \phi_{k_N}, \phi_{l_N} \rangle \end{pmatrix}.$$

for each $\mathbf{k}, \mathbf{l} \in \Lambda_n^A$. Then each entry of S and P^{ne} can be computed as

$$\sum_{i=1}^{N} \langle \partial_{x_1} \Phi_{\mathbf{k}}^A, \partial_{x_i} \Phi_{\mathbf{l}}^A \rangle = \frac{1}{N!} \sum_{i=1}^{N} \det(\tilde{S}_{\mathbf{k},\mathbf{l},i}), \tag{33}$$

$$\sum_{i=1}^{N} \langle |x_i| \Phi_{\mathbf{k}}^A, \Phi_{\mathbf{l}}^A \rangle = \frac{1}{N!} \sum_{i=1}^{N} \det(\tilde{P}^{ne}_{\mathbf{k},\mathbf{l},i}). \tag{34}$$

- For the electron-electron interaction potential matrix P^{ee}, we use the methodology proposed by G.Beylkin [2]. To show the idea, we need more notations.

$$\Phi_{\mathbf{k}}(\mathbf{x}) = \prod_{i=1}^{N} \phi_{k_i}(x_i), \qquad \Phi_{\mathbf{k}}(\mathbf{x}) = \begin{pmatrix} \phi_{k_1}(x_1) \\ \phi_{k_2}(x_2) \\ \cdots \\ \phi_{k_N}(x_N) \end{pmatrix}, \tag{35}$$

$$\Phi_{\mathbf{l}}(\mathbf{x}) = \prod_{i=1}^{N} \phi_{l_i}(x_i), \qquad \Phi_{\mathbf{l}}(\mathbf{x}) = \begin{pmatrix} \phi_{l_1}(x_1) \\ \phi_{l_2}(x_2) \\ \cdots \\ \phi_{l_N}(x_N) \end{pmatrix}, \tag{36}$$

$$\Theta_{\mathbf{k},\mathbf{l}} = \tilde{M}_{\mathbf{k},\mathbf{l}}^{-1} \Phi_{\mathbf{k}} := \begin{pmatrix} \theta_1(x_1) \\ \theta_2(x_2) \\ \cdots \\ \theta_N(x_N) \end{pmatrix}. \tag{37}$$

Then the electron-electron inner products can be computed by

$$
\langle \sum_{i=1}^{N} \sum_{j>i} W(x_i, x_j) \Phi_{\mathbf{k}}^{A}, \Phi_{\mathbf{l}}^{A} \rangle =
$$

$$
\frac{\det(\tilde{M}_{\mathbf{k},\mathbf{l}})}{2N!} \sum_{i \neq j} \int W(x_i, x_j) \Phi_{l_i}(x_i) \Phi_{l_j}(x_j) \det(\Theta_{\mathbf{k},\mathbf{l}}^{i,j}) \mathrm{d}x_i \mathrm{d}x_j, \tag{38}
$$

where the weight $W(x_i, x_j) = |x_i - x_j|$ and

$$
\Theta_{\mathbf{k},\mathbf{l}}^{i,j} = \begin{pmatrix} \theta_i(x_i) & \theta_i(x_j) \\ \theta_j(x_i) & \theta_j(x_j) \end{pmatrix}.
$$

The formula (38) should be very efficient for large N. However, for many cases, the matrix $\tilde{M}$ is singular, that is to say we need to redefine the $\Theta_{\mathbf{k},\mathbf{l}}$ in (37) and $\det(\tilde{M})$. The detailed discussion can be found in [2]. We omit the details here for simplicity.

The mass and stiffness and matrices M, S based on antisymmetric Legendre and Laguerre bases for 4 electrons with $n = 8$ are shown in Figs. 4 and 5. All of this matrices are symmetric and positive definite.

4   Numerical Results

It is well known that the performance of spectral methods in unbounded domains can be significantly enhanced by choosing a proper truncation or scaling parameter such that the extreme collocation points are at or close to the endpoints of the effective interval (outside of which the solution is essentially zero). For the mapped Legendre method, the scaling parameter is the parameter ξ in Eq. (6). For the Laguerre method, one usually needs to determine a suitable scaling parameter ζ [16, 24] and then make a coordinate transform $y = x/\zeta$. That is to say the basis function for problem (5) should be chosen as

$$
\phi_{k,\zeta}^{u}(x) := \phi_{k}^{u}(x/\zeta), \quad \zeta > 0, \tag{39}
$$

where ϕ_k^u is defined in (14).

We apply the efficient spectral methods proposed in the previous section to the ESE (1). More precisely, the methods used in this section are

- Antisymmetric full grids based on Legendre basis ('**Leg-FA**') with parameters ξ and Laguerre basis ('**Lag-FA**') with parameters ζ;
- Antisymmetric sparse grids based on Legendre basis ('**Leg-SA**') with parameters ξ and Laguerre basis ('**Lag-SA**') with parameters ζ.

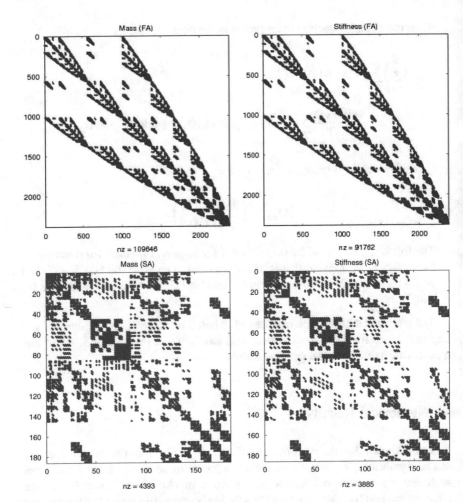

Fig. 4 Mass/stiffness matrices for Legendre basis in full/sparse antisymmetric grids: $N = 4$, $n = 8$

Besides, we make the following notations: *DoF* means the total number of degrees of freedom; E is the numerical estimates of first eigenvalue and ΔE denotes the relative difference between the two successive values of E. Moreover, the number with footnote in the following tables means the number with negative exponent, e.g., $4.49_{04} = 4.49 \times 10^{-04}$.

The numerical results for $N = 1$ are shown in Tables 2 and 3. The exponential rates of convergence could be observed both in Legendre and Laguerre basis, since the singularity in the V_{ne} has been taken care of (However, this exponential convergence does not extend to the more electron cases, see below). Further, the best choice for parameters is $\xi = 4$ and $\zeta = 0.1$ for $N = 1$. Note that the Legendre method with $\xi = 2$ has a faster convergence rate than Legendre methods with

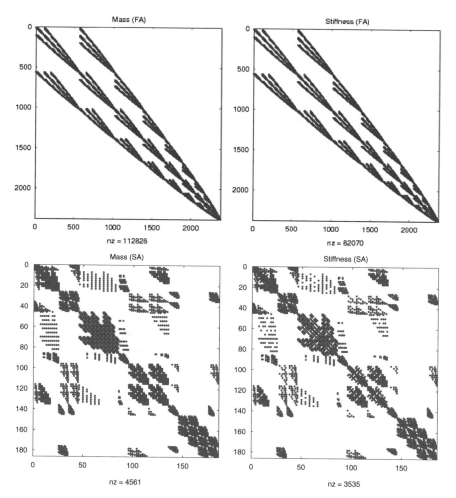

Fig. 5 Mass/stiffness matrices for Laguerre basis in full/sparse antisymmetric grids: $N = 4$, $n = 8$

larger ξ values, but it converges to an energy that is too far from the exact one. In Fig. 6, we plot the convergence curves of our Legendre and Laguerre methods together with the results of **'Fourier'** method (hyperbolic cross sparse grid method based on Fourier basis, proposed in [8]). The significant advantages of the bases proposed here over the Fourier bases demonstrates the importance of handling the nucleus-electron singularity. From Fig. 6, we also see that the Laguerre method is more sensitive to the scaling parameter than the Legendre method, although that an optimal scaled Laguerre method seems gives better solution than Legendre method.

The numerical results of **'Leg-FA'**, **'Lag-FA'**, **'Leg-SA'** and **'Lag-SA'** for $N = 2, 4, 6, 8$ are shown in Tables 4, 5, 6, and 7 respectively. We see that the advantages of sparse grids over full grids is not significant for small N ($N \le 4$). The

Table 2 First eigenvalues: Legendre basis for $N = 1$. Note that $4.49_{04} = 4.49 \times 10^{-04}$

	$\xi = 2$		$\xi = 4$		$\xi = 6$		$\xi = 8$	
DoF	E	ΔE	E	ΔE	E	ΔE	E	ΔE
9	0.80898847		0.85120259		0.88298404		0.88823511	
17	0.80862555	4.49_{04}	0.80875522	4.24_{02}	0.80937787	9.09_{02}	0.82066506	6.76_{02}
25	0.80862554	8.19_{09}	0.80861663	1.39_{04}	0.80863141	9.23_{04}	0.80885868	1.18_{02}
33	0.80862554	6.97_{13}	0.80861652	1.17_{07}	0.80861657	1.84_{05}	0.80861710	2.42_{04}
41	0.80862554	2.06_{15}	0.80861652	3.57_{11}	0.80861652	6.61_{08}	0.80861652	5.82_{07}
49	0.80862554	2.75_{16}	0.80861652	3.44_{15}	0.80861652	7.90_{11}	0.80861652	2.48_{09}
57	0.80862554	1.37_{16}	0.80861652	5.55_{16}	0.80861652	4.39_{14}	0.80861652	1.69_{11}
65	0.80862554	1.51_{15}	0.80861652	1.11_{16}	0.80861652	5.49_{16}	0.80861652	3.42_{14}
73	0.80862554	1.51_{15}	0.80861652	4.44_{16}	0.80861652	6.86_{16}	0.80861652	3.33_{15}
81	0.80862554	8.24_{16}	0.80861652	5.55_{16}	0.80861652	1.78_{15}	0.80861652	1.11_{16}

Table 3 First eigenvalues: Laguerre basis for $N = 1$

	$\zeta = 0.05$		$\zeta = 0.1$		$\zeta = 0.5$		$\zeta = 1$	
DoF	E	ΔE	E	ΔE	E	ΔE	E	ΔE
9	1.53707272		0.83910765		0.81007589		0.85464218	
17	0.87424212	7.58_{01}	0.80862774	3.05_{02}	0.80862429	1.80_{03}	0.80943446	4.52_{02}
25	0.81259311	7.59_{02}	0.80861652	1.12_{05}	0.80861676	9.31_{06}	0.80898253	4.52_{04}
33	0.80873830	4.77_{03}	0.80861652	1.30_{11}	0.80861653	2.83_{07}	0.80862929	3.53_{04}
41	0.80861843	1.48_{04}	0.80861652	2.22_{18}	0.80861652	1.67_{08}	0.80861978	9.51_{06}
49	0.80861653	2.35_{06}	0.80861652	3.33_{16}	0.80861652	1.51_{10}	0.80861711	2.67_{06}
57	0.80861652	2.01_{08}	0.80861652	2.22_{16}	0.80861652	8.05_{12}	0.80861654	5.74_{07}
65	0.80861652	9.65_{11}	0.80861652	4.44_{16}	0.80861652	4.20_{13}	0.80861653	8.02_{09}
73	0.80861652	2.63_{13}	0.80861652	4.44_{16}	0.80861652	4.39_{15}	0.80861652	9.20_{09}
81	0.80861652	1.10_{15}	0.80861652	1.11_{16}	0.80861652	4.12_{16}	0.80861652	1.88_{09}

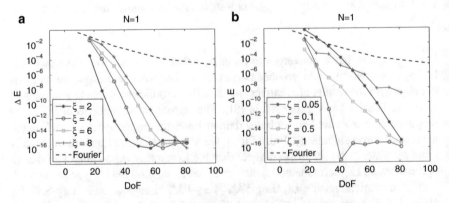

Fig. 6 Convergence rates for different scaling parameters ξ and ζ (one electron case). (**a**) Legendre basis. (**b**) Laguerre basis

Table 4 First eigenvalues: **'Leg-FA'** method for $N = 2, 4, 6, 8$

N	DoF	$\xi = 2$		$\xi = 4$		$\xi = 6$		$\xi = 8$	
		E	ΔE	E	ΔE	E	ΔE	E	ΔE
2	36	2.77064844		2.83934161		3.23853921		3.83477463	
2	136	2.75897331	4.23_{03}	2.75962610	7.97_{02}	2.78271494	1.64_{01}	2.79743867	1.04_{00}
2	300	2.75896782	1.99_{06}	2.75853265	1.09_{03}	2.75867580	8.71_{03}	2.76007806	3.74_{02}
2	528	2.75896776	2.02_{08}	2.75852536	7.29_{06}	2.75852924	5.31_{05}	2.75856947	1.51_{03}
2	820	2.75896776	1.98_{09}	2.75852515	2.03_{07}	2.75852540	1.39_{06}	2.75852700	4.25_{05}
2	1176	2.75896775	3.41_{10}	2.75852513	2.53_{08}	2.75852517	8.17_{08}	2.75852535	1.65_{06}
2	1596	2.75896775	8.62_{11}	2.75852512	5.58_{09}	2.75852513	1.34_{08}	2.75852518	1.75_{07}
2	2080	2.75896775	2.63_{11}	2.75852512	1.65_{09}	2.75852513	3.56_{09}	2.75852514	3.72_{08}
2	2628	2.75896775	3.59_{11}	2.75852512	8.05_{10}	2.75852513	4.76_{09}	2.75852513	1.77_{08}
2	3916	2.75896775	5.44_{12}	2.75852512	1.15_{10}	2.75852512	6.47_{10}	2.75852512	2.22_{09}
2	5460	2.75896775	1.09_{12}	2.75852512	2.47_{11}	2.75852512	1.32_{10}	2.75852512	4.35_{10}
2	7260	2.75896775	3.62_{13}	2.75852512	6.53_{12}	2.75852512	3.55_{11}	2.75852512	1.14_{10}
2	9316	2.75896775	3.62_{13}	2.75852512	2.17_{12}	2.75852512	1.16_{11}	2.75852512	3.66_{11}
4	35	11.3744684		14.6183222		19.5555592		25.2272328	
4	126	11.1529293	1.99_{02}	12.5401722	2.08_{00}	15.4303761	2.67_{01}	19.0645225	6.16_{00}
4	330	11.0906509	5.62_{03}	11.4897741	1.05_{00}	13.2777107	1.62_{01}	15.5602812	3.50_{00}
4	715	11.0387961	4.70_{03}	11.1928180	2.97_{01}	12.0181657	1.05_{01}	13.5778359	1.98_{00}
4	1365	11.0311564	6.93_{04}	11.1503562	4.25_{02}	11.3807587	5.60_{02}	12.3383459	1.24_{00}
4	2380	11.0309216	2.13_{05}	11.0667169	8.36_{02}	11.2139303	1.49_{02}	11.5916733	7.47_{01}
4	3876	11.0306460	2.50_{05}	11.0150199	5.17_{02}	11.1756559	3.42_{03}	11.2803427	3.11_{01}
4	5985	11.0305710	6.80_{06}	11.0034337	1.16_{02}	11.0953952	7.23_{03}	11.2180070	6.23_{02}

(continued)

Table 4 (continued)

N	DoF	$\xi = 2$		$\xi = 4$		$\xi = 6$		$\xi = 8$	
		E	ΔE	E	ΔE	E	ΔE	E	ΔE
4	8855	11.0305636	6.76_{07}	11.0025213	8.29_{05}	11.0312699	5.81_{03}	11.1718388	4.13_{03}
4	12,650	11.0305614	2.00_{07}	11.0020105	4.64_{05}	11.0076249	2.15_{03}	11.0932904	7.08_{03}
6	7	30.4774572		41.7709865		59.0079666		77.4736697	
6	84	27.7191179	9.95_{02}	34.4566265	2.12_{01}	45.3580818	3.01_{01}	58.2795553	3.29_{01}
6	462	27.3018885	1.53_{02}	30.7355475	1.21_{01}	37.404467	2.13_{01}	46.198215	2.62_{01}
6	1716	27.0408367	9.65_{03}	28.3071847	8.58_{02}	33.0501800	1.32_{01}	38.7185569	1.93_{01}
6	5005	26.8872665	5.71_{03}	27.3234899	3.60_{02}	30.2367754	9.30_{02}	34.2981982	1.29_{01}
6	12,376	26.8720859	5.65_{04}	27.1394812	6.78_{03}	28.3094643	6.81_{02}	31.4814019	8.95_{02}
8	9	58.97684862		75.48658360		104.9615300		137.2152993	
8	165	53.64284607	9.94_{02}	65.09745212	1.60_{01}	84.71660799	2.39_{01}	108.4869011	2.65_{01}
8	1287	52.91599468	1.37_{02}	59.33803295	9.71_{02}	72.01714565	1.76_{01}	89.02662541	2.19_{01}
8	6435	52.15277285	1.46_{02}	54.85127080	8.18_{02}	64.67355647	1.14_{01}	76.12430898	1.69_{01}
8	24,310	51.83345874	6.16_{03}	52.69663222	4.09_{02}	59.72645435	8.28_{02}	68.00062085	1.19_{01}

Table 5 First eigenvalues: '**Lag-FA**' method $N = 2, 4, 6, 8$

N	DoF	$\zeta = 0.05$		$\zeta = 0.1$		$\zeta = 0.5$		$\zeta = 1$	
		E	ΔE	E	ΔE	E	ΔE	E	ΔE
2	36	7.30379716		3.12261380		2.77777209		2.85513015	
2	136	3.34341896	1.18_{00}	2.75950375	3.63_{01}	2.75896094	6.82_{03}	2.78956672	6.56_{02}
2	300	2.82490601	1.84_{01}	2.75852513	9.79_{04}	2.75854050	1.52_{04}	2.75986845	2.97_{02}
2	528	2.76277165	2.25_{02}	2.75852512	6.59_{09}	2.75852626	5.16_{06}	2.75897551	8.93_{04}
2	820	2.75865751	1.49_{03}	2.75852512	8.16_{10}	2.75852547	2.85_{07}	2.75858518	3.90_{04}
2	1176	2.75852721	4.72_{05}	2.75852512	2.76_{10}	2.75852526	7.83_{08}	2.75853935	4.58_{05}
2	1596	2.75852514	7.51_{07}	2.75852512	1.15_{10}	2.75852519	2.53_{08}	2.75852947	9.88_{06}
2	2080	2.75852512	6.37_{09}	2.75852512	5.55_{11}	2.75852516	1.09_{08}	2.75852635	3.12_{06}
2	2628	2.75852512	6.41_{09}	2.75852512	3.12_{11}	2.75852514	1.64_{08}	2.75852581	1.33_{06}
2	3916	2.75852512	1.09_{12}	2.75852512	9.79_{12}	2.75852513	4.71_{09}	2.75852535	1.65_{07}
2	5460	2.75852512	3.63_{13}	2.75852512	3.99_{12}	2.75852512	1.79_{09}	2.75852522	4.68_{08}
2	7260	2.75852512	3.62_{13}	2.75852512	1.81_{12}	2.75852512	8.05_{10}	2.75852517	1.74_{08}
2	9316	2.75852512	3.64_{13}	2.75852512	1.09_{12}	2.75852512	4.09_{10}	2.75852515	8.05_{09}
4	35	84.5261141		24.0010489		11.4543515		14.4696514	
4	126	48.1712017	7.55_{01}	15.9511257	8.05_{00}	11.3868658	5.93_{03}	12.9426380	1.53_{00}
4	330	32.5685850	4.79_{01}	12.9989513	2.95_{00}	11.2011807	1.66_{02}	12.0343856	9.08_{01}
4	715	24.4059885	3.34_{01}	11.7815358	1.22_{00}	11.0620997	1.26_{02}	11.5559439	4.78_{01}
4	1365	19.6523245	2.42_{01}	11.2772003	5.04_{01}	11.0168962	4.10_{03}	11.3619089	1.94_{01}
4	2380	16.6988902	1.77_{01}	11.0837922	1.93_{01}	11.0106396	5.68_{04}	11.3091692	5.27_{02}
4	3876	14.7880278	1.29_{01}	11.0207238	6.31_{02}	11.0084099	2.03_{04}	11.2814318	2.77_{02}
4	5985	13.5204842	9.37_{02}	11.0046752	1.60_{02}	11.0054217	2.72_{04}	11.2304460	5.10_{02}

(continued)

Table 5 (continued)

N	DoF	$\zeta = 0.05$		$\zeta = 0.1$		$\zeta = 0.5$		$\zeta = 1$	
		E	ΔE	E	ΔE	E	ΔE	E	ΔE
4	8855	12.6679704	6.73_{02}	11.0017149	2.69_{04}	11.0033635	1.87_{04}	11.1647721	5.88_{03}
4	12,650	12.0916290	4.77_{02}	11.0013445	3.37_{05}	11.0023306	9.39_{05}	11.1040989	5.46_{03}
6	7	488.968428		126.923387		30.8058940		44.1789048	
6	84	194.6512414	1.51_{00}	55.6839778	1.28_{00}	28.3194836	8.78_{02}	37.5755909	1.76_{01}
6	462	114.5606263	6.99_{01}	38.0266025	4.64_{01}	27.8526463	1.68_{02}	34.1104499	1.02_{01}
6	1716	79.2855548	4.45_{01}	31.4021247	2.11_{01}	27.5394558	1.14_{02}	31.6957497	7.62_{02}
6	5005	60.3655526	3.13_{01}	28.5962373	9.81_{02}	27.0880214	1.67_{02}	29.9054490	5.99_{02}
6	12,376	49.0792283	2.30_{01}	27.3804649	4.44_{02}	26.7879171	1.12_{02}	28.7231166	4.12_{02}
8	9	862.2460823		225.0434282		61.4221824		89.76996256	
8	165	351.8825621	1.45_{00}	101.6315296	1.21_{00}	56.01728167	9.65_{02}	77.38512422	1.60_{01}
8	1287	211.4500817	6.64_{01}	70.80521708	4.35_{01}	54.10871261	3.53_{02}	70.39360988	9.93_{02}
8	6435	148.7485015	4.22_{01}	59.17412032	1.97_{01}	53.50632043	1.13_{02}	65.88471983	6.84_{02}
8	24,310	114.6497701	2.97_{01}	54.24858141	9.08_{02}	52.65395877	1.62_{02}	62.16347168	5.99_{02}

Table 6 First eigenvalues: 'Leg-SA' method $N = 2, 4, 6, 8$

N	DoF	$\xi = 2$		$\xi = 4$		$\xi = 6$		$\xi = 8$	
		E	ΔE	E	ΔE	E	ΔE	E	ΔE
2	30	2.77117360		2.85154776		3.29333192		4.00360881	1.02_{00}
2	76	2.75907121	4.39_{03}	2.77186894	7.97_{02}	2.84618789	1.57_{01}	2.98567403	1.94_{01}
2	188	2.75898056	3.29_{05}	2.75912734	1.27_{02}	2.77101057	2.71_{02}	2.79193494	3.12_{02}
2	440	2.75896878	4.27_{06}	2.75856637	5.61_{04}	2.75897392	4.36_{03}	2.76075915	2.02_{03}
2	1016	2.75896792	3.11_{07}	2.75852925	3.71_{05}	2.75856020	1.50_{04}	2.75873647	7.43_{05}
2	2288	2.75896776	5.80_{08}	2.75852531	1.43_{06}	2.75852667	1.22_{05}	2.75853155	
4	45	11.2855573		13.1270840		17.5972705		22.7581385	5.55_{00}
4	185	11.0456316	2.17_{02}	11.8762275	1.25_{00}	14.1293533	2.45_{01}	17.2075738	3.18_{00}
4	685	11.0315868	1.27_{03}	11.2689744	6.07_{01}	12.3384003	1.45_{01}	14.0281403	1.88_{00}
4	2166	11.0306207	8.76_{05}	11.0607111	2.08_{01}	11.3668441	8.55_{02}	12.1527206	7.38_{01}
4	6438	11.0305668	4.89_{06}	11.0050047	5.57_{02}	11.0629403	2.75_{02}	11.4151839	2.78_{02}
4	18,070	11.0305602	5.95_{07}	11.0018297	2.89_{04}	11.0111051	4.71_{03}	11.1064143	
6	4	28.8667472		40.2289337		57.5510278		74.9994059	2.33_{01}
6	42	27.5001095	4.97_{02}	34.8584080	1.54_{01}	47.0333024	2.24_{01}	60.8190456	2.12_{01}
6	258	26.9802045	1.93_{02}	31.3576095	1.12_{01}	39.9539319	1.77_{01}	50.1906803	2.65_{01}
6	1240	26.8819947	3.65_{03}	28.4995437	1.00_{01}	32.8373715	2.17_{01}	39.6716608	1.37_{01}
6	4984	26.8682208	5.13_{04}	27.0984039	5.17_{02}	30.0298964	9.35_{02}	34.9065789	9.76_{02}
6	18,232	26.8675138	2.63_{05}	26.7326315	1.37_{02}	28.4165043	5.68_{02}	31.8035686	
8	4	59.40209940		76.68172385		107.3043916		140.4792744	3.54_{01}
8	57	54.08534117	9.83_{02}	62.51098600	2.27_{01}	81.15357811	3.22_{01}	103.7463664	1.34_{01}
8	425	52.17333953	3.66_{02}	57.86688923	8.03_{02}	73.18590963	1.09_{01}	91.50522209	1.15_{01}
8	2425	51.81422944	6.93_{03}	55.23428566	4.77_{02}	67.37989091	8.62_{02}	82.06933495	1.04_{01}
8	11,641	51.78818166	5.03_{04}	53.37488078	3.48_{02}	62.67868975	7.50_{02}	74.35006938	

Table 7 First eigenvalues: **'Lag-SA'** method $N = 2, 4, 6, 8$

N	DoF	$\zeta = 0.05$		$\zeta = 0.1$		$\zeta = 0.5$		$\zeta = 1$	
		E	ΔE	E	ΔE	E	ΔE	E	ΔE
2	30	8.67617797		3.34484794		2.77827775		2.86808325	
2	76	4.95071500	7.53_{01}	2.85047130	4.94_{01}	2.75910728	6.95_{03}	2.79510521	7.30_{02}
2	188	3.56995225	3.87_{01}	2.76696643	8.35_{02}	2.75857016	1.95_{04}	2.76032963	3.48_{02}
2	440	2.97004425	2.02_{01}	2.75854879	8.42_{03}	2.75853604	1.24_{05}	2.75890350	1.43_{03}
2	1016	2.78412765	6.68_{02}	2.75852514	2.36_{05}	2.75852721	3.20_{06}	2.75857819	3.25_{04}
2	2288	2.75876032	9.20_{03}	2.75852512	7.25_{09}	2.75852561	5.80_{07}	2.75853878	1.43_{05}
4	45	86.6917450		24.7000085		11.4797932		13.0563335	
4	185	47.4767386	8.26_{01}	15.7385648	8.96_{00}	11.0834163	3.58_{02}	11.9600581	1.10_{00}
4	685	32.5716096	4.58_{01}	12.9635893	2.77_{00}	11.0087891	6.78_{03}	11.3378162	6.22_{01}
4	2166	22.7567084	4.31_{01}	11.5596749	1.40_{00}	11.0027861	5.46_{04}	11.0910824	2.47_{01}
4	6438	16.9797442	3.40_{01}	11.0511325	5.09_{01}	11.0015878	1.09_{04}	11.0167019	7.44_{02}
4	18,070	14.1738357	1.98_{01}	11.0034033	4.34_{03}	11.0013648	2.03_{05}	11.0045689	1.10_{03}
6	4	952.185785		242.664898		31.7056642		40.9407131	
6	42	278.0484912	2.42_{00}	75.6539917	2.21_{00}	28.0048853	1.32_{01}	35.2221880	1.62_{01}
6	258	183.0644900	5.19_{01}	52.9850239	4.28_{01}	27.0791781	3.42_{02}	31.7607206	1.09_{01}
6	1240	118.8419089	5.40_{01}	39.4694998	3.42_{01}	26.7470351	1.24_{02}	29.4671183	7.78_{02}
6	4984	88.6458674	3.41_{01}	33.3728721	1.83_{01}	26.6316558	4.33_{03}	28.1039769	4.85_{02}
6	18,232	62.0557435	4.28_{01}	28.4594806	1.73_{01}	26.6028587	1.08_{03}	27.2593918	3.10_{02}
8	4	1397.855236		358.1470801		63.17782097		90.38022455	
8	57	739.6581056	8.90_{01}	194.9633013	8.37_{01}	57.62889533	9.63_{02}	69.29072364	3.04_{01}
8	425	439.0533644	6.85_{01}	123.6510866	5.77_{01}	53.87219761	6.97_{02}	64.08200887	8.13_{02}
8	2425	291.0249117	5.09_{01}	89.23827767	3.86_{01}	51.68828658	4.23_{02}	59.76006788	7.23_{02}
8	11,641	202.0154616	4.41_{01}	69.31440411	2.87_{01}	50.95865047	1.43_{02}	57.05869814	4.73_{02}

Table 8 Best results of first eigenvalues achieved by various methods for $N = 2, 4, 6, 8$

Methods	$N = 2$		$N = 4$		$N = 6$		$N = 8$	
	DoF	E	DoF	E	DoF	E	DoF	E
'Leg-FA'	9316	2.75852512	12,650	11.0020105	12,376	26.872086	24,310	51.83345874
'Leg-SA'	2288	2.75852531	18,070	11.0018297	18,232	26.867514	11,641	51.78818166
'Lag-FA'	9316	2.75852512	12,650	11.0013445	12,376	26.787917	24,310	52.65395877
'Lag-SA'	2288	2.75852512	18,070	11.0013648	18,232	26.602859	11,641	50.95865047
'Fourier'	3409	2.758536	79,498	11.011562	297,605	27.571226	215,864	60.838970

reason might be that the sparse grids based on hyperbolic cross allow to treat the nucleus-electron cusps properly which are aligned to the particle coordinate axes of the system while does not fit well to the "diagonal" directions of the electron-electron cusps. However the sparse grids based on hyperbolic cross for six and eight dimensional case do give better results than the full grid cases. We observe also that carefully choice of parameters ξ and ζ is needed to obtain a decent accuracy for E. As showed in the one-dimensional case, the Legendre method is not very sensitive to the scaling parameter comparing to the Laguerre method. The results show that all our numerical methods have monotonic convergence property, which might be used to determine optimal scaling in practice through multiple runs.

As a comparison with published results using a Fourier method in [8], we list in Table 8 our results with the "best" parameters and the corresponding results in [8]. We observe that our method gives much better results with significant less number of unknowns.

5 Concluding Remarks

We developed in this paper efficient spectral-element methods with Legendre and Laguerre basis sets for ESE in one spatial dimension. To achieve high-order approximation to the nucleus-electron cusps, we construct the basis sets in spectral-element type. For the system with N electrons, we proposed to use sparse grids of hyperbolic cross type to deal with high dimensionality.

We also presented efficient procedure to enforce the antisymmetry using Slater determinants which reflect the Pauli principle, and lead to antisymmetric basis sets for full/sparse grid spaces with a substantially reduced amount of degree of freedoms. We performed numerical experiments which showed that our methods enjoy exponential convergence rate for the one electron case, and for multi-electron cases, can lead to a target accuracy with significantly fewer number of unknowns than other approaches.

We only presented some preliminary numerical results with one-dimensional particles here. We believe that these preliminary results are very encouraging, and many techniques developed in this paper can be extended to solving ESE in two and three spatial dimensions. For example, we can construct special basis functions of spectral-element type that take care of the nuclei-electron singularities like $1/|x|$ in $\mathbb{R}^3$ and $\log(|x|)$ in $\mathbb{R}^2$. Such consideration and other issues are currently under investigation.

Acknowledgements The work of J. Shen is partially supported by an NSF grant DMS-1419053. Y. Wang gratefully acknowledge the support of ICERM during the preparation of this manuscript. The work of H. Yu is partially supported by the National Science Foundation of China (NSFC 11101413 and NSFC 11371358). The authors also thank the anonymous referees for valuable comments and suggestions which led to a significant improvement of the presentation.

References

1. S. Agmon, *Lectures on Exponential Decay of Solutions of Second-Order Elliptic Equations: Bounds on Eigenfunctions of N-Body Schrödinger Operations*, vol. 29 (Princeton University Press, Princeton, 1982)
2. G. Beylkin, M.J. Mohlenkamp, F. Pérez, Approximating a wavefunction as an unconstrained sum of slater determinants. J. Math. Phys. **49**(3), 032107 (2008)
3. H. Bungartz, M. Griebel, Sparse grids. Acta Numer. **13**, 147–269 (2004)
4. C. Canuto, M.Y. Hussaini, A. Quarteroni, T.A. Zang, *Spectral Methods: Fundamentals in Single Domains* (Springer, Berlin/New York, 2006)
5. D.B. Cook, *Handbook of Computational Quantum Chemistry* (Courier Dover, 2012)
6. A. Gordon, C. Jirauschek, F.X. Kärtner, Numerical solver of the time-dependent Schrödinger equation with Coulomb singularities. Phys. Rev. A **73**, 042505 (2006)
7. M. Griebel, J. Hamaekers, A wavelet based sparse grid method for the electronic Schrödinger equation, in *Proceedings of the International Congress of Mathematicians*, vol. 3, Madrid (2006), pp. 1473–1506
8. M. Griebel, J. Hamaekers, Sparse grids for the Schrödinger equation. ESAIM: Math. Model. Numer. Anal. **41**(2), 215–247 (2007)
9. M. Griebel, J. Hamaekers, Tensor product multiscale many-particle spaces with finite-order weights for the electronic Schrödinger equation. Int. J. Res. Phys. Chem. Chem. Phys. **224**(4), 291–708 (2010)
10. B.-y. Guo, J. Shen, C.-l. Xu, Spectral and pseudospectral approximations using Hermite functions: application to the Dirac equation. Adv. Comput. Math. **19**(1–3), 35–55 (2003)
11. R. Pauncz, *The Symmetric Group in Quantum Chemistry* (CRC, Boca Raton, 1995)
12. M. Rizea, V. Ledoux, M. Van Daele, G.V. Berghe, N. Carjan, Finite difference approach for the two-dimensional schrödinger equation with application to scission-neutron emission. Comput. Phys. Commun. **179**(7), 466–478 (2008)
13. Y. Saad, J.R. Chelikowsky, S.M. Shontz, Numerical methods for electronic structure calculations of materials. SIAM Rev. **52**(1), 3–54 (2010)
14. J. Shen, Efficient spectral-Galerkin method I. Direct solvers for second- and fourth-order equations using Legendre polynomials. SIAM J. Sci. Comput. **15**, 1489–1505 (1994)
15. J. Shen, Efficient spectral-Galerkin method II: direct solvers for second- and fourth-order equations using Chebyshev polynomials. SIAM J. Sci. Comput. **16**, 74–87 (1995)
16. J. Shen, A new fast Chebyshev-Fourier algorithm for the Poisson-type equations in polar geometries. Appl. Numer. Math. **33**, 183–190 (2000)
17. J. Shen, T. Tang, L.-L. Wang, *Spectral Methods: Algorithms, Analysis and Applications*. Springer Series in Computational Mathematics, vol. 41 (Springer, Berlin/New York, 2011)
18. J. Shen, L.-L. Wang, Some recent advances on spectral methods for unbounded domains. Commun. Comput. Phys. **5**(2–4), 195–241 (2009)

19. J. Shen, L.-L. Wang, Sparse spectral approximations of high-dimensional problems based on hyperbolic cross. SIAM J. Numer. Anal. **48**, 1087–1109 (2010)
20. J. Shen, L.-L. Wang, H. Yu, Approximations by orthonormal mapped Chebyshev functions for higher-dimensional problems in unbounded domains. J. Comput. Appl. Math. **265**, 264–275 (2014)
21. J. Shen, H. Yu, Efficient spectral sparse grid methods and applications to high dimensional elliptic problems. SIAM J. Sci. Comput. **32**, 3228–3250 (2010)
22. J. Shen, H. Yu, Efficient spectral sparse grid methods and applications to high dimensional elliptic problems II: unbounded domains. SIAM J. Sci. Comput. **34**, A1141–A1164 (2012)
23. A. Szabo, N.S. Ostlund, *Modern Quantum Chemistry: Introduction to Advanced Electronic Structure Theory* (Dover, Mineola/New York, 1996)
24. T. Tang, The Hermite spectral method for Gaussian-type functions. SIAM J. Sci. Comput. **14**, 594–606 (1993)
25. H. Yserentant, Sparse grid spaces for the numerical solution of the electronic Schrödinger equation. Numer. Math. **101**(2), 381–389 (2005)
26. H. Yserentant, The hyperbolic cross space approximation of electronic wavefunctions. Numer. Math. **105**(4), 659–690 (2007)

A Sparse Grid Method for Bayesian Uncertainty Quantification with Application to Large Eddy Simulation Turbulence Models

Hoang Tran, Clayton G. Webster, and Guannan Zhang

Abstract There is wide agreement that the accuracy of turbulence models suffer from their sensitivity with respect to physical input data, the uncertainties of user-elected parameters, as well as the model inadequacy. However, the application of Bayesian inference to systematically quantify the uncertainties in parameters, by means of exploring posterior probability density functions (PPDFs), has been hindered by the prohibitively daunting computational cost associated with the large number of model executions, in addition to daunting computation time per one turbulence simulation. In this effort, we perform in this paper an *adaptive hierarchical sparse grid* surrogate modeling approach to Bayesian inference of large eddy simulation (LES). First, an adaptive hierarchical sparse grid surrogate for the output of forward models is constructed using a relatively small number of model executions. Using such surrogate, the likelihood function can be rapidly evaluated at any point in the parameter space without simulating the computationally expensive LES model. This method is essentially similar to those developed in Zhang et al. (Water Resour Res 49:6871–6892, 2013) for geophysical and groundwater models, but is adjusted and applied here for a much more challenging problem of uncertainty quantification of turbulence models. Through a numerical demonstration of the Smagorinsky model of two-dimensional flow around a cylinder at sub-critical Reynolds number, our approach is proven to significantly reduce the number of costly LES executions without losing much accuracy in the posterior probability estimation. Here, the model parameters are calibrated against synthetic data related

This material is based upon work supported in part by the U.S. Air Force of Scientific Research under grant numbers 1854-V521-12; by the U.S. Department of Energy, Office of Science, Office of Advanced Scientific Computing Research, Applied Mathematics program under contract numbers ERKJ259, ERKJE45; and by the Laboratory Directed Research and Development program at the Oak Ridge National Laboratory, which is operated by UT-Battelle, LLC., for the U.S. Department of Energy under Contract DE-AC05-00OR22725.

H. Tran • C.G. Webster (✉) • G. Zhang
Computer Science and Mathematics Division, Oak Ridge National Laboratory, Oak Ridge, TN 37831-6164, USA
e-mail: tranha@ornl.gov; webstercg@ornl.gov; zhangg@ornl.gov

© Springer International Publishing Switzerland 2016
J. Garcke, D. Pflüger (eds.), *Sparse Grids and Applications – Stuttgart 2014*,
Lecture Notes in Computational Science and Engineering 109,
DOI 10.1007/978-3-319-28262-6_12

291

to the mean flow velocity and Reynolds stresses at different locations in the flow wake. The influence of the user-elected LES parameters on the quality of output data will be discussed.

1 Introduction

For most turbulent flows encountered in industrial applications, the cost of direct numerical simulation (DNS) would exceed the capacity of current computational resource (and possibly continue to do so for the foreseeable future). As a result, many important decisions affecting our daily lives (such as climate policy, biomedical device design, pollution dispersal and energy efficiency improvement) are informed from simulations of turbulent flows by various models of turbulence. The accuracy of estimated quantities of interest (QoIs) by such models, however, frequently suffers from the uncertainties on the physical input data, user-chosen model parameters and the subgrid model. It is ideal to be able to incorporate these uncertainties in the predictions of QoIs.

The basic approach used for approximating turbulent flows has been to compute the time- and space-filtered velocity and pressure, which are less computationally demanding and of main technical interest, instead of solving for the pointwise velocity and pressure prescribed by the standard Navier-Stokes equations. The use of turbulence models leads to a level of uncertainty in the performance and inaccuracy in the simulation results, due to user-chosen model parameters whose true or optimal values are not well-known *(parametric uncertainty)*, or the inherent inability of the model to reproduce reality *(structural uncertainty)*. With the fast growth in available computational power, the literature on uncertainty quantification for fluid mechanics modeling has grown extensively recently. Many stochastic numerical methods have been developed, analyzed and tested for simulations of fluid flows with uncertain physical and model parameters, see, e.g., [18, 20, 53, 57, 61, 62]. Sensitivity analysis of LES to parametric uncertainty was conducted in [35]. Statistical methods to capture structural uncertainties in turbulence models were presented in [17, 23, 24]. For inverse uncertainty quantification, we refer to [13, 45] (Bayesian inference for Reynolds-averaged Navier Stokes (RANS) models) and [16] (adjoint based inverse modeling).

Bayesian inference has become a valuable tool for estimation of parametric and structural uncertainties of physical systems constrained by differential equations. Sampling techniques, such as Markov chain Monte Carlo (MCMC), have frequently been employed in Bayesian inference [19, 33, 50]. However, MCMC methods [28, 59, 60] are, in general, computationally expensive, because a large number of forward model simulations is needed to estimate the PPDF and sample from it. Given the fact that one solution of turbulence models easily takes thousands of computing hours, MCMC simulations in many CFD applications would require prohibitively large computational budgets. Perhaps due to this demand, efforts on model calibration up until now have been limited on the least expensive turbulence model − RANS equations [13, 45]. To make Bayesian inference tractable for other types of closure models, including LES, it is essential to perform the MCMC sampling in a time and cost effective manner.

A strategy to improve the efficiency of MCMC simulations is *surrogate modeling*, which has been developed in a wide variety of contexts and disciplines, see [49] and the reference therein. Surrogate modeling practice seeks to approximate the response of an original function (model outputs or the PPDF in this work), which is typically computationally expensive, by a cheaper-to-run surrogate. The PPDF can then be evaluated by sampling the surrogate directly without forward model executions. Compared to conventional MCMC algorithms, this approach is advantageous that it significantly reduces the number of forward model executions at a desired accuracy and allows sampling the PPDF in parallel. Several methods can be employed to construct the surrogate systems, including polynomial chaos expansion [22], stochastic Galerkin [3], stochastic collocation [2], and polynomial dimensional decomposition [48], to list a few. For problems where the quantities of interest have irregular dependence with respect to the random parameters, such as those studied herein, it should be noted that approximation approaches that use global polynomials are generally less effective than those allowing for multi-level, multi-scale decomposition. In this direction, one can develop multi-level hierarchical subspaces and employ adaptive grid refinement to concentrate grid points on the subdomains with a locally high variation of solutions, resulting in a significant reduction in the number of grid points.

In this paper, we present an adaptive hierarchical sparse grid (AHSG) surrogate modeling approach to Bayesian inference of turbulence models, in particularly LES. The key idea is to place a grid in the parameter space with sparse parameter samples, and the forward model is solved only for these samples. Compared to the regular full grid approach, sparse grid preserves the high level of accuracy with less computational work, see [4, 20, 26, 27, 42, 43]. As sparse grid methods require the bounded mixed derivative property, which is open for the solutions of Navier-Stokes equations and turbulence models in general, a locally adaptive refinement method, guided by hierarchical surpluses, is employed to extend sparse grid approach to possible non-smooth solutions. This refinement strategy is different from dimension-adaptive refinement [21], which puts more points in dimensions of higher relevance and more in line with those in [25, 34, 46]. Although similar surrogate methods has been studied in [36, 63] for geophysical and groundwater models, we tackle here a more challenging problem of uncertainty quantification of turbulence models. Indeed, turbulent flows are notorious for their extremely complex nature and the non-smoothness of the surface of LES output data may weaken the accuracy of the surrogate. The applicability of surrogate modeling techniques to LES therefore needs thorough investigation. In this work, we will demonstrate the accuracy and efficiency of the surrogate model through a numerical example of the classical Smagorinsky closure model of turbulent flow around a circular cylinder at a sub-critical Reynolds number ($Re = 500$), which is a benchmark test case for LES. The computation will be conducted for the two-dimensional flow, whose outputs have similar patterns as three-dimensional simulation, but which is significantly less demanding in computing budget. The synthetic data of velocity and Reynolds stresses at different locations in the flow wake are utilized for the calibration.

This work is only one piece in the complete process of calibration and validation of LES models to issue predictions of QoIs with quantified uncertainties, and many open questions remain. We do not attempt to fit the numerical solutions with physical data herein, as the two-dimensional model has been known to show remarkable discrepancy with the experiment results. Applying our framework to the three-dimensional simulation for parameter calibration against real-world data would be the next logical step. Another important problem is to evaluate and compare the performance of our AHSG with other surrogate methods (including some listed above) in this process. This would be conducted in future research. Also, characterization and quantification of the structural inadequacy and comparison of different competing LES models are beyond the scope of this study.

The rest of the paper is organized as follows. The Bayesian framework and the adaptive hierarchical sparse grid method of constructing the surrogate system are described in Sect. 2. In Sect. 3, we give a detailed description of the Smagorinsky model of sub-critical flow around a cylinder. The performance of surrogate modeling approach and results of the Bayesian analysis are presented in Sect. 3.4. Finally, discussions and conclusions appear in Sect. 4.

2 Adaptive Hierarchical Sparse Grid Methods for Surrogate Modeling in Bayesian Inference

2.1 Bayesian Inference

Consider the Bayesian inference problem for a turbulence model

$$d = f(\theta) + \varepsilon, \tag{1}$$

where $d = (d_1, \ldots, d_{N_d})$ is a vector of N_d reference data, $\theta = (\theta_1, \ldots, \theta_{N_\theta})$ is a vector of N_θ model parameters, $f(\theta)$ is the forward model, e.g., Smagorinsky model (see Sect. 3), with N_θ inputs and N_d outputs, and ε is a vector of residuals, including measurement, model parametric and structural errors. (Nonlinear model $d = \Xi(f, \theta, \varepsilon)$ can be considered as well, but leads to more complicated likelihood functions, as $\varepsilon = \Xi^{-1}(f, \theta)(d)$).

The posterior distribution $P(\theta|d)$ of the model parameters θ, given the data d, can be estimated using the Bayes' theorem [10] via

$$P(\theta|d) = \frac{L(\theta|d)P(\theta)}{\int L(\theta|d)P(\theta)d\theta}, \tag{2}$$

where $P(\boldsymbol{\theta})$ is the prior distribution and $L(\boldsymbol{\theta}|\boldsymbol{d})$ is the likelihood function that measure "goodness-of-fit" between model simulations and observations. In parametric uncertainty quantification, the denominator of the Bayes' formula in Eq. (2) is a normalization constant that does not affect the shape of the PPDF. As such, in the hereafter discussion concerning building surrogate systems, the notation $P(\boldsymbol{\theta}|\boldsymbol{d})$ or the terminology PPDF will only refer to the product $L(\boldsymbol{\theta}|\boldsymbol{d})P(\boldsymbol{\theta})$. The prior distribution represents knowledge of the parameter values before the data $\boldsymbol{d}$ is available. When prior information is lacking, a common practice is to assume uniform distributions with parameter ranges large enough to contain all plausible values of parameters.

Selection of appropriate likelihood functions for a specific turbulence simulation is an open question. A commonly used *formal* likelihood function is based on the simplistic assumption that the residual term $\boldsymbol{\varepsilon}$ in (1) follows a multivariate Gaussian distribution with mean zero and prescribed standard deviations, which leads to the Gaussian likelihood function:

$$L(\boldsymbol{\theta}|\boldsymbol{d}) = \exp\left[-\frac{1}{2}(\boldsymbol{d}-\boldsymbol{f}(\boldsymbol{\theta}))^{\top}\Sigma^{-1}(\boldsymbol{d}-\boldsymbol{f}(\boldsymbol{\theta}))\right]. \tag{MVN}$$

In this paper, we assume that the residual errors are independent, i.e., the covariance matrix Σ is diagonal. To describe the correlation of the errors or the inadequacy of turbulence models, other covariance matrices can also be used (and lead to inconsistent results) [13, 45]. In general, the formal approach has been criticized for relying heavily on residual error assumptions that do not hold. Alternatively, *informal* likelihood functions are proposed as a pragmatic approach to implicitly account for errors in measurements, model inputs and model structure and to avoid over-fitting to reference data [8]. Definition of informal likelihood functions is problem specific in nature, and there has been no consensus on which informal likelihood functions outperforms others. For the sake of illustration, in Sect. 3.4, the exponential informal likelihood function is used for the numerical example (together with (MVN)). It reads:

$$L(\boldsymbol{\theta}|\boldsymbol{d}) = \exp\left(-\zeta \cdot \frac{\sum_{i=1}^{N_d}\left((d_i - f_i) - (\bar{d} - \bar{f})\right)^2}{\sum_{i=1}^{N_d}\left(d_i - \bar{d}\right)^2}\right), \tag{EXP}$$

where $\bar{d}$ is the mean of observations, $\bar{f}$ is the mean of the outputs of forward model, and ζ is a scaling constant. For some other widely used informal likelihood functions, see [55].

2.2 Adaptive Hierarchical Sparse Grid Methods for Construction of the Surrogate PPDF

The central task of Bayesian inference is to estimate the posterior distribution $P(\theta|d)$. It is often difficult to draw samples from the PPDF directly, so the MCMC methods, such as the Metropolis-Hastings (M-H) algorithm [19] and its variants, are normally used for the sampling process. In practice, the convergence of MCMC methods is often slow, leading to a large number of model simulations. To tackle this challenge, surrogate modeling approaches seek to build an approximation (called the surrogate system) for $P(\theta|d)$, then the MCMC algorithm draws samples from it directly without executing the forward model. With this approach, the main computational cost for evaluating the PPDF is now transferred to the surrogate construction step. Naturally, an approximation method which requires minimal number of grid points in the parameter space, while not surrendering much accuracy is desired. The methodology we utilize to construct the surrogate system, presented in this subsection, is similar to the method introduced in [63]. Since the method can be applied to functions governed by partial differential equations, not limited to $P(\theta|d)$ or $f(\theta)$, a generic notation $\eta(\theta) : \Omega \to \mathbb{R}$ is used for the description. Recall the following assumptions are generally needed for sparse grid methods:

(a) The domain Ω is a rectangle, i.e., $\Omega = \Omega_1 \times \ldots \times \Omega_{N_\theta}$, where $\Omega_n \subset \mathbb{R}$, $n = 1, \ldots, N_\theta$.

(b) The joint probability density function $\rho(\theta)$ is of product-type:

$$\rho(\theta) = \prod_{n=1}^{N_\theta} \rho_n(\theta_n),$$

where $\rho_n : \Omega_n \to \mathbb{R}$ are univariate density functions.

(c) The univariate domains and density functions are identical:

$$\Omega_1 = \ldots = \Omega_{N_\theta}; \; \rho_1 = \ldots = \rho_{N_\theta},$$

yielding the same i-level univariate quadrature rules

$$\mathcal{Q}_i^{(1)}[\cdot] = \ldots = \mathcal{Q}_i^{(N_\theta)}[\cdot] =: \mathcal{Q}_i[\cdot].$$

(d) The univariate quadrature rules are nested.

In this setting, we can treat θ as a parametric variable and the probability density function $\rho(\theta)$, consequently, as uniform. Assumptions (c) and (d) will be imposed via our construction of quadrature points. It is possible that the plausible domain for θ (corresponding to positive $P(\theta|d)$) is far from rectangle. In these cases, the domain can be enclosed in a rectangle and as we shall see, the adaptive procedure will generate the grid points only on the plausible regions, with the exception of the starting level. Isoprobabilistic transformations to map the function into a unit cube, such as Rosenblatt transformation [52], can also be considered.

2.2.1 Adaptive Sparse Grid Interpolation

The basis of constructing the sparse grid approximation in the multi-dimensional setting is the one-dimensional (1-D) hierarchical interpolation. Consider a function $\eta(\theta) : [0, 1] \rightarrow \mathbb{R}$. The 1-D hierarchical Lagrange interpolation formula is defined by

$$\mathcal{U}_K[\eta](\theta) := \sum_{i=0}^{K} \Delta\mathcal{U}_i[\eta](\theta), \tag{3}$$

where K is the resolution level, and the incremental interpolation operator $\Delta\mathcal{U}_i[\eta]$ is given as

$$\Delta\mathcal{U}_i[\eta](\theta) := \sum_{j=1}^{m_i} c_{i,j}\phi_{i,j}(\theta), \quad i = 0, \ldots, K. \tag{4}$$

For $j = 1, \ldots, m_i$, $\phi_j^i(\theta)$ and $c_{i,j}$ in (4) are the piecewise hierarchical basis functions [12, 63] and the interpolation coefficients for $\Delta\mathcal{U}_i[\eta]$, respectively. For $i = 0, \ldots, K$, the integer m_i in (4) is the number of interpolation points involved in $\Delta\mathcal{U}_i[\eta]$, which is defined by

$$m_0 = 1, \quad m_1 = 2, \quad \text{and} \quad m_i = 2^{i-1} \text{ for } i \geq 2.$$

A uniform grid, denoted by $\Delta\mathcal{X}_i = \{\theta_{i,j}\}_{j=1}^{m_i}$, can be utilized for the incremental interpolant $\Delta\mathcal{U}_i[\eta]$. The abscissas of $\Delta\mathcal{X}_i$ are defined by

$$\theta_{0,1} = 0.5, \quad \theta_{1,1} = 0, \quad \theta_{1,2} = 1, \quad \text{and} \quad \theta_{i,j} = \frac{2j-1}{\sum_{k=0}^{i} m_k - 1} \text{ for } j = 1 \ldots, m_i, \ i \geq 2.$$

Then, the hierarchical grid for $\mathcal{U}_K[\eta](\theta)$ is defined by $\mathcal{X}_K = \cup_{i=0}^{K}\Delta\mathcal{X}_i$.

Based on the one-dimensional hierarchical interpolation, we can construct an approximation for a multivariate function $\eta(\boldsymbol{\theta}) : [0, 1]^{N_\theta} \rightarrow \mathbb{R}$, where $\boldsymbol{\theta} = (\theta_1, \ldots, \theta_{N_\theta})$, by hierarchical interpolation formula as

$$\mathcal{I}_K[\eta](\boldsymbol{\theta}) := \sum_{|\mathbf{i}| \leq K} \Delta_{\mathbf{i}}[\eta](\boldsymbol{\theta}) \tag{5}$$

and the multi-dimensional incremental interpolation operator $\Delta_{\mathbf{i}}[\eta]$ is defined by

$$\Delta_{\mathbf{i}}[\eta](\boldsymbol{\theta}) := \Delta\mathcal{U}_{i_1} \otimes \cdots \otimes \Delta\mathcal{U}_{i_{N_\theta}}[\eta](\boldsymbol{\theta}) = \sum_{\mathbf{j}\in B_{\mathbf{i}}} c_{\mathbf{i},\mathbf{j}} \boldsymbol{\phi}_{\mathbf{i},\mathbf{j}}(\boldsymbol{\theta}),$$

where $\mathbf{i} := (i_1, \ldots, i_{N_\theta})$ is a multi-index indicating the resolution level of $\Delta_{\mathbf{i}}[\eta]$, $|\mathbf{i}| = i_1 + \cdots + i_{N_\theta}$, $\boldsymbol{\phi}_{\mathbf{i},\mathbf{j}}(\boldsymbol{\theta}) := \prod_{n=1}^{N_\theta} \phi_{i_n,j_n}(\theta_n)$, and the multi-index set $B_{\mathbf{i}}$ is defined by $B_{\mathbf{i}} = \{\mathbf{j} \in \mathbb{N}^{N_\theta} \mid j_n = 1, \ldots, m_{i_n}, n = 1, \ldots, N_\theta\}$. As such, the grids for $\Delta_{\mathbf{i}}[\eta]$ and $\mathcal{I}_K[\eta]$ are defined by $\Delta\mathcal{H}_{\mathbf{i}} := \Delta\mathcal{X}_{i_1} \times \cdots \times \Delta\mathcal{X}_{i_{N_\theta}}$ and $\mathcal{H}_K := \cup_{|\mathbf{i}|\leq K} \Delta\mathcal{H}_{\mathbf{i}}$.

In this paper, we employ the piecewise linear hierarchical basis [12, 63] and the surplus $c_{\mathbf{i},\mathbf{j}}$ can be explicitly computed as

$$c_{0,1} = \Delta_0[\eta](\boldsymbol{\theta}_{0,1}) = \mathcal{I}_0[\eta](\boldsymbol{\theta}_{0,1}) = \eta(\boldsymbol{\theta}_{0,1}),$$

$$c_{\mathbf{i},\mathbf{j}} = \Delta_{\mathbf{i}}[\eta](\boldsymbol{\theta}_{\mathbf{i},\mathbf{j}}) = \eta(\boldsymbol{\theta}_{\mathbf{i},\mathbf{j}}) - \mathcal{I}_{K-1}[\eta](\boldsymbol{\theta}_{\mathbf{i},\mathbf{j}}) \quad \text{for } |\mathbf{i}| = K > 0,$$

as the supports of basis functions are mutually disjoint on each subspace. As discussed in [12], when the function $\eta(\boldsymbol{\theta})$ is smooth with respect to $\boldsymbol{\theta}$, the magnitude of the surplus $c_{\mathbf{i},\mathbf{j}}$ will approach to zero as the resolution level K increases. Therefore, the surplus can be used as an error indicator for the interpolant $\mathcal{I}_K[\eta]$ in order to detect the smoothness of the target function and guide the sparse grid refinement. In particular, each point $\boldsymbol{\theta}_{\mathbf{i},\mathbf{j}}$ of the isotropic level-K sparse grid $\mathcal{H}_K$ is assigned two children in each n-th direction, represented by

$$C_1^n(\boldsymbol{\theta}_{\mathbf{i},\mathbf{j}}) = \left(\theta_{i_1,j_1}, \ldots, \theta_{i_{n-1},j_{n-1}}, \theta_{i_n+1,2j_n-1}, \theta_{i_{n+1},j_{n+1}}, \ldots, \theta_{i_{N_\theta},j_{N_\theta}} \right),$$

$$C_2^n(\boldsymbol{\theta}_{\mathbf{i},\mathbf{j}}) = \left(\theta_{i_1,j_1}, \ldots, \theta_{i_{n-1},j_{n-1}}, \theta_{i_n+1,2j_n}, \theta_{i_{n+1},j_{n+1}}, \ldots, \theta_{i_{N_\theta},j_{N_\theta}} \right), \tag{6}$$

for $n = 1, \ldots, N_\theta$. Note that the children of each sparse grid point on level $|\mathbf{i}|$ belong to the sparse grid point set of level $|\mathbf{i}| + 1$. The basic idea of adaptivity is as follows: for each point whose magnitude of the surplus is larger than the prescribed error tolerance, we refine the grid by adding its children on the next level. More rigorously, for an error tolerance α, the adaptive sparse grid interpolant is defined on each successive interpolation level as

$$\mathcal{I}_{K,\alpha}[\eta](\boldsymbol{\theta}) := \sum_{|\mathbf{i}|\leq K} \sum_{\mathbf{j}\in B_{\mathbf{i}}^\alpha} c_{\mathbf{i},\mathbf{j}} \boldsymbol{\phi}_{\mathbf{i},\mathbf{j}}(\boldsymbol{\theta}), \tag{7}$$

where the multi-index set $B_{\mathbf{i}}^\alpha$ is defined by modifying the multi-index set $B_{\mathbf{i}}$, i.e., $B_{\mathbf{i}}^\alpha = \{\mathbf{j} \in B_{\mathbf{i}} \mid |c_{\mathbf{j}}^{\mathbf{i}}| > \alpha\}$. The corresponding adaptive sparse grid is a sub-grid of the level-K isotropic sparse grid $\mathcal{H}_K$, with the grid points becoming concentrated in the non-smooth region. In the region where $\eta(\boldsymbol{\theta})$ is very smooth, this approach saves a significant number of grid points but still achieves the prescribed accuracy.

2.2.2   Algorithm for Constructing the Surrogate PPDF

In the forthcoming numerical illustration, a surrogate PPDF will be constructed based on the sparse grid method, discussed above, with the use of the following procedure.

Algorithm 2.1

- *STEP 1: Determine the maximum allowable resolution K of the sparse grid by analyzing the trade off between the interpolation error and computational cost. Determine the error tolerance α.*
- *STEP 2: Generate the isotropic sparse grid at some starting coarse level ℓ. Until the maximum level K is reached or the magnitudes of all surpluses on the last level are smaller than α, do the following iteratively:*

  - *Step 2.1: Simulate the turbulence model $f(\theta)$ at each grid point $\theta_{i,j} \in \mathcal{H}_\ell$.*
  - *Step 2.2: Construct the sparse grid interpolant $\mathcal{I}_{\ell,\alpha}[f](\theta)$ based on formula (7).*
  - *Step 2.3: Generate the adaptive sparse grid for the next level based on the obtained surpluses. Set $\ell := \ell + 1$ and go back to Step 2.1.*

- *STEP 3: Construct an approximate likelihood function, denoted by $\tilde{L}(\theta\,|d)$, by substituting $\mathcal{I}_{\ell,\alpha}[f]$ for f into the likelihood formula using, e.g., (MVN) or (EXP).*
- *STEP 4: Construct the surrogate PPDF $\tilde{P}(\theta\,|d)$ via*

$$\tilde{P}(\theta\,|d) \propto \tilde{L}(\theta\,|d)P(\theta).$$

After the surrogate is constructed, an MCMC simulation is used to explore $\tilde{P}(\theta\,|d)$. Using our approach, drawing the parameter samples does not require any model executions but negligible computational time for polynomial evaluation using the surrogate system. The improvement of computational efficiency by using surrogate PPDF is more impressive when increased samples are drawn in the MCMC simulation.

Finally, it is worth discussing the flexibility of grid adaptive refinement strategies. It is known that in calibration problems of turbulence models, different likelihood models could lead to conflicting posterior distributions [13, 45]. Moreover, for a flow problem, experimental data given by different authors is sometimes inconsistent. There is also a wide variation of the physical quantities to be measured and recorded. Naturally, one would desire a surrogate modeling method that allows for the use of a variety of likelihood functions and data sets, at little cost, once the surrogate system has been built. An adaptive refinement strategy based on the smoothness of the likelihood functions [63] is obviously the least flexible, since the grid is likelihood-function-specific. The approach we apply in this work, i.e., an adaptive method that is guided by the smoothness of output interpolant, allows the use of an universal surrogate of the output, for different choices of likelihood functions and data of the *same* physical quantities. The surrogate for the output is, however, more expensive than that built directly for the likelihood function in the former approach, since grid points may be generated in the low density region of the likelihood where the forward simulations are wasteful. The most versatile method is certainly the non-adaptive, full sparse grid method, but the surrogate is also constructed with highest cost in this case. To this end, one has to sacrifice the flexibility of the sparse grid surrogate to improve the efficiency. The demand of investigating posterior distribution over different likelihood functions and data sets and the computational budget need to be balanced before an adaptive refinement strategy is determined.

3   Application to Large Eddy Simulation of Sub-critical Flow Around a Circular Cylinder

3.1   Parametric Uncertainty of Smagorinsky Model

In LES practice, the time dependent, incompressible Navier-Stokes equations are filtered by, e.g., box filter, Gaussian filter, differential filter and the governing equations are given by

$$\bar{u}_t + \nabla \cdot (\bar{u}\,\bar{u}) - \nu\Delta\bar{u} + \nabla\bar{p} - \nabla \cdot (2\nu_T\nabla^s\bar{u}) = \bar{f},$$
$$\nabla \cdot \bar{u} = 0, \tag{8}$$

where $\bar{u}$ is the velocity at the resolved scales, $\bar{p}$ is the corresponding pressure, $\nu_T \geq 0$ is the *eddy viscosity* and ∇^s is the symmetric part of ∇ operator, see [7].

The most common choice for ν_T, which is studied herein, is known in LES as the Smagorinsky model [41, 54] in which

$$\nu_T = \ell_S^2|\nabla^s\bar{u}|, \tag{9}$$

where $\ell_S = C_S\delta$ and $|\cdot| = \sqrt{2(\cdot)_{ij}(\cdot)_{ij}}$, ℓ_S is called the Smagorinsky lengthscale. There are two model calibration parameters in this term – the Smagorinsky constant C_S and the filter width δ. The pioneering analysis of Lilly [32], under some optimistic assumptions, proposed that C_S has a universal value 0.17 and is not a "tuning" constant. This universal value has been found later not the best choice for most LES computations and various different values ranging from 0.1 to 0.25 are usually selected leading to improved results, see, e.g., [1, 9, 14, 15, 37–40]. The optimal choice for C_S depends on the flow problems considered and even may be different for different regions in a flow field. Indeed, this poses a major drawback of the Smagorinsky model.

The second calibration parameter – the filter width δ-characterizes the short lengthscale fluctuations to be removed from the flow fields. Ideally, the filter width should be put at the smallest persistent, energetically significant scale (the flow microscale), which demarcates the deterministic large eddies and isotropic small eddies, [47]. Unfortunately, such a choice is infeasible, since the flow microscale is seldom estimated. Instead, due to the fact that LES requires the spatial resolution h to be proportional to δ, the usual practice is to specify the grid to be used in the computation, and then take the filter width according to the grid size. The specification of grid and filter without knowledge of the microscale could lead to poor simulation.

An additional calibration parameter involves in near wall treatment. The correct behavior of Smagorinsky eddy viscosity ν_T near the wall is $\nu_T \simeq 0$, since there is no turbulent fluctuation there. In contrast, the formulation (9) is nonzero and introduces large amounts of dissipation in the boundary layer. One approach to overcome this deficiency is to damp ℓ_S as the boundary is approached by the van Driest damping function [58]. The van Driest scaling reads:

$$\ell_S = C_S \delta \left(1 - e^{-y^{+n}/A^{+n}} \right)^p, \tag{10}$$

where y^+ is the distance from wall in wall units, A^+ is van Driest constant ascribed the value $A^+ = 25$. Various different values of (n, p) have been used – the most commonly chosen are $(1, 1)$ and $(3, 0.5)$, [51]. For simplicity, in this work, we fix $n = 1$ and treat p only as a calibration parameter. The variation of p alone can capture the full spectrum of near wall scaling: $p = 0$ means no damping function is applied, while a large p associates with fast damping. We call p van Driest damping parameter.

3.2 Sub-critical Flow Around a Circular Cylinder

The flow concerned in this study corresponds to a time-dependent flow through a channel around a cylinder. External flows past objects have been the subject of numerous theoretical, experimental and numerical investigations because of their many practical applications, see [5, 6, 44] and the reference therein. In the sub-critical Reynolds number range ($300 < Re < 2 \times 10^5$), these flows are characterized by turbulent vortex streets and transitioning free shear layers.

We consider the two-dimensional flow around a cylinder of diameter $D = 0.1$ in rectangular domain of size 2.2×1.4, consisting a $5D$ upstream, $17D$ downstream and $7D$ in lateral directions. We employ the finite element method with second order Taylor-Hood finite element and polygonal boundary approximation. Our computation is carried out on triangular meshes generated based on Delaunay-Voronoi algorithm and refined around the cylinder. The ratio of number of mesh points on the top/bottom boundaries, left/right boundaries and cylinder boundary is fixed at 3:2:4 (Fig. 1). As common practice, the filter width is chosen locally at each triangle as the size of the current triangle. Its value therefore varies throughout the domain, and is roughly 10 times smaller near the cylinder than that in the far field. Since the synthetic data will be taken in the near wake region, for simplicity, we characterize δ by the value of the filter width on the cylinder surface.

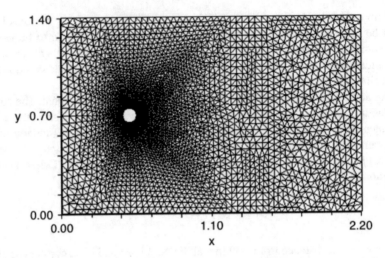

Fig. 1 A computational grid used in our study on LES of turbulent flow past a cylinder with $\delta = \pi/480$

The Smagorinsky model with van Driest damping (8), (9) and (10) is considered with $\nu = 2 \times 10^{-4}, f = 0, T = 12$ and $\Delta t = 0.01$. The statistics are compiled over the last 7 time units, equivalent to a period of ≈ 15 vortex shedding cycles. The inflow and outflow velocity is $(\frac{6}{1.4^2}y(1.4-y), 0)$. The mean velocity at the inlet is $U_0 = 1$. No-slip boundary conditions are prescribed along the top and bottom walls. Based on U_0 and the diameter of the cylinder D, the Reynolds number for this flow is $Re = 500$, in the sub-critical range. The temporal discretization applied in the computation is the Crank-Nicolson scheme. Denoting quantities at time level t_k by a subscript k, the time stepping scheme has the form:

$$\frac{\bar{u}_k - \bar{u}_{k-1}}{\Delta t} - \nu \Delta \frac{\bar{u}_k + \bar{u}_{k-1}}{2} + \frac{1}{2}(\bar{u}_k \cdot \nabla \bar{u}_k + \bar{u}_{k-1} \cdot \nabla \bar{u}_{k-1}) + \nabla \bar{p}_k$$
$$- (\nabla \cdot (\nu_T(\bar{u}_k)\nabla^s \bar{u}_k) + \nabla \cdot (\nu_T(\bar{u}_{k-1})\nabla^s \bar{u}_{k-1})) = 0, \tag{11}$$
$$\nabla \cdot \bar{u}_k = 0.$$

System (11) is reformulated as a nonlinear variational problem in time step t_k. This problem is solved iteratively by a fixed point iteration. Let $(\bar{u}_k^0, \bar{p}_k^0)$ be an initial guess. Given $(\bar{u}_k^m, \bar{p}_k^m)$, the iterate $(\bar{u}_k^{m+1}, \bar{p}_k^{m+1})$ is computed by solving

$$\frac{\bar{u}_k^{m+1} - \bar{u}_{k-1}}{\Delta t} - \nu \Delta \frac{\bar{u}_k^{m+1} + \bar{u}_{k-1}}{2} + \frac{1}{2}(\bar{u}_k^m \cdot \nabla \bar{u}_k^{m+1} + \bar{u}_{k-1} \cdot \nabla \bar{u}_{k-1}) + \nabla \bar{p}_k^{m+1}$$
$$- (\nabla \cdot (\nu_T(\bar{u}_k^m)\nabla^s \bar{u}_k^{m+1}) + \nabla \cdot (\nu_T(\bar{u}_{k-1})\nabla^s \bar{u}_{k-1})) = 0,$$
$$\nabla \cdot \bar{u}_k^{m+1} = 0. \tag{12}$$

The fixed point iteration in each time step is stopped if the Euclidean norm of the residual vector is less than 10^{-10}. The spatial and temporal discretizations we use herein are similar to [30, 31], in which they were applied to direct numerical simulations of flow around a cylinder at Reynolds number $Re = 100$.

3.3 The Prior PDF and Calibration Data

We will exploit Bayesian calibration for three model parameters C_S, p and δ. The uniform prior PDF of the uncertain parameters is assumed. The searching domains for C_S and p are $[0, 0.2]$ and $[0, 2]$ respectively, covering their plausible and commonly selected values. The range of the prior PDF of δ, on the other hand, would significantly affect the computational cost; since the filter width is proportional to the spatial resolution. Thus, to reduce the cost of flow simulations, the searching domain for δ is set to be $[\pi/600, \pi/200]$, corresponding to relatively coarse resolutions where the grid spacing on the cylinder surface ranges from ≈ 2 to 6 wall units. As we shall see, the response surfaces tend to be more complicated for the low-resolution simulation, possibly due to the non-physical oscillations in the underresolved solutions reflecting in the probability space. As a result, coarse grids pose a greater challenge for the surrogates to precisely describe the true outputs and are suitable for our purpose of verifying the accuracy of the surrogate modeling approach. Figure 2 shows the distribution of instantaneous vorticity at $t = 20$ in the near wake region for two different choices of turbulence parameters. We can see that the simulated flows display laminar vortex shedding, as expected for LES of flows past bluff bodies. The difference in phase of vortex shedding in two simulations is recognizable.

The synthetic data are generated by solving Smagorinsky model (8), (9) and (10) with $C_S = 0.15$, $p = 0.05$ and $\delta = \pi/480$. The data sets used for calibration process are taken at 11 stations in a distance of $\approx 1D$ downstream. Specifically, these points locate equidistantly on the vertical line $x = 0.65$ between $y = 0.6$ and $y = 0.8$. For each point, the data of average streamwise and vertical velocities,

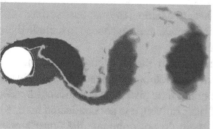 

Fig. 2 Intantaneous vorticity at $t = 20$ generated by two different choices of model parameters. *Left*: $C_s = 0.2$, $p = 0$, $\delta = \pi/480$. *Right*: $C_s = 0.05$, $p = 0$, $\delta = \pi/720$

Table 1 The true parameter values and the initial searching regions for model calibration

		True value	Γ
Smagorinsky constant	C_S	0.15	$[0, 0.2]$
van Driest parameter	p	0.5	$[0, 2]$
Filter width	δ	$\pi/480$	$[\pi/600, \pi/200]$

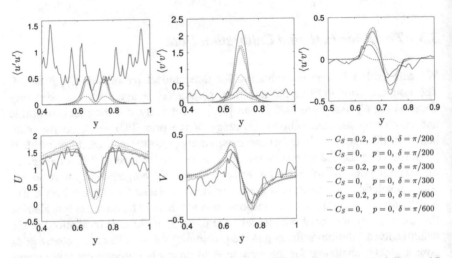

Fig. 3 Total resolved Reynolds stresses and average velocities along the vertical line at $1D$ downstream for some Smagorinsky models

denoted by U and V, as well as total streamwise, vertical and shear Reynolds stresses, i.e., $\langle u'u' \rangle$, $\langle v'v' \rangle$ and $\langle u'v' \rangle$, are selected, giving a total of 55 reference data. For clarity, the bounds of uniform prior PDFs and the true values of calibration parameters are listed in Table 1. In Fig. 3, the measurements of interested velocities and Reynolds stresses along $x = 0.65$ are plotted for some typical simulations. We observe that except for $\delta = \pi/200$, the approximated quantities are quite smooth and have expected patterns, see [5]. Certainly, the plots show significant differences among different models. In practice, LES models which give distinctly poor results such as those at $\delta = \pi/200$ could be immediately ruled out from the calibration process, informed by the fact that the wall-adjacent grid points lie outside the viscous sublayer. However, it is useful here to examine the response surfaces and the accuracy of the surrogate systems in these cases, and we choose to include these large filter widths in the surrogate domain instead.

Finally, it is worth mentioning that Smagorinsky model coefficients are not the only parameters that influence the quality of LES solutions. Indeed, other numerical parameters such as time step size and averaging time also have significant impacts, see, e.g., [11, 51]. While an estimation of their influence is not conducted here, we need to ensure that the errors caused by them do not dominate the uncertainties in the calibration parameters. A simple validation test is carried out on the flow statistics generated by Smagorinsky model of $C_S = 0$, $p = 0$ and $\delta = \pi/480$. The

Table 2 The maximum change in average velocity and Reynold stress profiles under the modifications of time step, averaging period and outflow BC

Component modified	U	V	$\langle u'u' \rangle$	$\langle v'v' \rangle$	$\langle u'v' \rangle$
Time step	0.0148	0.0213	0.0202	0.0371	0.0186
Averaging period	0.0048	0.0025	0.0014	0.0011	0.0045
Outflow BC	0.0022	0.0029	0.0016	0.0061	0.0020

flow simulation is replicated first with the temporal resolution refined by a factor of two, i.e., $\Delta t = 0.005$, and then with a doubled averaging period, i.e., by setting $T = 19$. We also conduct another simulation in which the zero gradient replaces Dirichlet outflow boundary condition to justify that the numerical oscillation at the downstream boundary does not disturb the inner domain. The maximum change in five velocity and Reynold stress profiles of interest in these modified models is presented in Table 2. We see that among three investigated source of numerical errors, the temporal resolution is the most prominent, as it makes up approximately 80 % of the change in all data. More importantly, Table 2 reveals that the total maximum change is approximately 0.05 in the vertical Reynolds stress data and 0.025 for other quantities. Numerical errors of the synthetic calibration data, as well as model outputs, are expected to be around these values. In the uncertainty analysis following, for the (MVN) likelihood model, we will assume that the reference data are corrupted with Gaussian random noise of 0.1.

3.4 Results and Discussions

This section justifies the accuracy and efficiency of the surrogate modeling method described in Sect. 2, when applied to the numerical example of two-dimensional flow around a cylinder specified in Sect. 3. We utilize the software package *FreeFem++* [29] in solving the Smagorinsky discretization scheme. The adaptive sparse grid interpolation and integration schemes are generated using functions in the TASMANIAN toolkit [56]. The DRAM algorithm [28] is chosen for MCMC sampling of the surrogate PPDF.

The surrogate system for outputs is constructed using the linear basis functions, first on the standard sparse grid of level 5, then the grids are refined adaptively up to level 8. The total numbers of model executions needed for the four interpolants are 177, 439, 1002 and 2190, respectively, which are also the number of points of the four corresponding adaptive sparse grids.

The accuracy of a surrogate modeling approach based on the AHSG method is largely determined by the smoothness of the surrogate system, so it is worth examining the surface of the output data in the parameter space. For brevity, we only plot here the vertical Reynolds stress data at the centerline, i.e., $\langle v'v' \rangle(0.65, 0.7)$, which is among the most fluctuating (See Fig. 3). Figure 4 represents some surfaces for

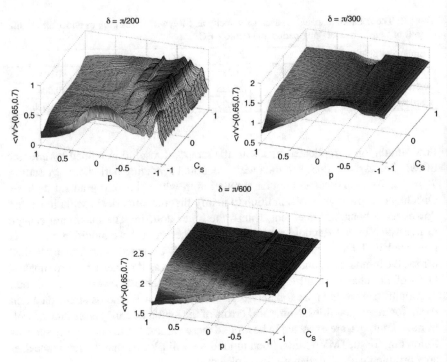

Fig. 4 Surfaces of the predicted vertical Reynolds stress data at $(0.65, 0.7)$ generated by the AHSG method at level 8. C_S and p are normalized such that their searching regions are $[-1, 1]$

typical values of filter width generated on level 8 grid. We observe that the surface according to $\delta = \pi/200$ differs from two other cases ($\delta = \pi/600$, $\delta = \pi/300$) that are remarkably rougher. This, together with Fig. 3, confirms the connection between the complexity of the output function in both the physical and parameter spaces. In Fig. 5, the scatter plots for the predicted outputs obtained with the surrogate system at level 7 are presented. The approximations show clear improvement in accuracy with $\delta \in [\pi/600, \pi/300]$, compared to those at larger values. While not considered herein, it is reasonable to expect that the surrogate outputs at least maintain the same accuracy for $\delta \leq \pi/600$, since more grid refinement will remove extra non-physical wiggles. In the next part, we justify that this level of accuracy is sufficient for our surrogate-based MCMC method. Although the surrogate systems show remarkable discrepancy for large δ, as previously mentioned, these values, leading to visibly inadequate outputs, should be excluded in practical calibration processes. While the original domain of δ is $[\pi/600, \pi/200]$, by choosing its true value as $\pi/480$, the effective searching region of δ is restricted to $[\pi/600, \pi/300]$.

To evaluate the accuracy and efficiency of our surrogate modeling approach, the DRAM-based MCMC simulations using the surrogate PPDF $\tilde{P}(\boldsymbol{\theta}|\boldsymbol{d})$ constructed in Algorithm 2.1 are conducted. Each MCMC simulation draws 60,000 parameter samples, the first 10,000 of which are discarded and the remaining 50,000 samples

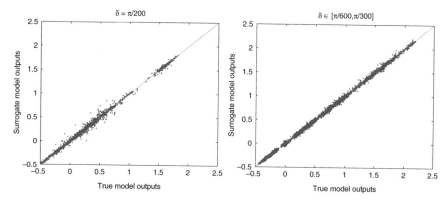

Fig. 5 Scatter plots for the prediction of the output data given by the surrogate system on level 7 sparse grid

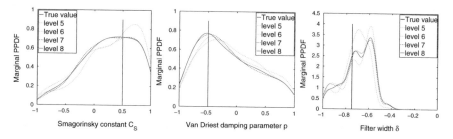

Fig. 6 Marginal posterior probability density functions of three Smagorinsky model parameters with (MVN) likelihood model estimated using the linear surrogate systems on level 5–8 adaptive sparse grids

are used for estimating the PPDF. For the first experiment, (MVN) likelihood function is employed; the data are corrupted by 10 % Gaussian random noise, treated as numerical errors. Figure 6 plots the marginal PPDFs where the three parameters are normalized such that the searching region is $[-1, 1]^3$. The black vertical lines represent the true values listed in Table 1. The red solid lines are the marginal PPDFs estimated by MCMC simulations based on the surrogate systems on level 8 grid, and the dashed lines represent those based on the surrogate systems on lower levels. The figure indicates that the MCMC results according to level 7 and level 8 sparse grids, which require 1002 and 2190 model executions correspondingly, are already close to each other. Thus, the surrogate PPDF on level 8 is accurate enough for MCMC simulations.

We proceed to compare the accuracy of the surrogate-based with the conventional MCMC with equal computational effort, i.e., same number of model executions. Due to the high computational cost, a proper conventional MCMC simulation is not conducted in this work. However, given the accuracy of the surrogate system, we expect that marginal PPDFs obtained from conventional MCMC are very close with those from surrogate-based MCMC on high-level grid and therefore, run the MCMC

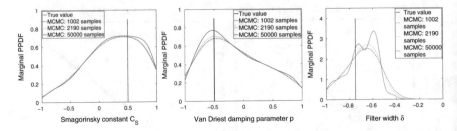

Fig. 7 Marginal posterior probability density functions of model parameters with (MVN) likelihood function estimated using the linear surrogate systems on level 8 adaptive sparse grids with 1002, 2190 and 50,000 samples (excluding 10,000 samples for burn-in period). These are the numbers of model executions that the conventional MCMC requires to obtain similar results

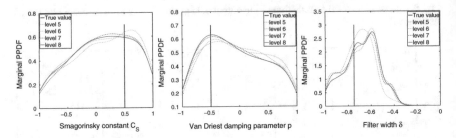

Fig. 8 Marginal posterior probability density functions of three Smagorinsky model parameters with (EXP) likelihood model estimated using the linear surrogate systems on level 5–8 adaptive sparse grids

simulation with samples drawn from level 8 surrogate. The first 10,000 samples are discarded to minimize the effect of initial values on the posterior inference. Figure 7 depicts the marginal PPDFs for model parameters obtained with 1002, 2190 and 50,000 samples after burn-in period. Let us remark that if conventional MCMC is employed, these are the numbers of *model executions* required to obtain similar results. Comparing Figs. 6 and 7 indicates that with the same number of model executions, the approximations using surrogate system are more accurate than those using conventional MCMC, highlighting the efficiency of our surrogate modeling method.

In order to demonstrate that our adaptive refinement strategy based on the smoothness of output data in probability space allows the change of likelihood models with minimal computational cost, we perform the above experiment with (EXP) likelihood function and $\xi = 500$ using the same surrogate of outputs. The marginal PPDFs of model parameters estimated using the linear surrogate systems are shown in Fig. 8. Again, they can be compared with marginal PPDFs estimated using conventional MCMC with the same number of model executions in Fig. 9. The plots confirm the accuracy of the surrogate PPDF for MCMC simulations and that surrogate-based MCMC requires less forward model executions than the conventional approach. On the other hand, it should be noted that some likelihood

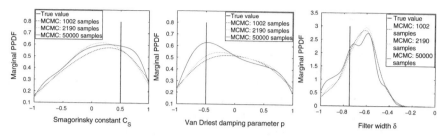

Fig. 9 Marginal posterior probability density functions of model parameters with (EXP) likelihood function estimated using the linear surrogate systems on level 8 adaptive sparse grids with 1002, 2190 and 50,000 samples (excluding 10,000 samples for burn-in period). These are the numbers of model executions that the conventional MCMC requires to obtain similar results

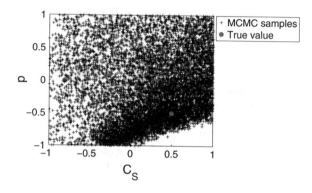

Fig. 10 Two-dimensional marginal posterior probability density function of C_S and p with (MVN) likelihood model. The MCMC samples are obtained using the linear surrogate system on level 8 sparse grid

models, especially those resulting in peaky PPDFs, may require a surrogate system more accurate than that on level 8 sparse grid. In those cases, the surrogate needs to be constructed on a grid of higher level.

The calibration results for both likelihood models show that the Smagorinsky constant C_S and van Driest damping parameter p have posterior maximizers near their true values, while smaller values are somewhat preferred for the filter width δ. Meanwhile, the posterior distribution of δ is peaky, indicating that the data depend on δ and the Smagorinsky models with our selections of filter width (and spatial resolution) are incomplete. Indeed, finer grids are needed to sufficiently resolve the energy. The plots also reveal that the data are significantly more sensitive with respect to δ than to other parameters. This elucidates why finding the optimal value for δ, i.e., determining the ideal place to truncate scale, is a very important issue in LES practice. Finally, the positive correlation between C_S and p can be observed in Fig. 10, in which the posterior samples projected on the (C_S, p)-plane are plotted. Given that our calibration data are extracted in near wake region, this correlation is expected. As larger value of C_S increases the Smagorinsky lengthscale ℓ_S, larger p would be needed for a stronger damping of ℓ_S near the boundary.

4   Conclusion and Future Works

In this paper, we present a surrogate modeling approach based on the AHSG method for Bayesian inference, with application to quantification of parametric uncertainty of LES turbulence models. The method is based on those developed in [62] for less complex geophysical and groundwater models, is model independent and can be flexibly used together with any MCMC algorithm and likelihood function. The accuracy and efficiency of our approach is illustrated by virtue of the numerical example consisting of the Smagorinsky model of two-dimensional flow around a cylinder. We combine the hierarchical linear basis and the local adaptive sparse grid technique to construct surrogate systems with a small number of model executions. Although the forward model investigated herein is highly nonlinear and more complicated than those in previous studies, our analysis indicates that the surrogate system is accurate for reasonable specifications of search regions. Compared to the conventional MCMC simulation, our surrogate-based approach requires significantly less model executions for estimating the parameter distribution and quantifying predictive uncertainty. Given the extremely high cost of turbulence simulations, this computational efficiency is critical for the feasibility of Bayesian inference in turbulence modeling.

While the performance of surrogate modeling method is evaluated in this work for a synthetic cylinder flow model on relatively coarse grids, we expect comparable results for practical, more complicated calibration and prediction problems using real-world data; since three-dimensional, more refined simulations and real experiments of these flows are known to produce similar patterns to the investigated physical outputs in this study. Still, a three-dimensional demonstration of our surrogate-based approach for these problems is irreplaceable and would be the next logical step. The framework presented here could be directly applied to other engineering flow models, as well as to the tasks of quantifying the structural uncertainties and comparing competing turbulence closure models. The accuracy of surrogate-based MCMC in these cases needs to be tested, but the verification, which is much less computational demanding than running the conventional MCMC, is possibly worthwhile. Finally, besides our AHSG, several other methods can be employed to construct the surrogate system. A thorough comparative assessment with those methods is essential to fully justify the efficiency of our approach in turbulence uncertainty quantification problems and would be considered in the future.

Concerning sparse grid interpolation methods, additional research in accelerating the convergence rate of the surrogate is necessary. One direction is high-order sparse grid methods, which utilize high-order (instead of linear) hierarchical polynomial basis and whose superior efficiency has been justified for uncertainty quantification of groundwater models [63]. On the other hand, given that the outputs and PPDFs do not experience same level of sensitivity to different calibration parameters, combining locally grid refinement strategy with dimension-adaptive sparse grid methods to further reduce the number of interpolation points is worth studying.

References

1. M. Antonopoulos-Domis, Large eddy simulation of a passive scalar in isotropic turbulent. J. Fluid Mech. **104**, 55–79 (1981)
2. I. Babuska, F. Nobile, R. Tempone, A stochastic collocation method for elliptic partial differential equations with random input data. SIAM J. Numer. Anal. **45**(3), 1005–1034 (2007) (electronic). MR MR2318799 (2008e:65372)
3. I. Babuska, F. Nobile, E. Zouraris, Galerkin finite element approximations of stochastic elliptic partial differential equations. SIAM J. Numer. Anal. **42**, 800–825 (2004)
4. F. Bao, Y. Cao, C.G. Webster, G. Zhang, A hybrid sparse grid approach for nonlinear filtering problems based on adaptive domain approximations of the Zakai equation. SIAM J. Uncertain. Quantif. **2**, 784–804 (2014)
5. P. Beaudan, P. Moin, Numerical experiments on the flow past a circular cylinder at sub-critical Reynolds number, Report No. TF-62, Thermosciences Division, Department of Mechanical Engineering, Stanford University, 1994
6. E. Berger, R. Wille, Periodic flow phenomena. Ann. Rev. Fluid Mech. **4**, 313–340 (1972)
7. L.C. Berselli, T. Iliescu, W. Layton, *Large Eddy Simulation* (Springer, Berlin, 2004)
8. K. Beven, A. Binley, The future of distributed models – model calibration and uncertainty prediction. Hydrol. Process. **6**(3), 279–298 (1992)
9. S. Biringen, W.C Reynolds, Large eddy simulation of the shear-free turbulent boundary layer. J. Fluid Mech. **103**, 53–63 (1981)
10. G. Box, G. Tiao, *Bayesian Inference in Statistical Analysis* (Wiley-Interscience, New York, 1992), p. 608
11. M. Breuer, Large eddy simulation of the subcritical flow past a circular cylinder: numerical and modeling aspects. Int. J. Numer. Methods Fluids **28**, 1281–1302 (1998)
12. H.J. Bungartz, M. Griebel, Sparse grids. Acta Numerica **13**, 1–123 (2004)
13. S.H. Cheung, T.A. Oliver, E.E. Prudencio, S. Prudhomme, R.D Moser, Bayesian inference with applications to turbulence modeling. Reliab. Eng. Syst. Saf. **96**, 1137–1149 (2011)
14. R.A. Clark, J.H. Ferziger, W.C. Reynolds, Evaluation of subgrid-scale models using an accurately simulated turbulent flow. J. Fluid Mech. **91**, 1–16 (1979)
15. J.W. Deardorff, A three-dimensional numerical study of turbulent channel flow at large Reynolds numbers. J. Fluid Mech. **41**, 453–480 (1970)
16. E. Dow, Q. Wang, Quantification of structural uncertainties in the $k - \omega$ turbulence model, AIAA Paper, 2011-1762 (2011)
17. M. Emory, J. Larsson, G. Iaccarino, Modeling of structural uncertainties in Reynolds-averaged Navier-Stokes closures. Phys. Fluids **25**, 110822 (2013)
18. J. Foo, X. Wan, G. Karniadakis, The multi-element probabilistic collocation method (ME-PCM): error analysis and applications. J. Comput. Phys. **227**, 9572–9595 (2008)
19. D. Gamerman, H. Lopes, *Markov Chain Monte Carlo: Stochastic Simulation for Bayesian Inference*, 2nd edn. (Chapman and Hall, London, 2006), p. 344
20. B. Ganapathysubramanian, N. Zabaras, Sparse grid collocation schemes for stochastic natural convection problems. J. Comput. Phys. **225**(1), 652–685 (2007)
21. T. Gerstner, M. Griebel, Dimension-adaptive tensor-product quadrature. Computing **71**, 65–87 (2003)
22. R.G. Ghanem, P.D. Spanos, *Stochastic Finite Elements: A Spectral Approach* (Springer, New York, 1991)
23. C. Gorlé, M. Emory, J. Larsson, G. Iaccarino, Epistemic uncertainty quantification for RANS modeling of the flow over a wavy wall, Center for Turbulence Research, Annual Research Briefs, 2012
24. C. Gorlé, G. Iaccarino, A framework for epistemic uncertainty quantification of turbulent scalar flux models for Reynolds-averaged Navier-Stokes simulations. Phys. Fluids **25**, 055105 (2013)

25. M. Griebel, Adaptive sparse grid multilevel methods for elliptic PDEs based on finite differences, Computing **61**(2), 151–179 (1998). doi:10.1007/BF02684411
26. M. Gunzburger, C.G. Webster, G. Zhang, An adaptive sparse grid iterative ensemble Kalman filter approach for parameter field estimation. Int. J. Comput. Math. **91**(4), 798–817 (2014)
27. M. Gunzburger, C.G. Webster, G. Zhang, Stochastic finite element methods for partial differential equations with random input data. Acta Numerica **23**, 521–650 (2014)
28. H. Haario, M. Laine, A. Mira, E. Saksman, DRAM: Efficient adaptive MCMC. Stat. Comput. **16**(4), 339–354 (2006). doi:10.1007/s11222-006-9438-0
29. F. Hecht, New development in freefem++. J. Numer. Math. **20**, 251–265 (2012)
30. V. John, Reference values for drag and lift of a two-dimensional time-dependent flow around a cylinder. Int. J. Numer. Math. Fluids **44**, 777–788 (2004)
31. V. John, G. Matthies, Higher-order finite element discretizations in a benchmark problem for incompressible flows. Int. J. Numer. Methods Fluids **37**, 885–903 (2001)
32. D.K. Lilly, The representation of small scale turbulence in numerical simulation experiments, in *Proceedings of IBM Scientific Computing Symposium on Environmental Sciences*, ed. by H.H. Goldstine (Yorktown Heights, New York, 1967), pp. 195–210
33. X. Liu, M.A. Cardiff, P.K. Kitanidis, Parameter estimation in nonlinear environmental problems. Stoch. Env. Res. Risk Assess. **24**(7), 1003–1022 (2010)
34. M. Liu, Z. Gao, J. Hesthaven, Adaptive sparse grid algorithms with applications to electromagnetic scattering under uncertainty. Appl. Numer. Math. **61**, 24–37 (2011)
35. D. Lucor, J. Meyers, P. Sagaut, Sensitivity analysis of large-eddy simulations to subgrid-scale-model parametric uncertainty using polynomial chaos. J. Fluid Mech. **585**, 255–279 (2007)
36. X. Ma, N. Zabaras, An efficient Bayesian inference approach to inverse problems based on an adaptive sparse grid collocation method. Inverse Probl. **25**(3), 035013 (2009)
37. P.J. Mason, Large eddy simulation of the convective atmospheric boundary layer. J. Atmos. Sci. **46**, 1492–1516 (1989)
38. P.J. Mason, S.H. Derbyshire, Large-eddy simulation of the stable-stratified atmospheric boundary layer. Boundary-layer Meteorol. **53**, 117–162 (1990)
39. P. Moin, J. Kim, Numerical investigation of turbulent channel flow. J. Fluid Mech. **18**, 341–377 (1982)
40. N.N. Monsour, P. Moin, W.C. Reynolds, J.H. Ferziger, Improved methods for large eddy simulations of turbulence, in F. Durst, B.E. Launder, F.W. Schmidt, J.H. Whitelaw (Eds.), Proceedings of the Turbulent Shear Flows I, Springer, Berlin, 1979, pp. 386–401
41. J. von Neumann, R.D. Richtmyer, A method for the numerical calculation of hydrodynamic shocks. J. Appl. Phys. **21**, 232–237 (1950)
42. F. Nobile, R. Tempone, C.G. Webster, A sparse grid stochastic collocation method for partial differential equations with random input data. SIAM J. Numer. Anal. **46**(5), 2309–2345 (2008)
43. F. Nobile, R. Tempone, C.G. Webster, An anisotropic sparse grid collocation method for elliptic partial differential equations with random input data. SIAM J. Numer. Anal. **46**(5), 2411–2442 (2008)
44. C. Norberg, Effects of Reynolds number and a low-intensity free-stream turbulence on the flow around a circular cylinder, Publication No. 87/2, Department of Applied Thermodynamics and Fluid Mechanics, Chalmer University of Technology, Gothenburg, 1987
45. T. Oliver, R. Moser, Bayesian uncertainty quantification applied to RANS turbulence models. J. Phys.: Conf. Ser. **318**, 042032 (2011)
46. D. Pflüger, Spatially adaptive sparse grids for high-dimensional problems, Ph.D. thesis, TU Munich, Munich, 2010
47. S. Pope, *Turbulent Flows* (Cambridge University Press, Cambridge/New York, 2000)
48. S. Rahman, A Polynomial Dimensional Decomposition for Stochastic Computing. Int. J. Numer. Methods Eng. **76**, 2091–2116 (2008)
49. S. Razavi, B.A. Tolson, D.H. Burn, Review of surrogate modeling in water resources. Water Resour. Res. **48**, W07401 (2012)
50. C. Robert, G. Casella, *Monte Carlo Statistical Methods*, 2nd edn. (Springer, New York, 2004)

51. W. Rodi, J.H. Ferziger, M. Breuer, M. Pourquié, Status of large eddy simulation: results of a workshop. Workshop on LES of Flows Past Bluff Bodies, Rottach-Egern, Tegernsee, 26–28 June 1995. J. Fluids Eng. **119**, 248–262 (1997)
52. M. Rosenblatt, Remarks on a multivariate transformation. Ann. Math. Statist. **23**, 470–472 (1952)
53. S. Sankaran, A. Marsden, A stochastic collocation method for uncertainty quantification and propagation in cardiovascular simulations. J. Biomech. Eng. **133**(3), 031001 (2011)
54. J.S. Smagorinsky, General circulation experiments with the primitive equations. Mon. Weather Rev. **91**, 99–164 (1963)
55. P. Smith, K.J. Beven, J.A. Tawn, Informal likelihood measures in model assessment: theoretic development and investigation. Adv. Water Resour. **31**, 1087–1100 (2008)
56. M. Stoyanov, User Manual: TASMANIAN sparse grid, ORNL Technical Report, 2013
57. H. Tran, C. Trenchea, C. Webster, A convergence analysis of stochastic collocation method for Navier-Stokes equations with random input data, ORNL Technical Report, 2014
58. E.R. van Driest, On turbulent flow near a wall. J. Aerosp. Sci. **23**, 1007–1011 (1956)
59. J.A. Vrugt, C.J.F. ter Braak, M.P. Clark, J.M. Hyman, B.A. Robinson, Treatment of input uncertainty in hydrologic modeling: doing hydrology backward with Markov chain Monte Carlo simulation. Water Resour. Res. **44**, W00B09 (2008)
60. J. Vrugt, C. Ter Braak, C. Diks, D. Higdon, B. Robinson, J. Hyman, Accelerating Markov chain Monte Carlo simulation by differential evolution with self-adaptive randomized subspace sampling. Int. J. Nonlinear Sci. Numer. Simul. **10**, 273–290 (2009)
61. X. Wan, G.E. Karniadakis, Long-term behavior of polynomial chaos in stochastic flow simulations. Comput. Methods Appl. Mech. Eng. **195**, 5582–5596 (2006)
62. J.A.S. Witteveen, G.J.A. Loeven, S. Sarkar, H. Bijl, Probabilistic collocation for period-1 limit cycle oscillations. J. Sound Vib. **311**(1–2), 421–439 (2008)
63. G. Zhang, D. Lu, M. Ye, M. Gunzburger, C.G. Webster, An adaptive sparse grid high-order stochastic collocation method for Bayesian inference in ground water reactive transport modeling. Water Resour. Res. **49**, 6871–6892 (2013)

Hierarchical Gradient-Based Optimization with B-Splines on Sparse Grids

Julian Valentin and Dirk Pflüger

Abstract Optimization algorithms typically perform a series of function evaluations to find an approximation of an optimal point of the objective function. Evaluations can be expensive, e.g., if they depend on the results of a complex simulation. When dealing with higher-dimensional functions, the curse of dimensionality increases the difficulty of the problem rapidly and prohibits a regular sampling. Instead of directly optimizing the objective function, we replace it with a sparse grid interpolant, saving valuable function evaluations. We generalize the standard piecewise linear basis to hierarchical B-splines, making the sparse grid surrogate smooth enough to enable gradient-based optimization methods. Also, we use an uncommon refinement criterion due to Novak and Ritter to generate an appropriate sparse grid adaptively. Finally, we evaluate the new method for various artificial and real-world examples.

1 Introduction

In this work, we want to solve optimization problems of the following form: Assume we are given a continuous function $f : [0, 1]^d \to \mathbb{R}$ (*objective function*). Our goal is to find a minimal point

$$\mathbf{x}_{\mathrm{opt}} = \arg\min_{\mathbf{x} \in [0,1]^d} f(\mathbf{x}) \,, \tag{1}$$

i.e., we want to solve a general, bound-constrained optimization problem. Optimization algorithms, whether gradient-free or gradient-based, usually perform a series of evaluations of f, its gradient, or its Hessian (if available) to find an approximation $\mathbf{x}_{\mathrm{opt}}^*$ of $\mathbf{x}_{\mathrm{opt}}$. As each evaluation can be expensive, e.g. by triggering a cascade of nested simulations, we want to use as few evaluations as possible. Of

J. Valentin (✉) • D. Pflüger

Institute for Parallel and Distributed Systems (IPVS), Universität Stuttgart, Universitätsstr. 38, 70569 Stuttgart, Germany

e-mail: julian.valentin@ipvs.uni-stuttgart.de; dirk.pflueger@ipvs.uni-stuttgart.de

© Springer International Publishing Switzerland 2016

J. Garcke, D. Pflüger (eds.), *Sparse Grids and Applications – Stuttgart 2014*,
Lecture Notes in Computational Science and Engineering 109,
DOI 10.1007/978-3-319-28262-6_13

course, for increasing d, the problem suffers from the curse of dimensionality, which obviously suggests the employment of sparse grids for the solution. Optimization with the aid of sparse grids was studied before, e.g. with additional constraints and piecewise linear functions [11] or with sparse grid surrogates defined via Lagrange polynomials on Chebyshev points [9, 10]. However, 1D Lagrange polynomials are asymmetrical, have global support [0, 1], and their degree 2^n is not tunable. In addition, polynomial interpolation prevents us from using equidistant grid points. We want to use B-splines as basis functions instead, as they do not have these drawbacks, but additionally feature many nice properties. B-splines have already been used in the context of sparse grids, e.g. for the purpose of data mining [27, 28] or quasi-interpolation [19]. The sufficient smoothness of B-splines allows us to use gradient-based optimization methods on the sparse grid interpolant efficiently, even if the gradient or Hessian of f are not available or costly to evaluate. Our optimization approach will be as follows:

1. Generate a spatially adaptive sparse grid $X = \{\mathbf{x}_k\}_k$ adapting to the peculiarities of f.
2. Interpolate f at X by an interpolant $\tilde{f}$ defined by a linear combination of B-splines on sparse grids.
3. Apply gradient-based optimization techniques to $\tilde{f}$ to get $\mathbf{x}_{\text{opt}}^*$.

In Sect. 2, we will define hierarchical B-splines and prove their linear independence in the univariate case, which generalizes to higher dimensionalities d. The B-splines will be modified to allow good approximations near the boundary of the domain $[0, 1]^d$. We will explain in Sect. 3 the refinement criterion by Novak and Ritter [26] we use to construct spatially adaptive sparse grids. A description of implementational details follows in Sect. 4. Finally, we evaluate our algorithm and compare it to established methods by studying various artificial and real-world examples in Sect. 5.

2   B-Splines on Sparse Grids

Conventional basis functions for sparse grids, including the piecewise polynomial functions by Bungartz [3], all share the shortcoming of not having globally continuous derivatives, hindering the use of gradient-based optimization. B-splines, which generalize the well-known hat functions, can tackle this problem. They were first studied by Schoenberg [34], who claimed that they were already known to Laplace [8]. But it was not until the 1960s when Schoenberg's results were rediscovered and the potential of B-splines for the emerging finite element method (FEM) was recognized. Important work was done by de Boor, who found simple B-spline algorithms [7]. B-splines have found application in a number of fields, e.g., for the aforementioned FEM [14], as non-uniform rational B-splines (NURBS) for geometric modeling [4, 15], for atomic and molecular physics [1, 23], and for financial mathematics [28], to name just a few examples. We will now repeat the definition of hierarchical B-Splines [28] and then prove their linear independence.

2.1 Cardinal B-Splines

The *cardinal B-spline* $b^p \colon \mathbb{R} \to \mathbb{R}$ of degree $p \in \mathbb{N}_0$ is defined by

$$b^0(x) = \chi_{[0,1)}(x) \,,$$

$$b^p(x) = \int_0^1 b^{p-1}(x-y)\mathrm{d}y \,, \quad p \geq 1 \,, \tag{2}$$

with the indicator function χ_A of $A \subset \mathbb{R}$, i.e., b^p is the convolution of b^{p-1} and b^0. This definition implies the following simple properties [15] (see Fig. 1, left). The support of b^p is $[0, p+1]$. On every interval $[k, k+1), k = 0, \ldots, p$ (*knot interval*), b^p is a non-negative polynomial of degree p. The B-spline is bounded by 1 and symmetric with respect to $x = \frac{p+1}{2}$. b^p is $(p-1)$ times continuously differentiable at $x = 0, \ldots, p+1$ (*knots*). By differentiation of (2) we get the simple identity

$$\frac{\mathrm{d}}{\mathrm{d}x}b^p(x) = b^{p-1}(x) - b^{p-1}(x-1) \,. \tag{3}$$

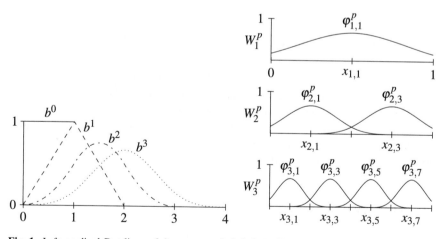

Fig. 1 *Left*: cardinal B-splines of degree $p = 0, 1, 2, 3$. *Right*: hierarchical B-splines of degree $p = 3$ and level $l = 1, 2, 3$

2.2 Hierarchical B-Splines

The *hierarchical B-spline* $\varphi_{l,i}^p : [0, 1] \rightarrow \mathbb{R}$ of level $l \in \mathbb{N}$ and index $i \in I_l :=$ $\{1, 3, 5, \ldots, 2^l - 1\}$ is defined by an affine parameter transformation [28],

$$\varphi_{l,i}^p(x) := b^p\left(\frac{x}{h_l} + \frac{p+1}{2} - i\right), \quad h_l := 2^{-l}.$$

$\varphi_{l,i}^p(x)$ has support $[0, 1] \cap (h_l \cdot [i \pm (p+1)/2])$ (see Fig. 1, right). For $p = 1$, we obtain the well-known piecewise linear hierarchical basis (hat functions). To simplify the next considerations, we only consider odd degree p, as the knots (where the B-spline is not infinitely many times differentiable) of $\varphi_{l,i}^p$ then coincide with the grid points

$$x_{l,i-(p+1)/2}, \quad \ldots, \quad x_{l,i}, \quad \ldots, \quad x_{l,i+(p+1)/2}$$

with $x_{l,i} := ih_l$. For even degree, the knots lie between the grid points, i.e.

$$x_{l,i-p/2} - \frac{h_l}{2}, \quad \ldots, \quad x_{l,i} - \frac{h_l}{2}, \quad x_{l,i} + \frac{h_l}{2}, \quad \ldots, \quad x_{l,i+p/2} + \frac{h_l}{2},$$

leading to slightly different, but related arguments. We can define the *nodal B-spline space* V_l^p and the *hierarchical B-spline subspace* W_l^p of level l by

$$V_l^p := \mathrm{span}\{\varphi_{l,i}^p \mid i = 1, \ldots, 2^l - 1\}, \quad W_l^p := \mathrm{span}\{\varphi_{l,i}^p \mid i \in I_l\}.$$

2.3 Linear Independence of Hierarchical B-Splines

In the piecewise linear case ($p = 1$), the relationship $V_n^1 = \bigoplus_{l=1}^n W_l^1$ can be seen easily. We prove that a similar relationship also holds for higher B-spline degrees. To this end, we first show the linear independence of the union $\{\varphi_{l,i}^p \mid l \leq n, i \in I_l\}$ of the hierarchical functions up to level n with the aid of B-splines on general knots.

Let $m, p \in \mathbb{N}_0$ and $\boldsymbol{\xi} = (\xi_0, \ldots, \xi_{m+p})$ be an increasing sequence of real numbers (knot sequence). Then for $k = 0, \ldots, m-1$ the B-splines $b_{k,\boldsymbol{\xi}}^p$ of degree p with knots $\boldsymbol{\xi}$ are defined by the Cox-de Boor recurrence [5, 7, 15]

$$b_{k,\boldsymbol{\xi}}^0 := \chi_{[\xi_k, \xi_{k+1})},$$

$$b_{k,\boldsymbol{\xi}}^p := \gamma_{k,\boldsymbol{\xi}}^p b_{k,\boldsymbol{\xi}}^{p-1} + (1 - \gamma_{k+1,\boldsymbol{\xi}}^p) b_{k+1,\boldsymbol{\xi}}^{p-1}, \quad \gamma_{k,\boldsymbol{\xi}}^p(x) := \frac{x - \xi_k}{\xi_{k+p} - \xi_k}, \quad p \geq 1.$$

For the special case of $\boldsymbol{\xi} = (0, 1, \ldots, p+1)$ and $k = 0$, we obtain the cardinal B-spline $b^p(x)$.

Proposition 1 *Let* $\boldsymbol{\xi} = (\xi_0, \ldots, \xi_{m+p})$ *be a knot sequence. Then the B-splines* $b_{k,\boldsymbol{\xi}}^p$, $k = 0, \ldots, m-1$, *form a basis of the spline space*

$$S_{\boldsymbol{\xi}}^p := \mathrm{span}\{b_{k,\boldsymbol{\xi}}^p \mid k = 0, \ldots, m-1\}. \tag{4}$$

$S_{\boldsymbol{\xi}}^p$ *contains exactly those functions which are continuous on* $D := [\xi_p, \xi_m]$, *polynomials of degree* $\leq p$ *on every knot interval* $[\xi_k, \xi_{k+1}]$ *in D and at least* $(p-1)$ *times continuously differentiable at every knot* ξ_k *in the interior of D.*

The proposition, a proof of which can be found in [15], implies linear independence of the nodal B-splines $\{\varphi_{n,i}^p \mid i = 1, \ldots, 2^n - 1\}$ of level $n \in \mathbb{N}$ by choosing

$$\xi_k := \left(k + 1 - \frac{p+1}{2}\right) h_n, \quad k = 0, \ldots, m+p, \quad m := 2^n - 1, \tag{5}$$

which leads to $\varphi_{n,i}^p = b_{i-1,\boldsymbol{\xi}}^p$ for $i = 1, \ldots, m$, i.e. $S_{\boldsymbol{\xi}}^p = V_n^p$ when restricting all B-splines to $D = [\xi_p, \xi_m]$. In particular, this means $\{\varphi_{n,i}^p \mid i \in I_n\}$ is a basis of W_n^p.

Proposition 2 *For every* $n \in \mathbb{N}$, *the hierarchical B-splines* $\{\varphi_{l,i}^p \mid l \leq n, i \in I_l\}$ *are linearly independent, i.e., the sum* $\bigoplus_{l=1}^n W_l^p$ *is indeed direct.*

Proof We prove the assertion by induction over n for the most common degrees $p \in \{1, 3, 5, 7\}$. For rather uncommon higher degrees, the proof can be viewed as a sketch. For $n = 1$, only one function exists. To proceed from $n-1$ to n, we assume that $\{\varphi_{l,i}^p \mid l \leq n-1, i \in I_l\}$ is linearly independent, so its span $\bigoplus_{l=1}^{n-1} W_l^p$ is a direct sum of hierarchical subspaces. Because both sets $\{\varphi_{n,i}^p \mid i \in I_n\}$ and $\{\varphi_{l,i}^p \mid l \leq n-1, i \in I_l\}$ are linearly independent, it is necessary and sufficient to show that $\mathrm{span}\{\varphi_{n,i}^p \mid i \in I_n\} \cap \bigoplus_{l=1}^{n-1} W_l^p = \{0\}$. Let $f_1 \in \mathrm{span}\{\varphi_{n,i}^p \mid i \in I_n\}$ and $f_2 \in \bigoplus_{l=1}^{n-1} W_l^p$ with $f_1 = f_2$. Then coefficients $c_{n,i}, c_{l,i} \in \mathbb{R}$ exist such that

$$\sum_{i \in I_n} c_{n,i} \varphi_{n,i}^p = f_1 = f_2 = \sum_{l=1}^{n-1} \sum_{i \in I_l} c_{l,i} \varphi_{l,i}^p.$$

The right-hand side is smooth in every grid point $x_{n,j}$, $j \in I_n$, of level n, as these grid points are not knots of the B-splines of level $< n$. So the left-hand side must be smooth there as well, i.e.

$$\partial_-^p f_1(x_{n,j}) = \partial_+^p f_1(x_{n,j}), \tag{6}$$

denoting with ∂_-^p and ∂_+^p the left and right derivative of order p, respectively. Now we use the combinatorial identity

$$\partial_+^p b^p(k) = (-1)^k \binom{p}{k} = \partial_-^p b^p(k+1), \quad k \in \mathbb{Z},$$

setting $\binom{p}{k} := 0$ for $k < 0$ or $k > p$. The identity stems from the repeated application of relation (3) (cf. [15]). Calculating the left and the right derivative in $x_{n,j}$ of each summand of f_1 respectively, we obtain from (6)

$$\sum_{i\in I_n} c_{n,i}(-1)^{k-1}\binom{p}{k-1} = \sum_{i\in I_n} c_{n,i}(-1)^k\binom{p}{k}, \quad k := k(i,j) = j - i + \frac{p+1}{2},$$

due to $\varphi_{n,i}^p(x_{n,j}) = b^p(k)$. The inner derivative $1/h_l^p$ canceled out from both sides. Using the relation $\binom{p}{k-1} + \binom{p}{k} = \binom{p+1}{k}$, we get

$$\sum_{i\in I_n} c_{n,i}(-1)^k\binom{p+1}{k} = 0, \quad j \in I_n. \tag{7}$$

As k is always odd or always even (for fixed j), we get

$$\sum_{i\in I_n} c_{n,i}\binom{p+1}{j-i+\frac{p+1}{2}} = 0, \quad j \in I_n,$$

by multiplying (7) by -1 if k is odd. This is a linear system with variables $c_{n,i}$, whose sparsity pattern depends on p. The corresponding matrix $A = A(p)$ is a symmetric $(2^{n-1} \times 2^{n-1})$ Toeplitz matrix with bandwidth $\lceil \frac{p-1}{4} \rceil$. For example, we obtain tridiagonal matrices for $p = 3$ or $p = 5$:

$$A(3) = \begin{pmatrix} 6 & 1 & & \\ 1 & \ddots & \ddots & \\ & \ddots & \ddots & 1 \\ & & 1 & 6 \end{pmatrix}, \quad A(5) = \begin{pmatrix} 20 & 6 & & \\ 6 & \ddots & \ddots & \\ & \ddots & \ddots & 6 \\ & & 6 & 20 \end{pmatrix}.$$

$A(p)$ is strictly diagonally dominant for $p = 1, 3, 5, 7$ and therefore invertible. For higher degrees, the regularity of $A(p)$ must be shown differently. If $A(p)$ is regular, we infer $c_{n,i} = 0$ for all $i \in I_n$, implying $f_2 = f_1 = 0$, which completes the proof for the common cases $p \in \{1, 3, 5, 7\}$. □

Proposition 3 *Let $n \in \mathbb{N}$. If we choose ξ as in (5) and restrict all functions involved to $D = [\xi_p, \xi_m]$, then $\bigoplus_{l=1}^n W_l^p = S_\xi^p = V_n^p$.*

Proof We already mentioned that $W_n^p \subseteq S_\xi^p = V_n^p$ holds. When restricting all of the basis functions $\varphi_{l,i}^p$ to $D = [\xi_p, \xi_m]$, $W_l^p \subseteq S_\xi^p$ also holds for smaller levels $l < n$: Each basis function $\varphi_{l,i}^p$, $i \in I_l$, is continuous on D, a polynomial of degree $\leq p$ on every knot interval of ξ (due to p odd) and at the knots themselves at least $(p-1)$ times continuously differentiable. From proposition 1 it follows $\varphi_{l,i}^p \in S_\xi^p$ and hence

$W_l^p \subseteq S_\xi^p$ for $l \leq n$. Consequently, $\bigoplus_{l=1}^n W_l^p \subseteq S_\xi^p$ and with a dimension argument,

$$\dim \bigoplus_{l=1}^n W_l^p = \sum_{l=1}^n |I_l| = \sum_{l=1}^n 2^{l-1} = 2^n - 1 = m = \dim S_\xi^p,$$

we obtain $\bigoplus_{l=1}^n W_l^p = S_\xi^p = V_n^p$, which proves the proposition. $\qquad\qquad\square$

2.4 Modified and Multivariate Hierarchical B-Splines

In $[0, 1] \setminus D$, linear combinations of hierarchical B-splines experience an unnatural decay towards the boundary of $[0, 1]$. As a side effect, this can result in overshoots of the linear combinations even when interpolating simple polynomials (see Fig. 2, left). Note that Fig. 2 does not contradict Prop. 3, as the $(p-1)/2$ leftmost and the $(p-1)/2$ rightmost grid points, where compliance with the interpolation condition is enforced, do not lie in $D = [\xi_p, \xi_m]$.

Grids with boundary points can help, but they spend proportionally too few points in the interior, most notably in higher dimensions. To overcome this difficulty,

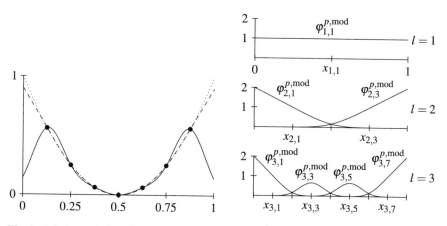

Fig. 2 *Left*: interpolation of the parabola $y = 4(x - 0.5)^2$ (*dotted line*) with unmodified (*solid*) and modified B-splines (*dashed*) for $p = 3$. *Right*: modified hierarchical B-splines of degree $p = 3$ and level $l = 1, 2, 3$

we modified the B-spline of level 1 and the first and last B-splines of higher levels [28],

$$\varphi_{l,i}^{p,\text{mod}}(x) := \begin{cases} 1 & \text{if } l = 1, i = 1, \\ \psi_l^p(x) & \text{if } l > 1, i = 1, \\ \psi_l^p(1-x) & \text{if } l > 1, i = 2^l - 1, \\ \varphi_{l,i}^p(x) & \text{otherwise,} \end{cases} \qquad \psi_l^p := \sum_{k=0}^{\lceil (p+1)/2 \rceil} (k+1)\varphi_{l,1-k}^p \,,$$

adding B-splines which have their maximum outside of $[0, 1]$ (see Fig. 2, right). Due to the relation

$$x = \sum_{k \in \mathbb{Z}} \left(k + \frac{p+1}{2} \right) b^p(x-k) \,, \quad x \in \mathbb{R} \,,$$

which can be proven with Marsden's identity [15], we infer for degree $1 \leq p \leq 4$

$$\psi_l^p(x) = 2 - \frac{x}{h_l} \,, \quad x \in \left[0, \frac{5-p}{2} h_l\right] \,.$$

In other words, modified B-splines with index $i \in \{1, 2^l - 1\}$ extrapolate linearly towards the boundary of $[0, 1]$, providing meaningful values for linear combinations near the boundary. For higher degrees $p > 4$, the deviation from $2 - x/h_l$ is hardly visible, as the second derivative at the boundary is numerically small.

Hierarchical B-splines of one dimension are generalized to the d-dimensional case as usual by a tensor product approach,

$$\varphi_{\mathbf{l,i}}^{\mathbf{p}}(\mathbf{x}) := \prod_{t=1}^{d} \varphi_{l_t,i_t}^{p_t}(x_t) \,, \quad \mathbf{x_{l,i}} := (x_{l_1,i_1}, \ldots, x_{l_d,i_d}) \,,$$

using multi-indices $\mathbf{l}, \mathbf{i} \in \mathbb{N}^d$, $\mathbf{p} \in \mathbb{N}_0^d$, and $\mathbf{x} \in [0, 1]^d$. We define d-variate nodal and hierarchical subspaces by

$$V_{\mathbf{l}}^{\mathbf{p}} := \text{span}\{\varphi_{\mathbf{l,i}}^{\mathbf{p}} \mid \forall_{t=1,\ldots,d} : i_t = 1, \ldots, 2^{l_t} - 1\} \,,$$

$$W_{\mathbf{l}}^{\mathbf{p}} := \text{span}\{\varphi_{\mathbf{l,i}}^{\mathbf{p}} \mid \mathbf{i} \in I_{\mathbf{l}}\} \,, \quad I_{\mathbf{l}} := I_{l_1} \times \cdots \times I_{l_d} \,,$$

Tensor products of linearly independent functions are linearly independent, i.e. the generating sets of $V_{\mathbf{l}}^{\mathbf{p}}$ and $W_{\mathbf{l}}^{\mathbf{p}}$ are their bases, respectively. By using an analogous

d-variate formulation of Prop. 1 (defining S_ξ^p appropriately), it follows as above that

$$V_n^p = S_\xi^p = \bigoplus_{l \leq n} W_l^p ,$$

if we choose the d knot sequences $\xi = (\xi_1, \ldots, \xi_d)$, $\xi_t = (\xi_{t,0}, \ldots, \xi_{t,m_t+p_t})$, accordingly to (5) and restrict all functions to $D = [\xi_{1,p_1}, \xi_{1,m_1}] \times \cdots \times [\xi_{d,p_d}, \xi_{d,m_d}]$. The sparse grid space $V_n^{p,s}$ of level n can now be constructed as usual by

$$V_n^{p,s} := \bigoplus_{\|l\|_1 \leq n+d-1} W_l^p .$$

We get the familiar piecewise linear sparse grid space with $\mathbf{p} = \mathbf{1} := (1, \ldots, 1)$. Sparse grid spaces consisting of modified B-splines are defined similarly.

3 Adaptive Grid Generation

The surrogate, which replaces the objective function f to be minimized, is defined as the interpolant on an adaptively generated sparse grid. The most widespread method is the refinement of the grid points whose hierarchical surpluses $\alpha_{l,i}$ (in the piecewise linear basis) have the highest absolute value [29]. However, this approach does not generate more points close to minima than elsewhere. Instead we want to use a slightly modified version of the Novak-Ritter refinement criterion [11, 16, 26] which was specifically made for optimization (initially for hyperbolic cross points, which are closely related to sparse grids).

The method works iteratively: We start with an initial regular sparse grid, e.g. of level 3. Let $X = \{\mathbf{x}_k := \mathbf{x}_{l_k,i_k} \mid k = 1, \ldots, N\} \subset \mathbb{R}^d$ be the current sparse grid at the beginning of an iteration. The Novak-Ritter criterion selects one point $\mathbf{x}_{k*}$ of X, which is then refined by inserting the $2d$ neighbors into the grid. The neighbors of $\mathbf{x}_{l,i}$ in the t-th dimension have level $l_t + 1$ and index $2i_t \pm 1$ in dimension t and the same level and index in all other dimensions. If one of the neighbors already exists in X, then the first higher-order neighbor, which is not in X, is inserted instead. The neighbor of order m has level $l_t + m$ and index $2^m i_t \pm 1$ in the t-th dimension. Therefore, in each iteration exactly $2d$ points are inserted. The grid generation is completed when a specific number $N \in \mathbb{N}$ of grid points, which is due to the overall effort that can be invested, has been reached.

The Novak-Ritter criterion determines $\mathbf{x}_{k*}$ as follows: Associate with each grid point $\mathbf{x}_k = \mathbf{x}_{l_k,i_k}$ three scalars $\|l_k\|_1$, d_k and r_k. $\|l_k\|_1$ is the sum of the levels, as usual. d_k represents the *degree* of $\mathbf{x}_k$, the number of times the point was already selected (initially 0). r_k is the *rank* of $\mathbf{x}_k$ defined by $r_k := |\{k' \mid f(\mathbf{x}_{k'}) \leq f(\mathbf{x}_k)\}|$, i.e., the point with the smallest objective function value gets rank 1, the next bigger

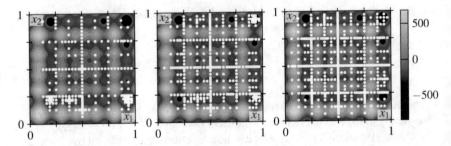

Fig. 3 Adaptive grid generation of $N = 500$ points with Novak-Ritter's refinement criterion for the Schwefel 2D test function with $\gamma = 0.6$ (*left*), $\gamma = 0.8$ (*center*), and $\gamma = 0.95$ (*right*). The global minimum lies in the upper right corner

one gets rank 2 etc. Now, k^* is selected as the index for which the *quality* β_{k^*} is minimal:

$$\beta_k := (\|\mathbf{l}_k\|_1 + d_k + 1)^\gamma \cdot r_k^{1-\gamma} .$$

We added 1 to the base of the first factor to prevent ambiguities if levels and degree sum up to zero (possible when working with boundary grids). $\gamma \in [0, 1]$ is the adaptivity parameter with $\gamma = 0$ meaning pure adaptivity and $\gamma = 1$ leading to an unadaptive algorithm with the function values being irrelevant. γ must be chosen carefully to allow the algorithm to explore the whole domain $[0, 1]^d$, while refining in promising regions sufficiently well to increase the accuracy of the sparse grid interpolant (see Fig. 3 for an example). Its best choice depends a lot on the characteristics of the objective function at hand. As a compromise, we choose a priori $\gamma = 0.85$ for all applications. Note that for γ large enough, the set X of generated grid points gets dense in $[0, 1]^d$ in the limit $N \to \infty$, implying that for arbitrary objective functions f, a global optimum will be found eventually (if N and γ are chosen large enough).

4 Implementation

After adaptively generating the grid as the first step, we replace the objective function f by the sparse grid interpolant $\tilde{f}$ and then apply existing optimization algorithms to $\tilde{f}$. In this section, we want to elaborate on the two remaining steps.

4.1 Hierarchization

The interpolant $\tilde{f}$ on the sparse grid $X = \{\mathbf{x}_1, \ldots, \mathbf{x}_N\} \subset \mathbb{R}^d$ is defined by the linear combination of the basis functions $\varphi_k := \varphi_{\mathbf{l}_k, \mathbf{i}_k}^{\mathbf{p}}$ (either modified or not) interpolating f in the grid points $\mathbf{x}_k := \mathbf{x}_{\mathbf{l}_k, \mathbf{i}_k}$. This leads to a linear system with the variables $\alpha_1, \ldots, \alpha_N \in \mathbb{R}$ (*hierarchical surpluses*):

$$\tilde{f}(\mathbf{x}) := \sum_{k=1}^{N} \alpha_k \varphi_k(\mathbf{x}), \quad \tilde{f}(\mathbf{x}_j) = f_j := f(\mathbf{x}_j), \quad j = 1, \ldots, N. \tag{8}$$

The basis transform $\mathbf{f} \mapsto \boldsymbol{\alpha}$ is usually called *hierarchization*. For $\mathbf{p} = \mathbf{1}$, the linear system can efficiently be solved via the unidirectional principle [28]: It suffices to apply one-dimensional hierarchization operators to all one-dimensional subgrids (so-called poles) of X in each dimension, working with updated values. However, the principle only works if every pole is a proper 1D sparse grid: Every hierarchical ancestor of a grid point of X must be in X, too. This requirement has severe effects, because every grid point insertion by Novak-Ritter's algorithm in the grid generation phase implies the recursive insertion of all (indirect) hierarchical ancestors. The number of the ancestors to be inserted grows rapidly with the number d of dimensions. For example, performing Novak-Ritter's grid generation for the well-known Rosenbrock function and $\gamma = 0.8$ leads to 1128, 223, 61, 33, 16 refinement iterations for $d = 2, 3, 4, 5, 10$ respectively, stopping when $N = 10,000$ points have been generated. As a result, for $d = 2$ only 45% of the maximum possible number $N/(2d)$ of iterations has been exploited, for $d = 3$ only 13% and for $d \geq 4$ less than 5%. The ancestors often lie at uninteresting places, wasting valuable evaluations of the objective function.

For higher B-spline degrees $\mathbf{p} > \mathbf{1}$, the unidirectional principle is in general not applicable anyway. This is due to the fact that in this case basis functions do not vanish at all grid points of coarser levels (unlike in the piecewise linear case). For our purposes with limited overall effort N, it is sufficient to solve the linear system (8) directly or iteratively. We thus do not have to generate additional hierarchical ancestors, allowing to exhaust the full number $N/(2d)$ of iterations in the grid generation phase. In general, the linear system is asymmetric and its sparsity structure depends on how many grid points are contained in the supports of the basis functions. For lower numbers d of dimensions and lower B-spline degrees $\mathbf{p}$, the system is sparse, which allows a solution by adequate solvers in reasonable time.[1]

---

[1]We used Gmm++ ([31], GMRES) and UMFPACK ([6], LU factorization) for sparse systems and Armadillo ([33], LU factorization) and Eigen ([13], QR Householder factorization) for full systems.

4.2 Global Optimization

The constructed sparse grid B-spline interpolant $\tilde{f}$ is $(p_t - 1)$ times partially continuously differentiable in dimension $t = 1, \ldots, d$. For $p_t > 1$, we can apply gradient-based optimization methods to $\tilde{f}$ without having to evaluate f additional times. We used local gradient-based algorithms [25], particularly the gradient descent method, the nonlinear conjugate gradient method with Polak-Ribière coefficients (NLCG, [30]), Newton's method, BFGS, and Rprop [32], in addition to the local gradient-free Nelder-Mead algorithm (NM, [24]). We also used Storn's and Price's Differential Evolution (DE, [35]), using a population size of $10d$, as a non-local gradient-free method. To prevent being stuck in local minima, we globalized all mentioned local algorithms by using a multi-start approach with $m := \min(10d, 100)$ uniformly random starting points (i.e. m parallel calls of the local algorithm, each with $1/m$ of the permitted function evaluations). The gradient-free techniques NM and DE are not only used for the global optimization of the surrogate $\tilde{f}$, but also to directly optimize the objective function and the standard piecewise linear interpolant (case of $\mathbf{p} = \mathbf{1}$), as we will explain later.

Our optimization algorithm to solve problem (1) for a given objective function $f: [0, 1]^d \rightarrow \mathbb{R}$ works as follows, assuming that the adaptivity parameter $\gamma \in [0, 1]$ of the grid generation and the maximal number $N \in \mathbb{N}$ of evaluations of f is given:

1. Generate the grid $X = \{\mathbf{x}_1, \ldots, \mathbf{x}_n\}$, $n \leq N$, using the adaptive Novak-Ritter method. This requires to evaluate the objective function n times, obtaining $f_j = f(\mathbf{x}_j)$.
2. Solve the linear system (8) to get the interpolant $\tilde{f}: [0, 1]^d \rightarrow \mathbb{R}$.
3. Optimize the interpolant: First, find $\mathbf{y}_0 := \mathbf{x}_{j^*}$ with $j^* := \arg\min_j f_j$. Then apply all gradient-based methods to $\tilde{f}$ with $\mathbf{y}_0$ as starting point. Let $\mathbf{y}_1$ be the resulting point with the minimal objective function value. Now use the globalized local algorithms and DE applied to $\tilde{f}$; let $\mathbf{y}_2$ be the best (i.e. in terms of the f value) point of the results. Take the point of $\{\mathbf{y}_0, \mathbf{y}_1, \mathbf{y}_2\}$ with the smallest f value as approximation $\mathbf{x}_{opt}^*$ to the optimum $\mathbf{x}_{opt}$ of f.

The third step requires (beyond the n evaluations during grid generation) some, say c, additional evaluations of f. Thus, a total of up to $N + c$ evaluations have to be performed during the algorithm. To keep the overall effort to at most N one can enforce $n \leq N - c$ in step 1. Because $\mathbf{y}_0$ is taken into account when determining $\mathbf{x}_{opt}^*$, the returned optimum is the point with the smallest objective function value of all points where f was evaluated during the algorithm.

We compared our optimization algorithm to the following common optimization techniques:

- Optimization of the piecewise linear sparse grid interpolant. Therefore we proceed as above with B-spline degree $\mathbf{p} = \mathbf{1}$, using only the gradient-free methods NM (globalized) and DE to optimize $\tilde{f}$. The best of the two results is called $\mathbf{x}_{opt}'$.

- Direct optimization of the objective function f (with globalized NM) without using sparse grids, but with only N evaluations permitted. The resulting optimum is called $\mathbf{x}''_{\mathrm{opt}}$.

5 Numerical Results and Applications

In this section, we want to review our optimization method with the aid of artificial test functions and real-world applications. As standard parameters, we used modified B-splines of degree $\mathbf{p} = \mathbf{5}$ as basis functions and $\gamma = 0.85$ as adaptivity.

5.1 Test Functions

We studied a wide variety of test functions for different dimensionalities [36]. In the following, we present three functions for two dimensions and three functions defined in arbitrary dimensions. The domain of each function is transformed to the unit hypercube $[0, 1]^d$ by an affine transformation. Additionally, some of the domains were translated and/or scaled first (when compared to the literature) to make sure that the optimum does not lie at the center of the domain. Otherwise sparse grid approaches would have been in advantage, because they spend proportionally few points near the corners of $[0, 1]^d$. In Table 1 we give the domains, minimal points, and corresponding function values (all before parameter scaling and translation). The two-dimensional test functions are shown in Fig. 4. All functions were perturbed in the parameter domain by a small pseudo-random normally distributed translation (standard deviation 0.01), while making sure that the optima of the perturbed functions still lie in the original domains. To increase the validity of our results, all results shown are the mean of five passes with different perturbations.

The plots depicted in Figs. 5, 6, and 8 show the difference between approximated and true minimal value of the objective function over the number N of evaluations of f. Each test function is associated with three lines: The solid lines represent the

Table 1 Employed test functions in two and arbitrary dimensions with abbreviations in bold

Name	Domain	$\mathbf{x}_{\mathrm{opt}}$	$f(\mathbf{x}_{\mathrm{opt}})$	Reference
Branin	$[-5, 10] \times [0, 15]$	$(-\pi, 12.275), (\pi, 2.275),$ $(9.42478, 2.475)$	0.397887	[18, Branin RCOS]
Eggholder	$[-512, 512]^2$	$(512, 404.2319)$	-959.6407	[37, $F101$]
Rosenbrock	$[-5, 10]^2$	$(1, 1)$	0	[38]
Ackley	$[-1, 9]^d$	$\mathbf{0}$	0	[38]
Rastrigin	$[-2, 8]^d$	$\mathbf{0}$	0	[38]
Schwefel	$[-500, 500]^d$	$420.9687 \cdot \mathbf{1}$	$-418.9829d$	[38]

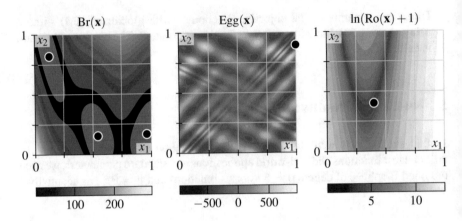

Fig. 4 Bivariate test functions with location of the minimal points (after normalization of the domain to $[0, 1]^2$)

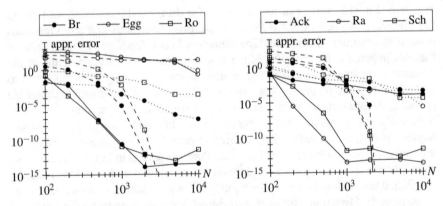

Fig. 5 Approximation errors $f(\mathbf{x}^*_{\mathrm{opt}}) - f(\mathbf{x}_{\mathrm{opt}})$ (*solid lines*), $f(\mathbf{x}'_{\mathrm{opt}}) - f(\mathbf{x}_{\mathrm{opt}})$ (*dotted*), and $f(\mathbf{x}''_{\mathrm{opt}}) - f(\mathbf{x}_{\mathrm{opt}})$ (*dashed*) over the number N of evaluations for different test functions with $d = 2$ variables

performance of our optimization algorithm with result $\mathbf{x}^*_{\mathrm{opt}}$, the dotted lines display the performance of the optimization of the piecewise linear sparse grid interpolant with Nelder-Mead (NM) and Differential Evolution with result $\mathbf{x}'_{\mathrm{opt}}$, and the dashed lines show the optimization of the objective function using globalized NM with result $\mathbf{x}''_{\mathrm{opt}}$. Note that in the notation of the last section, we have $f(\mathbf{x}'_{\mathrm{opt}}) \leq f(\mathbf{y}_0)$, implying that the gain of our method compared to the best Ritter-Novak grid point $\mathbf{y}_0$ is at least $f(\mathbf{x}'_{\mathrm{opt}}) - f(\mathbf{x}^*_{\mathrm{opt}})$ (difference between solid and dotted lines).

As can be seen in Fig. 5, functions like Branin and Rosenbrock are generally easier to optimize since they feature few local minima and few oscillations. Such functions can be approximated by B-spline linear combinations very well, which leads to a considerable advantage for the B-splines compared to the standard piecewise linear basis. Our method even beats the globalized NM method (dashed

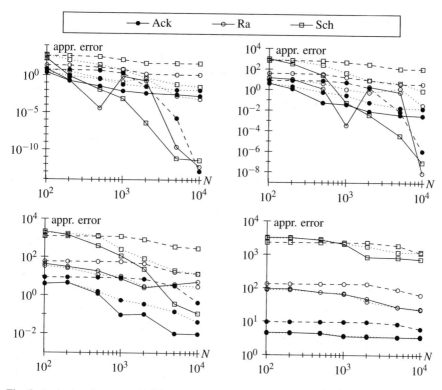

Fig. 6 Approximation errors $f(\mathbf{x}^*_{\mathrm{opt}}) - f(\mathbf{x}_{\mathrm{opt}})$ (*solid lines*), $f(\mathbf{x}'_{\mathrm{opt}}) - f(\mathbf{x}_{\mathrm{opt}})$ (*dotted*), and $f(\mathbf{x}''_{\mathrm{opt}}) - f(\mathbf{x}_{\mathrm{opt}})$ (*dashed*) over the number N of evaluations for different test functions with $d = 3$, $d = 4$, $d = 6$, and $d = 10$ variables (*from top left to bottom right*)

lines) for most of the test functions. For the Eggholder function, fast convergence of all methods is impeded not only by high oscillations and many local minima, but also by the fact that the global optimum lies on the boundary of the domain (before perturbing).

Figure 6 shows that for higher-dimensional functions, the problem of optimization rapidly becomes very difficult. With increasing d, the rate of convergence becomes substantially slower. We note that for moderate dimensions $d \in \{3, 4\}$, B-splines can provide a significant boost in the performance compared to the piecewise linear basis. For higher dimensions $d \geq 6$, both sparse grid approaches (B-splines and piecewise linear) perform better in general in comparison to the globalized NM optimization technique. It can be seen that Rastrigin is a very tough function as it exhibits numerous local minima in a neighborhood of the global minimum, all with a similar function value. This can lead to a non-monotonous error decay of our method as seen in the plots for $d \in \{3, 4, 6\}$, since the global minimum $\mathbf{x}^*_{\mathrm{opt}}$ of the interpolant occasionally does not match with the actual optimum $\mathbf{x}_{\mathrm{opt}}$ of the objective function.

5.2 Model of a DC Motor

As an example application we study an inverse problem of a simple DC (direct current) motor. If we denote with θ and ω the angular position and velocity in rad and rad/s, respectively, then an idealized model (with zero disturbance and torque) of the motor can be deduced [22], obtaining the linear state-space representation

$$\dot{\theta}(t) = \omega(t) , \quad \dot{\omega}(t) = -\frac{1}{\tau}\omega(t) + \frac{k}{\tau}U(t) , \quad \theta(0) = \theta_0 , \quad \omega(0) = \omega_0 , \quad (9)$$

with (θ, ω) as both state and output, input voltage U, and motor-dependent constants τ (*time constant*) and k (*steady-state gain*). It can easily be seen that for constant inputs $U \equiv U_0$, ω then satisfies $\omega(t) = (\omega_0 - kU_0)e^{-t/\tau} + kU_0$.

We have generated artificial data for the motor sampled at t_j with 10 Hz over a time span of 60 s (see Fig. 7). The sampled data (θ_j, ω_j) was generated by adding an artificial Gaussian noise with standard deviation 0.1 rad and 0.1 rad/s to the solution (θ, ω) of (9) for the generated voltage data U_j, respectively. Our goal is now to determine $(\tau, k, \theta_0, \omega_0)$ so that the resulting solution (θ, ω) of (9) minimizes the ℓ^2 norm $(\sum_j (\omega(t_j) - \omega_j)^2)^{1/2}$ of the difference of experimental and simulated angular velocity. The error functional does not need to take θ into account, as $\dot{\theta} = \omega$ should imply a good match of θ and θ_j and including θ in the error functional would lead to worse results due to overfitting. In total, this leads to a 4D optimization problem.

Before we can start optimizing, we need to determine reasonable parameter intervals. Looking at the data, we guess $\theta_0 \in [-2\,\text{rad}, 2\,\text{rad}]$ and $\omega_0 \in [-2\,\text{rad/s}, 0\,\text{rad/s}]$. We guess τ by looking at the half-life period $\tau \ln 2$ of the transient response after a change in input voltage polarity. This period roughly equals 0.3 s which would imply $\tau \approx 0.43$ s, justifying the assumption of $\tau \in [0.2\,\text{s}, 0.6\,\text{s}]$. For the interval of k, we look at the steady-state angular velocity kU_0, which is around 1 rad/s, leading

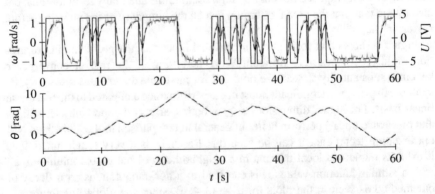

Fig. 7 Input voltage U (*solid black line*), artificial motor data (θ_j, ω_j) (*solid gray*), and simulated data (θ, ω) of optimal model (*dashed*)

Fig. 8 Approximation errors $f(\mathbf{x}_{\text{opt}}^*) - f(\mathbf{x}_{\text{opt}})$ (*solid line*), $f(\mathbf{x}_{\text{opt}}') - f(\mathbf{x}_{\text{opt}})$ (*dotted*), and $f(\mathbf{x}_{\text{opt}}'') - f(\mathbf{x}_{\text{opt}})$ (*dashed*) over the number N of evaluations for the objective function resulting from the DC motor

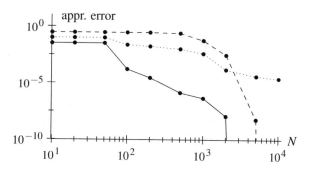

with $U_0 = 5\,\text{V}$ to $k \approx 0.2\,\text{rad/(Vs)}$. Therefore, we generously set the interval to $k \in [0.1\,\text{rad/(Vs)}, 0.4\,\text{rad/(Vs)}]$.

Figure 8 (log-log plot as above) shows the performance of our optimization method as well as the performance of the piecewise linear basis and the direct optimization of the objective function with the globalized Nelder-Mead method (NM). We see that the objective function is sufficiently smooth to allow good B-spline interpolation, leading to a faster convergence compared to the piecewise linear basis and to the classical NM technique, with convergence beginning at $N = 100$ grid points. Our method exactly finds the optimal parameters $\tau_{\text{opt}} = 0.496\,\text{s}$, $k_{\text{opt}} = 0.214\,\text{rad/(Vs)}$, $\theta_{0,\text{opt}} = -0.198\,\text{rad}$, $\omega_{0,\text{opt}} = -0.411\,\text{rad/s}$ with error functional value 2.693 using just $N = 1000$ evaluations.

5.3 Shape Optimization with Homogenization

As another application, we have also employed B-splines on sparse grids in a two-scale shape optimization setting [17]. Classical approaches in shape optimization share the drawback of severely restricting the set of feasible topologies before starting to optimize, which leads to homeomorphic results. However, one often does not know the topology of the optimal shape beforehand. For example, if we take the cantilever in Fig. 9 (left), which is fixed at one side and deformed by a force $\mathbf{F}$ at the other side, and want to determine the cantilever shape which minimizes displacement, we do not know if we should use one, two, or even more crossbars, if the cantilever is only allowed to take up a specific volume.

In the homogenization approach [2], the whole domain Ω is potentially filled with material with varying density $\varrho(\mathbf{x}) \in [0, 1]$. For a fixed force $\mathbf{F}$, we want ϱ to minimize the *compliance function* $J(\mathbf{u}_\varrho)$ with a volume constraint,

$$\min_\varrho J(\mathbf{u}_\varrho) \quad \text{s.t.} \quad \frac{1}{|\Omega|} \int_\Omega \varrho(\mathbf{x}) \, d\mathbf{x} \le \varrho_{\max}, \quad J(\mathbf{u}_\varrho) := \int_\Omega \mathbf{F} \cdot \mathbf{u}_\varrho(\mathbf{x}) \, d\mathbf{x}, \tag{10}$$

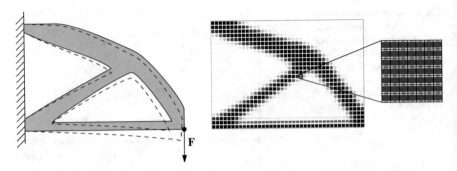

Fig. 9 *Left*: shape optimization of a 2D cantilever fixed at the left side with force **F** resulting in a deformation. *Right*: optimized model consisting of macro cells (compounds of micro cells)

where $\varrho_{max} \in [0, 1]$ is the maximal total density and $\mathbf{u}_\varrho(\mathbf{x})$ is the displacement in $\mathbf{x} \in \Omega$ (depending on the density ϱ), which can be determined by the finite element method (FEM, [14, 17]). In the following, we restrict ourselves to the two-dimensional case, but the 3D case could also be handled in an analogous way.

We choose a two-scale approach as Hübner [17]: First, we discretize the domain in $(N_1 \times N_2)$ *macro cells*. Each macro cell k is a compound of tiny periodic and identical *micro cells* (see Fig. 9), which are formed by shearing an axis-parallel cross with thicknesses a_k, b_k by an angle $\varphi_k \in (-\pi/2, \pi/2)$, resulting in parallelogram-shaped micro cells. If a_k, b_k, φ_k are known, one can compute the symmetrical elasticity tensor $E_k = (E_{k,i,j})_{1 \le i,j \le 3} \in \mathbb{R}^{3 \times 3}$ of macro cell k by the FEM (*micro problem*). When we know all the E_k, we can compute $\mathbf{u}_\varrho$ with the density ϱ given by the $3N_1N_2$ parameters $a_k, b_k, \varphi_k, k = 1, \ldots, N_1N_2$, again by solving a FEM problem (*macro problem*). Our goal is finding a combination of the $3N_1N_2$ parameters which solves (10). Because every evaluation of $J(\mathbf{u}_p)$ triggers the solution of a macro problem (which depends on N_1N_2 micro problems), a single evaluation is very expensive. As in [17], we use the FEM solver CFS++ [20] which provides interfaces to established optimizers like SNOPT [12], requiring gradients that have to be approximated by finite differences, since they are not available explicitly.

To increase performance, Hübner [17] precomputes values of the elasticity tensor $E: [0, 1]^3 \to \mathbb{R}^{3 \times 3}$ for different combinations of normalized parameters a, b, φ and replaces the task of solving micro problems by the evaluation of an interpolant $\tilde{E}: [0, 1]^3 \to \mathbb{R}^{3 \times 3}$ of E. Full grid interpolation approaches are possible, but more complex micro cell models (e.g. in 3D) will feature more parameters, which would imply a prohibitively large precomputational effort in terms of both computing time and storage space. In [17], also the suitability of sparse grid interpolation with piecewise linear functions was studied. However, these lead to problems because of the discontinuous derivatives calculated by the gradient-based optimizer. It seems natural to employ B-spline basis functions instead, as the function E to be interpolated is supposedly relatively smooth. Additionally, the optimizer can use exact derivatives of the interpolant.

As an example, we consider the cantilever in Fig. 9 ("example A" in [17]), where the domain consists of $40 \cdot 26$ macro cells. The emerging optimization problem with $3 \cdot 40 \cdot 26 = 3120$ variables is solved by CFS++/SNOPT and visualized as in Fig. 9 (right), where each macro cell is represented by a single micro cell cross. We compare the performance of B-splines to the piecewise linear basis and to piecewise tricubic interpolation [21] on the full grid of level 6.

First, we examine the B-spline sparse grid interpolation method when all $\varphi_k = 0$ are fixed and only a_k and b_k are optimized. In this case, four elasticity tensor entries $E_{1,3} = E_{2,3} = 0 = E_{3,2} = E_{3,1}$ vanish. This makes a significant difference in complexity as there are only four non-trivial entries of E left and only two variables per interpolant, resulting in an optimization problem of $2 \cdot 40 \cdot 26 = 2080$ variables. If we look at the results in Table 2 (top half), we observe that sparse grid interpolation produces similar objective function values as tricubic interpolation on the full grid. However, using a full grid interpolant leads to better convergence and faster termination. With sparse grid B-splines (see Fig. 10, left) we even get a smaller compliance function value than the full grid interpolation.

Second, we look at the general case where all $3N_1N_2$ are to be optimized. The optimization now takes much more time since there are more tensor entries to be interpolated, more partial derivatives to be evaluated, and more unknown

Table 2 Results for the two-parameter case (where $\varphi_k = 0$ is fixed) and the three-parameter case with optimal compliance function value, number of iterations, and time needed by SNOPT for optimization without precomputation of the elasticity tensors ($\rho_{max} = 0.5$)

#Param.	Grid	Basis	Obj. Fcn.	#Iter.	Time
2	Full, level 6	Piecewise tricubic interpolation	42.85	170	128 s
2	Sparse, level 7	Modified piecewise linear	43.00	704	546 s
2	Sparse, level 7	Modified cubic B-splines	42.80	377	299 s
3	Full, level 6	Piecewise tricubic interpolation	41.86	1307	27 min
3	Sparse, level 8	Modified piecewise linear	41.95	4203	163 min
3	Sparse, level 8	Modified cubic B-splines	41.26	1483	139 min

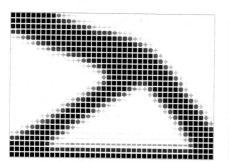

Fig. 10 Results with modified B-splines of degree $\mathbf{p} = 3$ and a material fraction of $\rho_{max} = 0.5$ for fixed $\varphi_k = 0$ with a regular sparse grid of level 7 with 2815 points (*left*) and optimized φ_k with a refined sparse grid with 4439 points (*right*)

optimization parameters. But as we have more degrees of freedom, the obtained objective function values (cf. Table 2, bottom half) are slightly better than in the case of $\varphi_k = 0$. If we compare the visualizations in Fig. 10, we note the exploitation of the additional degrees of freedom as the incline of the crossbar is more gentle in the case of three parameters per micro cell.

For the trivariate sparse grid interpolants, it is not sufficient to discretize $[0, 1]^3$ with a regular sparse grid of level 7 due to too many oscillations of the resulting $\tilde{E}$. Even for a regular sparse grid of level 8 and piecewise linear functions, the optimizer terminates early because of numerical difficulties, showing the problems introduced by approximating discontinuous derivatives by difference quotients. Again, B-splines perform quite well since they find the best parameter combination in terms of compliance function values compared to full tricubic or sparse piecewise linear interpolation, spending a multiple of the computational time of the full grid interpolation, though. This will, however, change as soon as we consider a more complicated micro cell model (e.g. in three dimensions), where we will not be able to employ full grid interpolation anymore.

Starting from the regular sparse grid of level 7, we also generated a spatially adaptive sparse grid, specifically tailored for this interpolation problem [36]. With only 4439 points (cf. 2815 and 7423 points of the regular grids of level 7 and 8, respectively), we get moderately worse results (objective function value of 42.14) for the modified cubic B-splines than for the regular sparse grid of level 8, but the optimization then only takes 84 min. Additionally, we have not taken into account the precomputation time to create the elasticity tensor data at the sparse grid interpolation points, which, accordingly to the smaller number of grid points, would be smaller, too (compared to the level 8 regular grid). Of course, the precomputation effort for the full grid of level 6 is much larger as it needs 258,048 data points.

6 Conclusion

We constructed a surrogate-based optimization approach using B-splines on sparse grids. After proving their linear independence and studying the direct sum of hierarchical subspaces, we used an adaptive grid generation method by Novak-Ritter to generate spatially adaptive sparse grids with adjustable adaptivity $\gamma \in [0, 1]$. Finally, we successfully employed our new optimization method to various artificial test functions and real-world examples. The new method works well for smooth, moderately dimensioned objective functions without high-frequency oscillations. We would like to mention we have applied our method to a lot more test functions (e.g. Beale, Goldstein-Price, Griewank) [36], and we picked a somewhat representative subset for this work. We also studied other sparse grid types like grids with boundary points or Clenshaw-Curtis grids with non-uniform Chebyshev points. However, the modified B-splines on the standard grid without boundary points seem to exhibit the best performance for a given number of grid points.

Certainly, there is room for improvement, as we used the same fixed B-spline degree $\mathbf{p} = p \cdot \mathbf{1}$ for all of the dimensions. We could start to use different degrees p_t depending on the dimensions t. Going one step further, we could even choose $\mathbf{p}$ adaptively depending on the objective function f, to adapt to discontinuities of f or its derivatives.

Acknowledgements This work was financially supported by the Juniorprofessurenprogramm of the Landesstiftung Baden-Württemberg.

References

1. H. Bachau, E. Cormier, P. Decleva, J.E. Hansen, F. Martín, Applications of B-splines in atomic and molecular physics. Rep. Prog. Phys. **64**(12), 1815–1942 (2001)
2. M.P. Bendsøe, N. Kikuchi, Generating optimal topologies in structural design using a homogenization method. Comput. Methods Appl. Mech. Eng. **71**(2) 197–224 (1988)
3. H.-J. Bungartz, Finite elements of higher order on sparse grids, Habilitationsschrift, Institut für Informatik, TU München, 1998
4. E. Cohen, R.F. Riesenfeld, G. Elber, *Geometric Modeling with Splines: An Introduction* (A K Peters, Natick, 2001)
5. M.G. Cox, The numerical evaluation of B-splines. IMA J. Appl. Math. **10**(2), 134–149 (1972)
6. T.A. Davis, Algorithm 832: UMFPACK V4.3-an unsymmetric-pattern multifrontal method. ACM Trans. Math. Softw. **30**(2), 196–199 (2004)
7. C. de Boor, On calculating with B-splines. J. Approx. Theory **6**(1), 50–62 (1972)
8. C. de Boor, Splines as linear combinations of B-splines. A survey, in *Approximation Theory II*, ed. by G.G. Lorentz, C.K. Chui, L.L. Schumaker (Academic, New York, 1976), pp. 1–47
9. F. Delbos, L. Dumas, E. Echagüe, Global optimization based on sparse grid surrogate models for black-box expensive functions, http://dumas.perso.math.cnrs.fr/JOGO.pdf, (2016)
10. M.M. Donahue, G.T. Buzzard, A.E. Rundell, Parameter identification with adaptive sparse grid-based optimization for models of cellular processes, in *Methods in Bioengineering: Systems Analysis of Biological Networks*, ed. by A. Jayaraman, J. Hahn (Artech House, Boston/London, 2009), pp. 211–232
11. I. Ferenczi, Globale Optimierung unter Nebenbedingungen mit dünnen Gittern, Diploma thesis, Department of Mathematics, TU München, 2005
12. P.E. Gill, W. Murray, M.A. Saunders, SNOPT: an SQP algorithm for large-scale constrained optimization. SIAM J. Optim. **12**(4), 979–1006 (2002)
13. G. Guennebaud, B. Jacob et al., Eigen, http://eigen.tuxfamily.org/, (2016)
14. K. Höllig, *Finite Element Methods with B-Splines* (SIAM, Philadelphia, 2003)
15. K. Höllig, J. Hörner, *Approximation and Modeling with B-Splines* (SIAM, Philadelphia, 2013)
16. Y.-K. Hu, Y.P. Hu, Global optimization in clustering using hyperbolic cross points. Pattern Recognit. **40**(6), 1722–1733 (2007)
17. D. Hübner, Mehrdimensionale Parametrisierung der Mikrozellen in der Zwei-Skalen-Optimierung, Master's thesis, Department of Mathematics, FAU Erlangen-Nürnberg, 2014
18. M. Jamil, X.-S. Yang, A literature survey of benchmark functions for global optimisation problems. Int. J. Math. Model. Numer. Optim. **4**(2), 150–194 (2013)
19. Y. Jiang, Y. Xu, B-spline quasi-interpolation on sparse grids. J. Complex. **27**(5), 466–488 (2011)
20. M. Kaltenbacher, Advanced simulation tool for the design of sensors and actuators, in *Proceedings of Eurosensors XXIV*, Linz, vol. 5, 2010, pp. 597–600

21. F. Lekien, J. Marsden, Tricubic interpolation in three dimensions. Int. J. Numer. Methods Eng. **63**(3), 455–471 (2005)
22. L. Ljung, *System Identification: Theory for the User*, 2nd edn. (Prentice Hall, Upper Saddle River, 1999)
23. C.W. McCurdy, F. Martín, Implementation of exterior complex scaling in B-splines to solve atomic and molecular collision problems. J. Phys. B: At. Mol. Opt. Phys. **37**(4), 917–936 (2004)
24. J.A. Nelder, R. Mead, A simplex method for function minimization. Comput. J. **7**(4), 308–313 (1965)
25. J. Nocedal, S.J. Wright, *Numerical Optimization* (Springer, New York, 1999)
26. E. Novak, K. Ritter, Global optimization using hyperbolic cross points, in *State of the Art in Global Optimization*, ed. by C.A. Floudas, P.M. Pardalos (Springer, Boston, 1996), pp. 19–33
27. D. Pandey, Regression with spatially adaptive sparse grids in financial applications, Master's thesis, Department of Informatics, TU München, 2008
28. D. Pflüger, *Spatially Adaptive Sparse Grids for High-Dimensional Problems* (Verlag Dr. Hut, München, 2010)
29. D. Pflüger, Spatially adaptive refinement, in *Sparse Grids and Applications*, ed. by J. Garcke, M. Griebel. Lecture Notes in Computational Science and Engineering (Springer, Berlin/Heidelberg, 2012), pp. 243–262
30. E. Polak, G. Ribière, Note sur la convergence de méthodes de directions conjuguées. Rev. Fr. Inf. Rech. Oper. **3**(1), 35–43 (1969)
31. Y. Renard, J. Pommier, Gmm++, http://download.gna.org/getfem/html/homepage/gmm/index.html, (2016)
32. M. Riedmiller, H. Braun, A direct adaptive method for faster backpropagation learning: the RPROP algorithm, in *Proceedings of 1993 IEEE International Conference on Neural Networks*, San Francisco, CA, vol. 1, 1993, pp. 586–591
33. C. Sanderson, Armadillo: an open source C++ linear algebra library for fast prototyping and computationally intensive experiments, Technical report, NICTA, 2010
34. I.J. Schoenberg, Contributions to the problem of approximation of equidistant data by analytic functions. Q. Appl. Math. **4**, 45–99, 112–141 (1946)
35. R. Storn, K. Price, Differential evolution – a simple and efficient heuristic for global optimization over continuous spaces. J. Glob. Optim. **11**(4), 341–359 (1997)
36. J. Valentin, Hierarchische Optimierung mit Gradientenverfahren auf Dünngitterfunktionen, Master's thesis, IPVS, Universität Stuttgart, 2014
37. D. Whitley, S. Rana, J. Dzubera, K.E. Mathias, Evaluating evolutionary algorithms. Artif. Intel. **85**(1–2), 245–276 (1996)
38. X.-S. Yang, *Engineering Optimization* (Wiley, Hoboken, 2010)

Editorial Policy

1. Volumes in the following three categories will be published in LNCSE:

i)   Research monographs
ii) Tutorials
iii) Conference proceedings

Those considering a book which might be suitable for the series are strongly advised to contact the publisher or the series editors at an early stage.

2. Categories i) and ii). Tutorials are lecture notes typically arising via summer schools or similar events, which are used to teach graduate students. These categories will be emphasized by Lecture Notes in Computational Science and Engineering. **Submissions by interdisciplinary teams of authors are encouraged.** The goal is to report new developments – quickly, informally, and in a way that will make them accessible to non-specialists. In the evaluation of submissions timeliness of the work is an important criterion. Texts should be well-rounded, well-written and reasonably self-contained. In most cases the work will contain results of others as well as those of the author(s). In each case the author(s) should provide sufficient motivation, examples, and applications. In this respect, Ph.D. theses will usually be deemed unsuitable for the Lecture Notes series. Proposals for volumes in these categories should be submitted either to one of the series editors or to Springer-Verlag, Heidelberg, and will be refereed. A provisional judgement on the acceptability of a project can be based on partial information about the work: a detailed outline describing the contents of each chapter, the estimated length, a bibliography, and one or two sample chapters – or a first draft. A final decision whether to accept will rest on an evaluation of the completed work which should include

- at least 100 pages of text;
- a table of contents;
- an informative introduction perhaps with some historical remarks which should be accessible to readers unfamiliar with the topic treated;
- a subject index.

3. Category iii). Conference proceedings will be considered for publication provided that they are both of exceptional interest and devoted to a single topic. One (or more) expert participants will act as the scientific editor(s) of the volume. They select the papers which are suitable for inclusion and have them individually refereed as for a journal. Papers not closely related to the central topic are to be excluded. Organizers should contact the Editor for CSE at Springer at the planning stage, see *Addresses* below.

In exceptional cases some other multi-author-volumes may be considered in this category.

4. Only works in English will be considered. For evaluation purposes, manuscripts may be submitted in print or electronic form, in the latter case, preferably as pdf- or zipped ps-files. Authors are requested to use the LaTeX style files available from Springer at http://www.springer.com/gp/authors-editors/book-authors-editors/manuscript-preparation/5636 (Click on LaTeX Template → monographs or contributed books).

For categories ii) and iii) we strongly recommend that all contributions in a volume be written in the same LaTeX version, preferably LaTeX2e. Electronic material can be included if appropriate. Please contact the publisher.

Careful preparation of the manuscripts will help keep production time short besides ensuring satisfactory appearance of the finished book in print and online.

5. The following terms and conditions hold. Categories i), ii) and iii):

Authors receive 50 free copies of their book. No royalty is paid.
Volume editors receive a total of 50 free copies of their volume to be shared with authors, but no royalties.

Authors and volume editors are entitled to a discount of 33.3 % on the price of Springer books purchased for their personal use, if ordering directly from Springer.

6. Springer secures the copyright for each volume.

Addresses:

Timothy J. Barth
NASA Ames Research Center
NAS Division
Moffett Field, CA 94035, USA
barth@nas.nasa.gov

Michael Griebel
Institut für Numerische Simulation
der Universität Bonn
Wegelerstr. 6
53115 Bonn, Germany
griebel@ins.uni-bonn.de

David E. Keyes
Mathematical and Computer Sciences
and Engineering
King Abdullah University of Science
and Technology
P.O. Box 55455
Jeddah 21534, Saudi Arabia
david.keyes@kaust.edu.sa

and

Department of Applied Physics
and Applied Mathematics
Columbia University
500 W. 120 th Street
New York, NY 10027, USA
kd2112@columbia.edu

Risto M. Nieminen
Department of Applied Physics
Aalto University School of Science
and Technology
00076 Aalto, Finland
risto.nieminen@aalto.fi

Dirk Roose
Department of Computer Science
Katholieke Universiteit Leuven
Celestijnenlaan 200A
3001 Leuven-Heverlee, Belgium
dirk.roose@cs.kuleuven.be

Tamar Schlick
Department of Chemistry
and Courant Institute
of Mathematical Sciences
New York University
251 Mercer Street
New York, NY 10012, USA
schlick@nyu.edu

Editor for Computational Science
and Engineering at Springer:
Martin Peters
Springer-Verlag
Mathematics Editorial IV
Tiergartenstrasse 17
69121 Heidelberg, Germany
martin.peters@springer.com

Lecture Notes
in Computational Science
and Engineering

76. J.S. Hesthaven, E.M. Rønquist (eds.), *Spectral and High Order Methods for Partial Differential Equations.*

77. M. Holtz, *Sparse Grid Quadrature in High Dimensions with Applications in Finance and Insurance.*

78. Y. Huang, R. Kornhuber, O.Widlund, J. Xu (eds.), *Domain Decomposition Methods in Science and Engineering XIX.*

79. M. Griebel, M.A. Schweitzer (eds.), *Meshfree Methods for Partial Differential Equations V.*

80. P.H. Lauritzen, C. Jablonowski, M.A. Taylor, R.D. Nair (eds.), *Numerical Techniques for Global Atmospheric Models.*

81. C. Clavero, J.L. Gracia, F.J. Lisbona (eds.), *BAIL 2010 – Boundary and Interior Layers, Computational and Asymptotic Methods.*

82. B. Engquist, O. Runborg, Y.R. Tsai (eds.), *Numerical Analysis and Multiscale Computations.*

83. I.G. Graham, T.Y. Hou, O. Lakkis, R. Scheichl (eds.), *Numerical Analysis of Multiscale Problems.*

84. A. Logg, K.-A. Mardal, G. Wells (eds.), *Automated Solution of Differential Equations by the Finite Element Method.*

85. J. Blowey, M. Jensen (eds.), *Frontiers in Numerical Analysis - Durham 2010.*

86. O. Kolditz, U.-J. Gorke, H. Shao, W. Wang (eds.), *Thermo-Hydro-Mechanical-Chemical Processes in Fractured Porous Media - Benchmarks and Examples.*

87. S. Forth, P. Hovland, E. Phipps, J. Utke, A. Walther (eds.), *Recent Advances in Algorithmic Differentiation.*

88. J. Garcke, M. Griebel (eds.), *Sparse Grids and Applications.*

89. M. Griebel, M.A. Schweitzer (eds.), *Meshfree Methods for Partial Differential Equations VI.*

90. C. Pechstein, *Finite and Boundary Element Tearing and Interconnecting Solvers for Multiscale Problems.*

91. R. Bank, M. Holst, O. Widlund, J. Xu (eds.), *Domain Decomposition Methods in Science and Engineering XX.*

92. H. Bijl, D. Lucor, S. Mishra, C. Schwab (eds.), *Uncertainty Quantification in Computational Fluid Dynamics.*

93. M. Bader, H.-J. Bungartz, T. Weinzierl (eds.), *Advanced Computing.*

94. M. Ehrhardt, T. Koprucki (eds.), *Advanced Mathematical Models and Numerical Techniques for Multi-Band Effective Mass Approximations.*

95. M. Azaïez, H. El Fekih, J.S. Hesthaven (eds.), *Spectral and High Order Methods for Partial Differential Equations ICOSAHOM 2012.*

96. F. Graziani, M.P. Desjarlais, R. Redmer, S.B. Trickey (eds.), *Frontiers and Challenges in Warm Dense Matter.*

97. J. Garcke, D. Pflüger (eds.), *Sparse Grids and Applications – Munich 2012.*

98. J. Erhel, M. Gander, L. Halpern, G. Pichot, T. Sassi, O. Widlund (eds.), *Domain Decomposition Methods in Science and Engineering XXI.*

99. R. Abgrall, H. Beaugendre, P.M. Congedo, C. Dobrzynski, V. Perrier, M. Ricchiuto (eds.), *High Order Nonlinear Numerical Methods for Evolutionary PDEs - HONOM 2013.*

100. M. Griebel, M.A. Schweitzer (eds.), *Meshfree Methods for Partial Differential Equations VII.*

101. R. Hoppe (ed.), *Optimization with PDE Constraints - OPTPDE 2014*.

102. S. Dahlke, W. Dahmen, M. Griebel, W. Hackbusch, K. Ritter, R. Schneider, C. Schwab, H. Yserentant (eds.), *Extraction of Quantifiable Information from Complex Systems*.

103. A. Abdulle, S. Deparis, D. Kressner, F. Nobile, M. Picasso (eds.), *Numerical Mathematics and Advanced Applications - ENUMATH 2013*.

104. T. Dickopf, M.J. Gander, L. Halpern, R. Krause, L.F. Pavarino (eds.), *Domain Decomposition Methods in Science and Engineering XXII*.

105. M. Mehl, M. Bischoff, M. Schäfer (eds.), *Recent Trends in Computational Engineering - CE2014*. Optimization, Uncertainty, Parallel Algorithms, Coupled and Complex Problems.

106. R.M. Kirby, M. Berzins, J.S. Hesthaven (eds.), *Spectral and High Order Methods for Partial Differential Equations - ICOSAHOM'14*.

107. B. Jüttler, B. Simeon (eds.), *Isogeometric Analysis and Applications 2014*.

108. P. Knobloch (ed.), *Boundary and Interior Layers, Computational and Asymptotic Methods – BAIL 2014*.

109. J. Garcke, D. Pflüger (eds.), *Sparse Grids and Applications – Stuttgart 2014*.

For further information on these books please have a look at our mathematics catalogue at the following URL: www.springer.com/series/3527

Monographs in Computational Science and Engineering

For further information on this book, please have a look at our mathematics catalogue at the following URL: www.springer.com/series/7417

Texts in Computational Science and Engineering

For further information on these books please have a look at our mathematics catalogue at the following URL: www.springer.com/series/5151

Printed in the United States
By Bookmasters